PROSPERITY IN THE FOSSIL-FREE ECONOMY

MELISSA K. SCANLAN

Prosperity in the Fossil-Free Economy

COOPERATIVES AND THE DESIGN OF SUSTAINABLE BUSINESSES

Yale

UNIVERSITY PRESS

NEW HAVEN & LONDON

Published with assistance from the foundation established in memory
of Amasa Stone Mather of the Class of 1907, Yale College.

Yale University Press books may be purchased in quantity for educational, business,
or promotional use. For information, please e-mail sales.press@yale.edu (U.S. office)
or sales@yaleup.co.uk (U.K. office).

Set in Scala and Scala Sans types by Newgen North America.
Printed in the United States of America.

Library of Congress Control Number: 2021930040
ISBN 978-0-300-25399-3 (hardcover : alk. paper)

A catalogue record for this book is available from the British Library.

This paper meets the requirements of ANSI/NISO z39.48–1992 (Permanence of Paper).

10 9 8 7 6 5 4 3 2 1

Like planting an acorn to grow into an oak, I wrote this book for today's generations, with aspirations and affection for my loved ones who will be living in 2100 and future generations I will never meet.

CONTENTS

PREFACE

AS WE ENTER THE 2020S, our world is marked by a remarkable global connectivity. We are able to communicate instantly, share information, and understand as an interconnected human community. Unlike at any other time in human history, we have platforms for multiple forms of collaborations globally, nationally, and locally. We have the tools to collectively see what we are doing to our life-support structure and assess the variable material wealth of people across the planet. And what many of us see is that the Earth faces an unprecedented climate emergency, mass extinctions, and disrupted ecological systems, while the wealth gap continues to widen.

Without hiding from the mounting horrors of rising temperatures and inequality, we can also see that we live in an incredibly exciting time of possibility. We have the knowledge, technology, and global connections to create a more just, democratic, and environmentally sustainable world. The massive change that occurred with the industrial revolution is in the process of changing again, and it could be toward a better political and economic system. The direction we ultimately take hangs in the balance as this book goes to press.

Nothing brought this lesson in global interconnectedness and disruption home more forcefully than the novel coronavirus, which emerged in January 2020 in China and rapidly spread throughout the world. People everywhere were forced to stay home, but through technology we were able to stay connected to each other and lessen our fear and isolation. Every country experienced loss of life, unemployment, and abruptly contracted economies. Governments everywhere were figuring out how to respond.

By the end of April 2020, the United States alone had already approved spending $3 trillion to get its economy moving, but by fall it was clear more was needed. The looming uncertainty is whether governments around the world will create economic stimulus programs that tackle the climate emergency. This is an opportunity to jump-start the economy in ways that promote ecological balance and healthy communities. The European Union and South Korea have endorsed passing or strengthening Green New Deals to put people back to work building renewable power generation, making buildings more energy-efficient, and growing diversified, sustainable agriculture.

We stand at a history-making moment. If governments fail to align economic stimulus with mitigating the climate emergency, climate risk and suffering will intensify. Going back to business as usual will mean that companies prioritize existing liabilities, payrolls, and shareholder payouts over investments in long-term sustainability. This will result in more climate-disrupting emissions and a failure to hold down temperature increases. On top of the huge stock market losses experienced during the early pandemic, a failure to transform the economy as we recover will lead to even greater risk of climate-related losses to life and business from flooding, sea-level rise, extreme temperatures, drought, productivity declines, and violent weather.

Moving the world off fossil fuels and powering our system on renewables requires governments to design economic stimulus programs that reject business as usual in our climate-changing emissions. If we also rethink how businesses can be organized to produce broadly shared wealth, we can prosper while we transform. Our choices will involve tradeoffs between the extent and speed of mitigating (reducing) climate change, adapting to disruptions, suffering, and prospering. The more aggressively we mitigate in this decade, the less we will need to adapt, and the less we will suffer. But what if our approach also meant we could prosper? This book explores how we can prosper while mitigating climate disruption, and envision how this intentional evolution will enable us to suffer less and have less need to adapt to extremes.

To prosper in the coming era will involve taking the democratic *political* experiment to the next level and building a democratic *economy*: one that is of the people, by the people, and for the people. Such an economy would involve creating businesses that reflect broad-based ownership and participation in decisions by workers, weaving sustainability and deep decarbonization into the core purpose of businesses, improving the communities where business is conducted, and sharing prosperity more equitably across nations, races, and genders. If we keep these goals in mind, the resulting system could be one that is democratic, just, and sustainable.

PROSPERITY IN THE FOSSIL-FREE ECONOMY

Introduction: The Dream from 2035, Looking Back on 2020

THE YEAR 2020 WAS AN INFLECTION point when the world changed dramatically and unexpectedly. A novel coronavirus spread from country to country and sent billions of people into homebound isolation. We buried so many. The virus showed us that no matter where we live or how famous we are, all are connected and vulnerable. It also revealed the disparities in our access to health care, as the poor, people of color, migrants, prisoners, and the elderly suffered the most.

We saw how the free market and individualism led to shortages in tests, protective gear, face masks, medicines, food, and even toilet paper. A few socially oriented companies quickly shifted production from things like snowboards to masks, which they donated to hospitals. Some governments flexed almost forgotten muscles and directed companies to produce ventilators and other essential supplies.

Some people hoarded food and supplies out of fear and a sense of disconnection. Many more started building new relationships among neighbors, volunteered to bring food and medicine to those most at risk, and planted gardens to keep people from going hungry. Nations closed borders, students from kindergarten to graduate school attended classes online, economies ground to a slow crawl, and the ranks of the unemployed swelled (in the first month alone, twenty million Americans lost their jobs). And yet, we remained home and headed off the worst-case scenario. In order to save lives, we slowed ourselves into the worst global economy since the Great Depression of the 1930s.

Although fighting the pandemic was in many ways horrific, as people stayed home and factories shuttered operations, ecosystems started to recover. The citizens of India could see blue skies and the Himalaya Mountains for the first time in thirty years. From Los Angeles and New York to Milan, and from New Delhi to Beijing and Seoul, the air was free of unhealthy smog and smoke. Urban rivers ran clear. Wild animals started to explore vacant city streets.

This unprecedented experience seeded myriad ideas about how to work with minimal travel, to power our lives without pollution, to feed ourselves from local farms and gardens, and to ensure everyone had access to basic health care and safe homes. Scientists eventually created a vaccine and the virus crisis passed, while the larger existential threat of climate change remained.

We emerged from our isolation changed, having learned the power of collective and cooperative action. Communities where people collaboratively engaged—stayed home, maintained physical distance when out, wore face masks, washed hands—flattened the exponential growth curve of the coronavirus most quickly. We then applied the knowledge we had gained to the challenge of reducing greenhouse gases (GHGs). The events of 2020 disrupted business as usual. It was the point where the upward trajectory of GHGs shifted dramatically downward. The community bonds we forged through singing together out our apartment windows, sharing supplies, and supporting health care workers gave us the surprising experience of social cohesion. When police brutality against an unarmed Black man—a far too common occurrence in the United States—resulted in his death, people rose up in protest. It turned out that cooperation was more contagious than isolation, and this powered a grassroots movement for greater ecological protection and equality when it came time to restart our economy.

This movement pushed governments to create economic stimulus programs, drawing on knowledge gained from the 2007–2008 global financial crisis. We avoided the past mistakes of implementing interventions that ended up primarily benefiting the wealthy while socializing financial risk. We knew we could not go back to economic growth that sacrificed ecological balance while sickening and dividing us. There were initial stumbles, with some leaders arguing that addressing environmental injustice would distract us from the task of jump-starting the economy. Ultimately, strong governments aligned economic stimulus with the epic challenge to mitigate climate change and ensure that the most marginal were not left behind.

India doubled down on its renewable energy target and incentivized co-operatives to build clean-energy infrastructure, which brought electricity to some people for the first time and paved the way for improving the overall quality of life. China used its central control to mandate an end to burning coal and innovated to electrify transportation. As envisioned by the Paris Agreement, the Green Climate Fund provided critical capital for the Global South. Although African nations have historically negligible carbon emissions, Morocco built the world's largest concentrated solar facility, while other countries created microgrid solutions and designed low-carbon and resilient infrastructure to reduce poverty and climate change. Deforestation in the Amazon came to a screeching halt as regenerative agriculture took root on degraded lands to draw down carbon; today the Amazon rain forest fuels a "green economy" where trees are worth more standing than fallen and people earn a living from forest-based products. The United States, the European Union, and South Korea implemented Green New Deals to put people back to work building renewable power generation, green infrastructure, and diversified, sustainable agriculture. When governments bailed out airlines, they required carbon neutrality by 2030. The United States stopped its $15 billion annual subsidy of oil, gas, and coal. Instead, as fossil fuel companies teetered on bankruptcy, Congress conditioned economic stimulus funds on the government acquiring controlling interests in the companies, with an explicit plan for a just transition to renewables and managed phase-out of fossil fuel production in short order.

The just transition to a new, resilient economy involved democratic input from workers; labor unions; impacted communities; governments; scientists; and others. The cooperation we learned during the pandemic was essential in the transition off fossil fuels. The path-breaking cooperatives featured in this book sparked ideas for how to structure social enterprises to lead the way in providing triple-bottom-line returns (financial, social, and environmental). We expanded public and employee ownership to bring the cooperative pathbreakers to scale, putting millions back to work. We grew these jobs in businesses structured to reinforce democratic practices. Enterprises now emphasize democratic ownership by employees, consumers, and producers, with cooperatives playing a substantial role. Wealth is shared more equitably, hours are reasonable, and we practice our democratic voice (one person, one vote) at our workplaces.

This democratic economy encourages people to spend more time outside of work engaging in political life. We passed laws that require public financing for elections, with a cap on overall spending and term limits to counter career politicians. We placed strict limits on the revolving doors connecting elected office, government agencies, and lobbying positions. We strengthened transparency and open meetings in governance at all levels.

Millions went back to work as owners in family-sustaining jobs in democratic enterprises that deeply decarbonized our economic activity. We worked to retrofit buildings, electrify transportation and heating, and diversify renewable energy sources and storage. We grew sustainable food, planted and managed trees to absorb carbon, and modified cities to adapt to rising seas and tumultuous weather. These actions drove down GHGs by the major emitters.

In 2035, our global GHG emissions are plunging toward zero, and it looks as if we'll reach that goal before 2050. The air we breathe isn't giving us asthma; we have more free time; and yes, we are happier than back in the 20-teens. We're in a green and affordable community where, by checking a smart phone, one can find a shared bike or e-scooter within a few blocks, call up an autonomous vehicle for a lift, or ride public transportation that is safe, clean, and reliable. With the rise of new mobility services, getting around is cheaper than when we depended on private cars. This means less debt and more financial security for our families. It enables the elderly to live independently longer. Congestion is reduced, and since all the vehicles are electric fleets, powered by renewables, it is incredibly quiet here! The sound of birds and smell of clean air are now common.

The children in our lives walk or bike to neighborhood schools that are high-quality and diverse, reflecting a housing policy that creates mixed-price, inclusive housing in each neighborhood and rewards valued teachers with middle-class salaries.

The pandemic forced universities to teach online and medical offices to see patients on video calls. It was difficult at first, but now it is common to gather online for everything from classes and public lectures to parent-teacher meetings and doctor visits. This has reduced the cost of higher education and health care, as well as making them more accessible.

Many people who first started gardening to enhance their food security during the pandemic continued to do so. We have fresh, locally produced food at multiple convenient neighborhood markets, and have developed relationships with the farmers who are producing our food. Single-use plastics have been banned for quite some time, and the stuff we buy lasts longer. Most nations promote a circular economy and require companies to take back their products at the end of their useful lives. This has led to companies designing products that are more durable and repairable.

After being physically isolated during the coronavirus crisis, we value our civic life more than ever. We don't get self-esteem from flamboyant consumerism, and we live in smaller, more energy-efficient homes near urban green spaces, which gives us greater opportunities to interact with our surrounding community. We now have more money to spend on services that make life enjoyable: hobbies, music and the arts, experiences, and restaurants. We have high levels of voter turnout, not as the culmination of our political involvement, but as one piece of our broader civic participation.

In the depths of despair in 2020, we saw another world was possible and changed the trajectory. As we left fossil fuels behind, we built a better life in this new democratic economy defined by shared prosperity. Beyond financial strength, our prosperity manifests in a high quality of life, health, happiness, strong relationships, trust in community, shared meaning and purpose, and satisfaction at work.

PART ONE

THEORIES, PRINCIPLES, AND LAWS

FOR SUSTAINABLE ENTERPRISES

Our Present Challenge

Current Path: Climate Disruption

We are at a critical juncture in our global development that calls for a fresh approach to cooperating for a livable planet. Because humans have not existed outside the relatively stable climate conditions that marked the Holocene period, anxieties are rising about our future. The international scientific consensus indicates we are on a deadly trajectory of our making because we are releasing increasing amounts of GHGs as a result of our economic activities. Atmospheric levels of carbon dioxide (CO_2) are at their highest levels in over 800,000 years and continue to rise. Global average temperatures are 1°C higher than pre-industrial temperatures, and the sea has been rising, warming, and becoming more acidic as it has absorbed CO_2 and heat.[1]

At the close of the 20-teens, we saw biodiversity worldwide in an alarming decline, as reported by the Intergovernmental Science-Policy Platform on Biodiversity and Ecosystem Services. The Amazon and Australia were engulfed in fires, releasing dangerous air pollution and reducing the capacity of established trees to absorb more carbon dioxide. To make matters worse, the global problems are broader than the environmental challenges. The United Nations' unanimously adopted Sustainable Development Goals of eradicating poverty and reducing wealth inequality, to name a couple of fundamental priorities, will be harder to reach due to climate disruption.[2]

If we continue business as usual, the path we're on leads to climate-related risks to health, livelihoods, food security, water supply, personal security, and economic growth. To put it more clearly, if we do not change course in this decade, within the next generation parts of Earth will be uninhabitable due to extreme heat, all the life-supporting coral in the oceans will die, and the Amazon and Australian fires of 2019 will be the norm.[3]

The government, the private sector, and individual consumers have brought the world to this climate emergency, but some are more complicit than others. Although in the 20-teens it has been commonplace to hear that every individual has contributed to the causes of climate disruption and that one can make a difference by individual actions, like installing more efficient lighting, these are partial truths. To the extent that this claim implies that we all share equal responsibility for the crisis and power for correcting it, it is dead wrong.

Climate disruption is a study in inequity. The United States and the European Union have cumulatively contributed the most climate-altering gases—25 percent and 22 percent, respectively. On an immediate annual basis, in 2019, China emitted the most (more than 25 percent), followed by the United States (15 percent) and the European Union (10 percent). By contrast, the world's poorest people and nations have contributed the least to this global problem (less than 1 percent of CO_2), but have the least amount of protection when hit by extreme weather and rising seas.[4]

The responsibility within the private sector is similarly concentrated. In 2013, the geographer Richard Heede's calculations revealed that just ninety companies are responsible for two-thirds of all the GHGs emitted between 1854 and 2010 when one considers all the carbon extracted and supplied. Despite increasing warnings about the link between GHGs and climate disruption, more than half those emissions have occurred since 1986. Further, based on 2016 air-emissions data, of the 100 largest U.S. power producers, a mere sixteen companies are responsible for 50 percent of the total CO_2 emissions for electricity. Yet, a targeted approach aimed at these entities has not been on the U.S. government's short-list agenda. Although temperatures and sea levels continue to rise, and people face greater extremes in droughts and floods, the government appears to be allowing business to continue as usual.[5]

Future Path: Deep Decarbonization and Sustainable Development

We not only know the causes and the culprits, but also the solutions to climate disruption. There is a growing awareness around the world that we must act decisively to move the global economy off fossil fuels immediately and rethink and redesign our relationship with natural resources. The nations and companies that are the largest sources of emissions, both from a cumulative and current annual accounting perspective, should take the lead in the transformation.

The United Nations has provided an international consensus about the system-changing sustainability the world needs to implement throughout the 2020s and beyond. The 2018 report of the UN Intergovernmental Panel on Climate Change (IPCC) urges swift decarbonization to get to net-zero CO_2 as soon as possible. The IPCC's recommendations and the Sustainable Development Goals are an international call for action.[6]

To deeply decarbonize, we need rapid transitions in energy, farming and forestry, transportation, urban infrastructure, and industrial systems. Such transitions depend on using fewer resources, creating greater energy efficiencies, and powering our energy needs with renewables. An essential aspect of the redesign is integrating the Sustainable Development Goals to address the socioeconomic implications of transitioning to a clean-energy economy. Fossil-fuel–based workers will be displaced, and a just transition is one that includes retraining those workers for the fossil-free future. The faster we can reach and sustain net-zero global CO_2 emissions before 2050, the greater chance we have to maintain a livable planet.[7]

Meeting this challenge will require transitions that, according to the IPCC, "are unprecedented in scale, but necessary in terms of speed, and imply deep emissions reductions in all sectors." Due to the coronavirus pandemic that spread across the world, economies abruptly slowed as people were forced to stay home. As of this writing, fighting the pandemic is ongoing and the death toll keeps climbing. Yet, there has been one positive outcome: as of April 2020, the International Energy Agency (IEA) expects an 8 percent drop in global CO_2 emissions, taking them to 2010 levels. This is a larger drop than occurred during the global financial crisis of

2007–2008. In order to continue on that trajectory and use this as a turning point for the world, investments to stimulate the economy need to be aligned with a clean-energy transition. The global Task Force on Climate-Related Financial Disclosures estimates $1 trillion per year for the foreseeable future should be invested to transition off fossil fuels. This is a matter of getting the investments not just into the right sectors, but into the optimal business structures. This book translates the sweeping climate and sustainability goals into a blueprint for designing businesses to deeply decarbonize economic activities and aim toward meeting the Sustainable Development Goals.[8]

We are entering a renewables revolution. Several announcements by business leaders in early 2020 show that the headwinds are shifting rapidly. BlackRock, the world's largest asset manager, with $7 trillion under management, is making sustainability integral to their risk management and exiting high-risk investments such as thermal coal. Tri-State, a U.S. power producer with some of the dirtiest power in terms of CO_2 per unit of electricity, is retiring coal facilities early and adding new renewable generation to supply 50 percent of its electricity with renewables by 2024. More and more businesses are setting goals to power their operations on 100 percent renewables in short order, such as Walmart, Apple, and Google. As we rethink business as usual in terms of meeting global energy and resource needs, we also need to rethink how we structure businesses to best accomplish the multiple co-benefits of building democratic participation, reducing poverty, eliminating hunger, and increasing economic equity.

How Did We Get Here?

Change on the scale and in the time frame needed depends on governments, businesses, and individuals making the shift to deep decarbonization and sustainable development a top priority. To understand how to move forward, and the best vehicles within the private sector for these system changes, we need to understand the path dependence arising from our political economy and the limits of its dominant business structure: the investor-owned corporation.

The rise of capitalism in seventeenth- and eighteenth-century Europe was a reaction to a concentration of wealth and exclusive property rights and political influence. Capitalism redefined ownership and power. In a move away from feudalism, people were able to rely more on ability, effort, and willingness to pool financial resources and take business risks, instead of the disempowered gamble of being born into a land-owning class. Capitalism, famously advocated by Adam Smith in *The Wealth of Nations* (1776), emphasized a system organized around self-interest and competition. Freeing people to pursue profit, as the theory goes, leads to investments in the most productive and useful activities. When businesses compete in this pursuit, society benefits through reduced prices, improved quality and choices, and growth, which in turn increase wages.[9]

Quality of Life and Economic Growth Diverge

This economic system set the world on a meandering path, with ups and downs, but trending toward greater levels of equality and quality of life until about the 1970s. After that time, people living in a wide variety of capitalist countries throughout the world started to experience a disconnect between quality of life and affluence. A 2013 study of seventeen countries, containing over half of the global population, by Ira Kubiszewski and others provides a lens to view the global economy and quality of life. They analyze gross domestic product (GDP) growth alongside twenty-four quality-of-life measures, called "Genuine Progress Indicators," such as crime, poverty, equity, health, and pollution. Their research shows quality of life stagnating and not increasing with GDP growth after the 1980s in multiple countries. Put more simply, the average person in the United States or the European Union in 2019 was not better off than in 1979, despite incredible growth in terms of the stock market and GDP.[10]

Wealth Inequality

So where are all the benefits of GDP growth and stock market value going? Thomas Picketty's *Capital in the Twenty-first Century* (2014) opens a breathtaking view of rising economic inequality as a global phenomenon. Material wealth in the period from the 1970s to the present has disproportionately gone to those who invest and own stocks, rather than to those who rely on their labor to earn money. Picketty tabulates shares of overall income and wealth over long time periods in a variety of countries. Looking at the United States over a 100-year period, from 1910 to 2010, the share of all income taken by the top 10 percent of households climbed steeply until the stock market crash of 1929. From the mid-1940s to the mid-1970s, the top 10 percent of Americans owned a smaller share of total income at a fairly stable level, but since that period inequality has accelerated. In 2007, the richest 10 percent took half of all the income. When narrowed to income generated by work alone, Piketty notes, "the level of inequality in the United States is 'probably higher than in any other society at any time in the past, anywhere in the world.'" While not as extremely unequal, charts of the same time period for South Africa, Argentina, Indonesia, India, Colombia, and China follow a similar trajectory, with the top 1 percent taking a growing share of income after 1970. In 2010, the top 1 percent in the United States and Europe owned a third and a quarter of all the wealth, respectively.[11]

Wealth in the United States is unequal based on race/ethnicity. The Urban Institute analyzed data on wealth from 1963 to 2016 and showed a gap during that entire period. In 2016 an average white family held seven times more wealth than a Black family and five times more than a Hispanic family.

These disparities are sticky. Wealth and income inequality are intertwined with social mobility. Research in 2017 by Raj Chetty and his colleagues concluded that the chance of moving from the bottom to the top income quintile are lower in the United States (7.5 percent) than in the United Kingdom (9.0 percent), Denmark (11.7 percent), or Canada (13.5 percent). In the United States, this research further showed that only half of the people born in 1980 have surpassed their parental family income, and one's address impacts mobility. So, where one lives and what one inherits matter in terms of chances of improving standard of living.[12]

Income Gaps Between Workers and CEOs

In another measure of inequality, the income gap between CEOs and workers has grown by an order of magnitude since 1978. In that year the ratio was 30 to 1, but by 2013 an average American CEO made about 300 times more than workers in the same company. As CEO pay has grown, workers' wages have stagnated, despite substantial increases in productivity.[13]

In some respects, we have returned to a form of feudalism. Three features stand out: 1) our economic activities are damaging the life-support structure of the Earth through climate disruption, species extinction, and pollution; 2) the countries most vulnerable to climate disasters are the poorest ones; and 3) the gap between the world's richest and poorest people is growing. We are once again living in a time when birth plays a significant role in success, access to opportunities, resources, and health.

Hyper-Capitalism and Investor-Owned Corporations

We are now functioning within a sort of genetically modified capitalism, a hyper-capitalism, that has come to shape the world. In this type of capitalism, producing financial returns for investors is the top priority; information, goods, and people move around the globe with a speed not previously known; business lacks a commitment to a specific place; consumerism orients people's lives and values; and governments have very limited will to regulate or tax the multinational corporations that shape this reality.[14]

The entity that dominates our hyper-capitalist existence is the multinational investor-owned corporation. The design of this type of enterprise reflects capital bias; it prioritizes and empowers those who provide capital, the shareholders. Distant shareholders can control investor-owned corporations, even those who speculate and own shares for moments in time, while the employees at those corporations and the communities around them are left to live with the consequences.

Corporate consolidation and globalization have increased corporate power and weakened the state's control. According to Global Justice's analysis, which compared 2017 revenues, sixty-nine of the top 100 economic entities in the world are corporations rather than governments. These corporations have been very effective at creating new products, encouraging consumer demand, and growing GDP and shareholder wealth, while holding down labor costs and externalizing environmental costs to the public.[15]

The hyper-capitalist system depends on policy to advance its capital bias. The concentration of wealth at the top is reinforced by overpowering the common good in policy decisions. We see this manifest in a variety of ways that further exaggerate income inequality and environmental degradation. For instance, in the United States, the government rewards capital with lower taxes on capital gains and punishes labor with relatively higher taxes on income from work. The law allows wages to stagnate while GDP grows. The government bailed out big banks while ordinary homeowners lost their homes to foreclosure during the 2007–2008 global financial crisis. Large-scale corporate agriculture secures tax benefits and subsidies from the federal Farm Bill, while small organic farmers fend for themselves. The government encourages, permits, and subsidizes continued extraction and use of fossil fuels while the world's climate becomes increasingly unstable.

Whereas increasing the minimum wage, taxing stock gains more than labor, and putting a price on carbon would be helpful in curbing some externalities, these measures only assist at the margins; these policies don't reach the fundamental design of business. Marjorie Kelly and Ted Howard correctly observe in *The Making of a Democratic Economy* (2019) that holding down wages and pushing environmental costs onto the public are not a glitch but a design of the current economic system and its primary institution, the investor-owned corporation.[16]

This is not to say corporations and sustainability are incompatible. When there is a strong business case for sustainability, we see corporate leadership. If multinational corporations commit to environmental sustainability and corporate social responsibility more broadly, the world will doubtless be a better place because of the sheer scale of their potential impact. It is essential for investor-owned corporations to lead when it comes to environmental sustainability, and it is encouraging to see more corporations moving in this direction. Fifteen years ago, in *Green to Gold* (2006), Daniel C. Esty and Andrew Winston argued that no company can afford to ignore environmental issues due to a variety of developments, including increasing clarity about environmental problems, chronic poverty in parts of the world, transparency, and rising levels of public expectations for corporate social responsibility. They asserted that major corporations must do more than comply with the law: industry leaders must incorporate environmental considerations into all their operations. That Walmart has committed to obtaining 100 percent of its electricity from renewable energy by 2035 (currently operating over 11,000 stores) and producing or procuring 7,000 GWh of renewable energy by the end of 2020 indicates the world is at a significant and exciting transition point. Even absent legal mandates, when there is a business case for environmental protection, as there clearly is for the private sector's embrace of renewables, investor-owned corporations can make strong advances to meet the goals.[17]

However, legal design and norms make this leadership and commitment far too infrequent and shallow. Investor-owned corporations do best what they are designed to do: maximize profits for shareholders. When this purpose conflicts with protecting the environment, paying workers better wages, involving stakeholders in democratic governance, and being a committed neighbor in the local community, those stakeholders tend to lose out to shareholder profit. The overarching profit goal has left the public to absorb many negative externalities: loss of life and property from climate disruption; shuttered factories in the search for bigger government incentives; and wealth inequalities resulting from stagnant wages and greater accumulation and concentration of wealth among a relatively small number of owners and investors. While sustainability reporting is now commonplace across the world's largest corporations, this practice has failed to alter the underlying business models, and we are seeing increases exactly where sustainability calls for decreases: in CO_2 emissions and widening wealth gaps. If we want to reorient the economy to the systemwide global priorities of rapidly decarbonizing and meeting the UN's Sustainable Development Goals, we need to redesign businesses to accomplish this shift.[18]

New Operating System

Transforming the corporation must be at the center of building the new operating system for a democratic, sustainable, and equitable economy. The United Nations has provided an international consensus about the system-changing sustainability the world needs to implement throughout the 2020s. In essence, businesses must be designed to serve the common good, which requires a broader engagement with stakeholders, prioritizing labor and giving workers more of a voice in governance, and dampening the current capital bias.

Creating a better future involves reimagining the role of corporations. The limitations of the current model become apparent when the business case for sustainability is weak, depends on long-term calculations, and conflicts with the priority to deliver quarterly profits for shareholders. This is where the social enterprise, with its legal design that advances the triple bottom line holds potential for greater and faster implementation of the Sustainable Development Goals while deeply decarbonizing. The ownership structure of businesses is a design choice and not an immutable fact. In *Frugal Value* (2017), Carina Millstone observes that the environmental movement has neglected ownership design, missing the important role it plays in driving corporate decisions about sustainability. Millstone argues for businesses that have fewer shareholders, but ones who are closer to the business and engaged and committed to the common good of a social and environmental mission.[19]

Cooperatives and Social Enterprise

It is an exciting time to envision future enterprises aligned with the climate imperative to transition the world off fossil fuels while distributing wealth more equitably. Ideas and examples exist for major shifts, including new corporate forms to create stakeholder enterprises that are community- and environmentally friendly, as well as democratic in the workplace. B-certified or benefit corporations started appearing in the last decade and hold promise; however, cooperatives have been around for centuries. They are one of the oldest and most enduring business forms. The iconic American, Benjamin Franklin, started a cooperative insurance company in Philadelphia in 1752, before the colonies broke away from Great Britain, and it still exists today. In a country fixated on Wall Street, people are often surprised to learn that today more people in the United States are members of cooperatives than participants in the stock market.[20]

Cooperatives are a form of enterprise distinct from investor-owned corporations in many ways. Cooperatives are enterprises with a purpose based on internationally shared ethics, values, and principles. Built to serve their members on the democratic foundation that each member has the same voting power, they operate in most sectors of the economy and in most countries in the world. This democratic control tamps down external control of a cooperative and ideally stimulates member participation, which has positive spillover effects for participating in the political and economic life of a community. Perhaps that's why the United Nations asserts that cooperative businesses serve an important purpose in economic and social development around the world. The combination of their legal structure, purpose, and values and principles set them apart. This is not to say that all cooperatives operate as designed. But with strong leadership committed to the Cooperative Values and Principles, as explained below, cooperatives have a tremendous potential to fundamentally alter the sustainability of how we do business and how much prosperity will be broadly shared as we deeply decarbonize.[21]

As public concern is rising about climate change and the inequality of wealth concentrated in accounts of the One Percent, cooperating is emerging as a unifying framework for the creation of sustainable, equitable, and democratic institutions and economies. We are at a time when organizations and entrepreneurs everywhere are increasingly looking for guidance and inspiration to establish and build social enterprises. Renewable energy, food and agriculture, water, finance, and trade are all areas where cooperatives could provide essential platforms for mass participation in deeply decarbonizing the economy. We need to approach legal design of ownership and governance with a clear understanding of social enterprise and how to maximize its co-benefits for democracy, society, and the environment.

Businesses, governments, and consumers want to participate in the shift to renewable energy, electric vehicles, locally and organically produced foods, reusable goods, and more. Importantly, the challenges before us call for every one of us to enthusiastically engage in the system-changing work the climate emergency and sustainable development require. As platforms for participation and resource sharing for greater efficiencies, cooperatives are a form of business organization that may be best positioned to facilitate broad engagement in rapid decarbonization and attaining the Sustainable Development Goals.

By design, cooperatives can facilitate a fairer division of wealth among producers and consumers, create more owners, and teach democratic practices among the members. They offer the opportunity to reinvest surplus revenue into the community instead of capturing it for shareholder profit. This is not to say cooperatives are a panacea. Like any business, they are composed of and run by imperfect people who possess all the failings of the human condition. Yet, with a strong legal design that reflects Cooperative Values and Principles, they have the flexibility to unapologetically pursue the triple bottom line free of short-term stock market pressure.

Terminology

Sustainability and *democracy* are contested terms that merit an explanation at the outset. As people grapple with how to respond to increasing awareness about the environment, they use "sustainability" ubiquitously. The term is so malleable that it can mean almost anything, from increasing profits to adding a recycling program at the office of a company whose primary business is oil production.

Sustainability is most profoundly understood as building the triple bottom line of protecting the environment, society, and the financial health of the enterprise into the overarching organizational purpose and day-to-day practices. The concept involves a long-term view and avoids harm to future generations. Creating a single organic product line or an employee transportation policy, while a step in the right direction, falls short of the system-changing sustainability the world needs. Instead, we need to make sustainability part of the DNA of the business enterprise, its core purpose, so it infuses the business model and affects every decision, employee, and aspect of day-to-day implementation. This understanding of sustainability is the one that will animate this book and be explored in greater depth in the upcoming chapters.

As discussed in this book, democracy is both an aspiration and a governance style. Facilitating broad participation in government or an enterprise can be direct or representative. Direct democracy utilizes the principle of one person, one vote on matters of shared importance. Representative democracy exists when one votes to select a representative of a group to have a voice on a smaller decision-making counsel, congress, and so on. A core feature of cooperatives is democratic governance among the members, whether they are workers, consumers, producers, or a combination. Typically, cooperatives use a combination of direct and representative democracy for workplace governance. They can be designed for different degrees of member involvement and control. Some are collectives where each member is involved in all decisions, but most use a hybrid direct/representative model of participating in broader policy choices. As cooperatives grow, members tend to elect a professional board and management team to make day-to-day decisions.

As David Orr contends in *Dangerous Years* (2018), the climate emergency "will require stable, effective, and truly democratic governments that protect natural systems and human rights alike, including those of future generations." He describes the erosion of democracy and governing capacity in the United States, and observes that the nation's founders could not have foreseen the tyranny of the corporation and its corrupting influence on politics, free press, and mass consumerism. Kelly and Howard argue the founders focused on government structures and functions, and that we now need to complete that revolution with a democratic economy. Some define a *democratic economy* as one where economic institutions are democratically controlled. Kelly and Howard go further and define a democratic economy as a system intended to serve the common good, whose fundamental design "aims to meet the essential needs of all of us, balance human consumption with the regenerative capacity of the earth, respond to the voices and concerns of regular people, and share prosperity without regard to race, gender, national origin or wealth."[22]

If we are to reinvigorate democracy in political life, we need business structures that teach democracy in work life. This book explores that and other notions about democracy and sustainability.

Core Questions and Research Design

Prosperity in the Fossil-Free Economy wrestles with core questions about enterprise design and sustainability.

What laws and systems of private governance act as incentives, mandates, or barriers to environmentally sustainable enterprises? How do sustainability and democracy pathbreakers build their purposes into the legal design of their businesses? Can the cooperative model of collaboration be harnessed and multiplied to transform and deeply decarbonize the economy on a short time scale?

The book uses case studies to explore these and related questions through a comparative analysis of sustainability leaders in cooperatives in Spain and the United States. Although my study centers on two countries, its findings and lessons are applicable in many jurisdictions around the world searching for enterprise design that is more democratic, equitable, and environmentally sustainable. In that spirit, the term "government" is intended to be broadly understood to describe any jurisdiction unless specified as the European Union, United States, or so on. The case studies show the complexity of real-world experiences, identify successes and failures, and extract lessons for a livable planet.

Spain and the United States are featured because each has thriving cooperative sectors, while adopting different legal approaches. Each country has sustainability leaders that offer lessons for the entire world. Spain is home to the largest worker-owned cooperative in the world (Mondragón), boasts the most organic acres under cultivation in Europe, and exhibits a strong shift to renewable energy. The United States has the most financially successful organic farming cooperative in the world, a long history of cooperatives delivering utility services like electricity and water, and numerous cooperative memberships, all coexisting in an economy dominated by the investor-owned corporation and Wall Street. In 2019, Spain had close to 19,000 cooperatives, serving approximately 6.8 million members and providing more than 300,000 jobs. In 2009, the United States had roughly 30,000 cooperatives, serving 350 million members, owning $3 trillion worth of assets, with revenues of nearly $654 billion, and providing 2 million jobs.[23]

In each of these countries, cooperatives are delivering the basic necessities of life, while strengthening social cohesion and democratic participation. The coopérative exemplars in the case studies discussed in Part II of this book operate in the key economic sectors we need to redesign to deeply decarbonize the global economy: energy, food and agriculture, water, trade, and finance. They provide a proof of concept for integrating sustainability in the purpose and day-to-day operations of social enterprises on the road to deeply decarbonizing.

The case studies highlight what distinguishes each cooperative from investor-owned corporations. They show the legal design of each of the businesses to distill how sustainability pathbreakers lock in their mission and purpose in legal documents, how they structure democratic participation among members, and how this influences their day-to-day sustainability practices. Some jurisdictions in which they are operating have stronger laws that support the establishment and development of sustainable cooperatives. Some cooperatives have better legal designs to lock in their environmental sustainability purpose and infuse it into their day-to-day operations through leadership changes. Some have impressive systems of democratic participation, across vast distances, that pair attention to personal relationships in small groups with the utilization of technology for virtual interactions with large groups.

These sustainability exemplars motivate and inspire others to accelerate the transition to deep decarbonization. Some readers of these stories will attempt to replicate the models, some will design further research, and some will advocate for law and policy changes to encourage their flourishing. Throughout this book, the question is how we might utilize cooperative ownership more effectively to transition the world to a new system that mitigates climate disruption and encourages prosperity, and that prioritizes environmental sustainability integrated with promoting democratic participation, employment, dispersed ownership, and equitable compensation.

Corporate Purpose and Governance
for a Livable Planet

CHAPTER 2 CONSIDERS THE ROLE of business law within the field of environmental law and makes the case that environmental law has largely ignored business law and economic system dynamics, at its peril. Integrating the work of business law scholars, it explains shareholder primacy as a law or norm of investor-owned corporations, giving arguments in favor and against. It discusses how the norm shapes corporate purpose and undermines Sustainable Development Goals and decarbonization. It then summarizes proposals to transform the corporation, including creating an environmental priority principle in law, defining corporate purpose to benefit stakeholders, and sparking social enterprise for innovative entrepreneurs and conversions of existing businesses.

Environmental Law Meets Business Law

The relationship between corporate purpose and sustainability is like the connection between a home's insulation and staying warm in winter: it is hidden from view, but essential for the overall result. For many involved in environmental law, corporate purpose and its relationship with environmental outcomes have been obscured.

Over the past forty years, the field of environmental law has focused primarily on laws directly related to controlling pollution or protecting natural

resources. This focus has the benefit of tailoring the law to reflect an expertise in highly technical and scientific questions about ecology and public
health. The government uses environmental law to set standards; require
permits or trade credits to pollute; and issue licenses to harvest, mine, or
use natural resources. The law tends to fall into various typologies: documenting and registering environmental harms, establishing important processes for public participation and transparency, setting limits, using market mechanisms or permits to achieve the goals, and mitigating the worst
excesses of economic activity. While an oversimplification, in general there
are chronic problems associated with the political will to set standards at the
right level and to allocate resources to enforce those standards.

Environmental law, as conventionally understood, corrects market failures but does not fundamentally transform the economic activity it regulates. This is at the heart of the all-too-frequent *jobs versus the environment*
scenarios that dominate environmental disputes. The government is in an
ongoing internal conflict revolving around the need to promote jobs while
protecting the public from the environmental consequences of those jobs.
Environmental law, cast as a bit part in this drama, is fighting a losing battle
against a growing population, driven by consumerism to fuel greater stock
market and GDP gains, and fed by expanding resource use.

Globally, the human appetite for consumption is voracious. Consumers
increased their buying from $1.5 trillion in 1900 to $24 trillion in 1998. In
terms of volume of raw materials used in the economy, in 2010 we used
65 billion tons, and by 2020 this figure is expected to grow to about 82 billion tons. As I documented in my earlier edited book *Law and Policy for a
New Economy: Sustainable, Just, and Democratic* (2017), a growing population
pursuing GDP-measured economic growth, powered by fossil fuels, is on
a collision course with maintaining a livable climate. According to the International Energy Agency, the combined growth in population (35 percent)
and per capita GDP (60 percent) led to a dramatic increase in global carbon
dioxide emissions (56 percent) between 1990 and 2013.[1]

Even the corporate leadership group of the Global Reporting Initiative,
the most widely used sustainability reporting platform in the world, worries
that despite almost universal adoption of sustainability reporting by major

corporations, fundamental business models have not changed. As a result, the group notes that sustainability reporting has not changed the increasing trend lines for:

- annual consumption of raw materials,
- pollution,
- global temperatures, and
- concentrations of wealth.

We need to fundamentally decouple or disconnect use of natural resources and GHG emissions from economic growth. In other words, meeting our current needs should not compromise the ability to meet future generations' needs. This is a redefinition of progress that, in John Dernbach's words, "calls on us to address issues such as climate change and high consumption of materials and energy that our environmental laws do not fully address." In order to pursue this new framework, sustainability, at its best, should be aimed to go beyond environmental law and function at an earlier stage in economic activity.[2]

Environmental outcomes do not only depend on the quality, clarity, and enforceability of environmental law. Perhaps more important are the motivations and incentives of the primary economic actors, both individuals and corporations. After all, they determine whether they will go beyond the letter of the law, merely comply, ignore, or bury litigants in obfuscation.

Sustainability should inform the fundamental purpose, mission, governance, norms, policies, and practices around which a business, its products, and services are designed. Yet, most U.S. environmental law scholarship ignores business law in favor of analysis of litigation over the meaning of environmental standards, federalism in environmental decision making, enforcement by the government or citizens, and debates about property rights. What we have missed, in focusing our attention on these assuredly important details, are the larger economic forces at play and the extent to which enterprises incorporate environmental considerations into their business model and day-to-day operations, beyond the minimal requirements of environmental law. Fundamentally, we have failed to recognize and address the conflict between the dominant corporate structure and positive environmental outcomes.

Some environmental scholars are starting to move beyond these environmental law boundaries. James Gustave Speth in *America the Possible* (2012) identifies a dozen features of the American political economy where transformative change is essential in order to prioritize improving conditions for the environment and people. *Law and Policy for a New Economy: Sustainable, Just, and Democratic* expanded the notion of environmental law to include laws that significantly impact environmental results. The authors included in the latter volume argued that the challenges are much broader than setting parameters around pollution, and go to the heart of the dominant global political economy. They explored the legal infrastructure needed to support the emergence of a new economy across a variety of major areas—from energy to food, common-pool natural resources, and shifting investments to capitalize locally connected and mission-driven businesses.

Sarah Light's article "The Law of the Corporation as Environmental Law" similarly broke new ground in 2019 by examining the role of U.S. securities, antitrust, and bankruptcy laws in impacting corporate environmental behavior. She discusses how business law shapes "norms, markets, and architecture" to impact corporate environmental decision making. According to her taxonomy, environmental laws can be grouped into categories based on whether they are incentives, disincentives, safe harbors, mandates, or prohibitions for businesses.[3]

The recent surge of climate litigation based on the public trust doctrine, constitutional, and common law tort claims is a response to the failures of statutory and regulatory controls, and a plea for courts to order the government to step into the breach. Our Children's Trust is a nonprofit law firm that has filed or supported lawsuits in a wide variety of courts throughout the world to pursue climate-change claims. In *Juliana v. United States*, they asked the court to order the federal government to prepare and implement an enforceable national remedial plan to phase out fossil fuel emissions and draw down excess atmospheric CO_2. While the *Juliana v. United States* case is laudable and important, without addressing larger political and economic dynamics, even a successful court order may not be enough. Like *Brown v. Board of Education*, which resulted in court-ordered desegregated schools that have not yet been fully realized in the United States generations later, these court actions are absolutely necessary, but perhaps will not be sufficient in themselves to change behavior.

A serious reckoning with the dynamics that shape corporate purposes and operations is overdue. Since corporations have such a significant impact on environmental and social outcomes, the legal foundations of corporations should be interrogated in the search for environmental sustainability. While all jurisdictions require corporations to have a lawful purpose, that does not mean they will comply with applicable laws. In practice, corporate boards ignore environmental compliance in pursuit of profits, asserts Norwegian corporate law professor Beate Sjåfjell. As U.S. corporate law professor Kent Greenfield writes in *The Failure of Corporate Law: Fundamental Flaws and Progressive Possibilities* (2006), there is not a single modern case in which a court held directors liable to shareholders because the directors broke the law; indeed, as long as the expected profits are higher than the penalties, there are incentives to break the law.[4]

Investor-Owned Corporations and Shareholder Primacy

The choice of legal structure for an enterprise impacts how decisions are made, what factors are considered, and what takes priority. This includes, for instance, whether an enterprise vests supervisory or strategy decisions in a board, relations between minority and majority shareholders, and relations between the corporation and its creditors. Legal structures also determine whether workers have a role in governing, managing, or owning the entity. Finally, founders decide whether the business is composed of multiple stakeholders with governance roles.[5]

Although the early forms of capitalism predate the corporation, its modern expression would not be possible without the state's sanction of the investor-owned corporation. Beate Sjåfjell contends that capitalism is grounded in laws that allowed the creation of the corporation as a legal entity and granted limited liability to its shareholders. Shareholders contribute their capital to fund the corporation and hire management. They not only have the right to elect directors, but the corporation then owes them fiduciary duties to act in their interest.[6]

Shareholder primacy in corporate law postulates that this fiduciary duty informs the central purpose of the corporation, which is to maximize the wealth of shareholders. *Dodge v. Ford Motor Co.* is routinely cited to endorse shareholder primacy. In that 1919 case, the Michigan Supreme Court ruled that Henry Ford could not operate his company in a charitable manner to benefit employees or customers instead of shareholders. This focus on maximizing returns for shareholders has come to inform most aspects of corporate and securities law.[7]

Proponents of shareholder primacy maintain that shareholder wealth maximization will in turn increase societal wealth, as measured in innovative research and improved consumer products as well as material wealth. Because shareholders want to increase their profits, they have incentives to closely monitor the corporation's directors and officers. Further, pursuing a singular purpose of wealth maximization is the clearest and most efficient way to run a business without confusing managers about additional goals.[8]

By contrast, a broader definition of corporate stakeholders could include workers, neighbors, the community where the business operates, customers, creditors, and the environment. Some argue that stakeholders who are not shareholders are already protected by regulations, such as environmental and labor laws, so corporate managers should not need to do more than comply with those regulations. Milton Friedman maintained that business managers who undertake socially responsible actions that fail to increase profits are imposing a "tax" on their shareholders and inappropriately engaging in public policymaking. In this vein of thinking, to the extent that protecting stakeholders supports the long-term economic interests of the corporation's shareholders, it can coexist with shareholder primacy. But when stakeholders' interests cost too much or the time period for realizing benefits is too long, shorter-term shareholders' interests will prevail.[9]

Proponents of shareholder primacy point out that other models have failed. As Carol Liao explains, some assert that the United States had a manager-oriented model from 1930 to 1960, Germany had a labor-oriented model that peaked in the 1970s, and Japan and France had state-oriented models after World War II. Yet, Liao argues, the model that has produced the best economic outcomes, measured in GDP growth, is the U.S. model that prevailed from 1970 to 2008, which is grounded in shareholder primacy.[10]

Lawyers behind the Purpose of the Corporation Project, an initiative of the European law firm Frank Bold, recount that the corporate focus on maximizing shareholder value started to dominate corporate governance in the 1970s. They claim that this focus is associated with negative externalities that are detrimental both to sustainability and to financial performance, because companies are now investing less in their employees and giving a greater percentage of profits to shareholders. Striking a similar chord, Greenfield contends that the "law has created an entity that is guaranteed to throw off as many costs and risks onto others as it can." It does this by "centralizing power in management, limiting the involvement of other stakeholders in corporate decision making, and imposing a requirement that the firm's management care about making money first and foremost."[11]

There is no dispute that the shareholder primacy model has produced growth in GDP and increased the value of the stock market in the United States. However, the time frame of shareholder primacy's ascendancy in the 1970s also correlates with a rise in GHGs and a decline in biodiversity, an increasing gap between the wealthiest and poorest, stagnation in the pay of labor, and an overall divergence between economic growth and quality of life. The research of Ira Kubiszewski and colleagues, noted above, shows quality of life stagnating and not increasing with GDP growth after the 1980s in multiple countries. This is also when CEO pay compared to worker's pay started to diverge dramatically. According to the Economic Policy Institute, under a conservative scenario, from 1978 to 2018, CEO pay grew 940 percent, while wages for the typical worker grew only 11.9 percent.[12]

Since the global financial crisis of 2007–2008 and the increasing understanding of the depth of the climate emergency, there has been a growing critique of corporations producing so many externalities in the pursuit of shareholder value. The economic efficiencies claimed by shareholder primacy advocates nowhere address the social and environmental harms that persist despite purported regulations to protect nonshareholder stakeholders.

In *Company Law and Sustainability*, Sjåfjell and her coauthors contend that the "main flaw" in the argument that shareholder primacy is best for society is "the idea that only shareholders have a residual interest, with the implication being that shareholders benignly further all interests in furthering their own." The argument also fails to recognize that shareholder value is not the best representative of the interests of the firm, since most shareholders care about their overall stock portfolios or retirement funds, not the performance of an individual company. The Purpose of the Corporation Project examines corporate purpose and critiques an exclusive focus on maximizing shareholder value as reflected in share price. Among other things, the lawyers behind the project argue that shareholder primacy has increased inequality over the last twenty years because increased corporate profits have gone to shareholders instead of increasing salaries across the companies.[13]

The Frank Bold lawyers claim some law and business schools are teaching an inaccurate version of the law that exaggerates the requirement to maximize shareholder value. A cohort of law professors appear to agree and argue that shareholder primacy is a norm and not a legal requirement. In 2016, the Cass Business School's Modern Corporation Project coordinated a statement of corporate law endorsed by more than fifty professors and lawyers in the United States, United Kingdom, and European Union. It includes this explanation of shareholder primacy:

> Contrary to widespread belief, corporate directors generally are not under a legal obligation to maximise profits for their shareholders. This is reflected in the acceptance in nearly all jurisdictions of some version of the business judgment rule, under which disinterested and informed directors have the discretion to act in what they believe to be in the best long-term interests of the company as a separate entity, even if this does not entail seeking to maximise short-term shareholder value. Where directors pursue the latter goal, it is usually a product not of legal obligation, but of the pressures imposed on them by financial markets, activist shareholders, the threat of a hostile takeover and/or stock-based compensation schemes.[14]

Sarah Light's work similarly identifies the business judgment rule as a safe harbor from shareholder claims that a director should not have prioritized environmental protection if it harmed short-term shareholder profit. She recognizes, however, that while this rule protects a director from liability, it does not tip the scales in favor of environmentally responsible behavior. Whether law or norm, the singular goal of maximizing shareholder value undisputedly continues to shape corporate behavior; and although the business judgment rule may provide a basis for a safe harbor, it does not influence the overarching purpose of an enterprise.[15]

Sjåfjell and her colleagues similarly observe that the norm of share-holder primacy creates an impediment to environmental sustainability. They say it impedes "exploration of how far boards—the strategy-setting, supervisory organ of companies—are legally permitted to go in shifting the management."[16]

Further, although shareholders have the ability to influence environmentally sustainable corporate behavior, their impact has been limited. Activist shareholders can target their investments to companies that are promoting environmental sustainability, but socially responsible investing has not picked up enough momentum.[17]

Corporate Reform Proposals

From a variety of disciplines and sectors, people have been critiquing the current role of investor-owned corporations globally and revisioning the future corporation. The reform proposals include creating an environmental-priority principle; requiring all corporations to have a social and environmental purpose; developing alternative business structures as social enterprises; and utilizing the enduring business form of coopera-tives, which, some scholars maintain, can more easily pursue social and environmental purposes because they do not need to focus on maximizing shareholder return.[18]

Create Environmental Priority Principle

Light calls for integrating an "environmental priority principle" into various aspects of business law (securities, antitrust, and bankruptcy), and provides a sketch of the mechanisms to accomplish this. As envisioned, corporations would have "first order" goals of environmental protection, alongside their traditional goals of efficiency, profit, and others. When in conflict, "environmental values [should] not be outweighed in the absence of a strong showing both that some other measure of justice will be com-promised, or a significant degree of efficiency or market functioning lost, and that the environmental values at stake are below some threshold. In other words, the principle would be limited by a degree of proportionality."[19]

Ideally, Light sees implementing this principle as a continuum, ranging from transforming prohibitions against environmental protection into safe harbors to moving safe harbors like the business judgment rule into incentives and turning incentives like a benefit corporation statute into a mandate. She explains in the United States this implementation would have to occur in all fifty states, with legislatures essentially requiring all corporations to become benefit corporations, and simultaneously strengthening benefit corporation legislation to the level of an environmental mandate.[20]

Define Corporate Purpose to Benefit Stakeholders

Corporate law scholars have been debating the philosophical and theoretical aspects of the purpose of corporations as separate legal entities with limited liabilities for more than 150 years. Without re-creating that debate here, suffice it to say that many of the arguments have focused on the interests of the company rather than the societal purpose of corporations. Yet, in recent years, a growing international cohort of corporate lawyers and professors have been advocating to reorient the corporation to meet global sustainable development challenges.[21]

The founders of a company define its purpose. This is the case regardless of corporate form, whether it is a cooperative, nonprofit, partnership, or an investor-owned corporation. Founders may choose any purpose that is legal. Once founders define the business purpose, this drives governance, goals, strategies, and the metrics they use to define success. There are no laws in the United States or European Union that require investor-owned corporations to have a social or environmental purpose. This is a primary point of contention. Leading corporate lawyers and scholars have been calling for corporate law reforms that would require the purpose of the corporation to be defined to serve society.[22]

The Future of the Corporation project of the British Academy, led by Oxford professor Colin Mayer, is revisioning the corporation with a group of academics and business leaders. The British Academy is the United Kingdom's national academy for the humanities and social sciences. After several years of convenings and research involving 100 people from business, academia, institutional investment offices, government, and think tanks, the academy released its conclusions in 2019. The report asserts that the climate crisis and urgency of meeting the United Nations' Sustainable Development Goals, among other factors, mean we must reform business around its purposes, trustworthiness, values, and culture. The report calls for new laws that require directors to state a purpose, governance that establishes accountability to a range of stakeholders, and special regulations requiring a high duty of care to the public interest for companies that perform particular public and social functions, such as utilities, banks, and companies with significant market power. The report concludes that "business as usual is no longer possible." At its core, the authors argue, the purpose of all business "is to profitably solve the problems of people and planet, and not profit from causing problems." Businesses should be required to articulate their purpose in their articles of incorporation and bylaws. They identify Delaware's public benefit corporation statute as an example of a law that enables purposeful business, and urge that this option should become the norm.[23]

Similarly, the Purpose of the Corporation Project focuses on shifting corporations from maximizing shareholder value to integrating societal interests in corporate purpose, governance, and decision making. They urge a paradigm shift that views corporate purpose as benefiting society. Recalling that corporations are creatures of law and that incorporation is a privilege, the project calls on civil society to articulate the purpose for which it is granting the privilege.[24]

The Purpose of the Corporation Project identifies a need to achieve transparency and accountability through accurate accounting of a corporation's environmental and social impacts. Moreover, boards of directors should consider the effect the business is having on the environment and society, systemic risks, and the ability of the company to achieve success in the long term. The project defines a "well-run company" as having "long-term plans charting their way towards environmental and economic sustainability."[25]

One step in the direction of transparency occurred in 2017, when the European Union's Non-Financial Reporting Directive became effective. This

law requires all publicly traded corporations and financial and insurance institutions in the union to report on environmental, social, and governance
metrics. Strategically increasing access to information holds promise because it moves corporate social responsibility out of the voluntary realm;
but early results already indicate areas where the law needs to be improved,
as discussed in Chapter 6.[26]

On the other side of the Atlantic, in 2019, the American Business Roundtable issued a "Statement on the Purpose of the Corporation." An association of CEOs of U.S. corporations, the Roundtable endorsed a stakeholder
commitment, breaking from its longstanding support for shareholder primacy. The newly articulated stakeholder purpose includes protecting the
environment and embracing sustainability. Businesses must "commit" to
"supporting the communities" where they operate, respecting people in
those communities, and "protect[ing] the environment by embracing sustainable practices" across their businesses. This shows a broadening of the
view of corporate directors to encompass the impacts companies have on
their communities.[27]

Sjåfjell and her coauthors go further and argue in favor of new corporate
law defining the societal purpose of companies as "to create sustainable
value within the planetary boundaries." They advocate for multi-stakeholder
corporations. Their alternative normative definition for a corporation's societal purpose is "to manage their businesses in a manner that creates benefits for investors, creditors, customers, employees, and local communities,
in a way that contributes to or at least does not harm overarching societal
goals such as sustainable development." This purpose would then set the
stage for orienting the duties of the board and all who manage and supervise it. Sustainability needs to be part of corporate decision making with an
imperative for actions geared toward the long term.[28]

The Sustainable Companies Project, which Sjåfjell led, is concerned that
nothing short of the above corporate reform will do more than nudge in-
cremental changes, which will not meet the climate imperative. They argue that sustainability cannot be left to voluntary action by corporations.
They call for legally enforceable standards related to the purpose of enterprises over the long term. Greenfield reaches an aligned, but more general,

conclusion that corporations should be required by law to consider the interests of multiple stakeholders.[29]

Spark Social Enterprise

All of the attention around reforming corporate purpose and support for it from leading CEOs may ultimately lead to the needed legal reforms to require it. While that aim should certainly be pursued, given the serious deficiencies in the state's willingness, ability, and capacity to reform corporate law along the lines discussed above and the weakness of environmental law as a counterforce, greater consideration should be given to social enterprise development that can accelerate in advance of public-law reforms. While it is only a partial solution to the overarching problems with investor-owned corporations, it is at least a solution that is readily available, and already under way.

An approach to reorienting corporations for a broader societal purpose is for business leaders to design new, and convert existing, small and mid-sized corporations into social enterprises. According to Bronwen Morgan, "social enterprise" is framed differently in the United States than in Europe, with an emphasis on individualistic social entrepreneurs in the United States and an emphasis on collective and participatory dimensions in Europe. Antonio Fici contends that in Europe "social enterprise" is associated with private organizations distinguished by the "purpose pursued, the activity conducted to pursue this purpose, and the structure of internal governance." A common core of European legislation requires that social enterprise's purpose be either exclusively or prevalently to serve the "community or general interest," with a constraint on profit distribution during the enterprise's life and dissolution. Further, in Europe, social enterprise "is subject to specific governance requirements, including the obligation to issue a social report, to involve its various stakeholders in the management of the enterprise, and to provide fair and equitable treatment of its workers, notably the disadvantaged ones."[30]

In October 2011, the European Commission launched a "Social Business Initiative" that defined social enterprise as having the "main objective" to have a "social impact rather than make a profit for their owners or shareholders." The social enterprise provides "goods and services for the market in an entrepreneurial and innovative fashion and uses its profits primarily to achieve social objectives." Social enterprises are managed in ways that are "open and responsible" and "involve employees, consumers and stakeholders affected by its commercial activities."[31]

EU governmental institutions celebrate and promote social enterprises as contributing to their shared socio-economic objectives, such as sustainable and inclusive growth, high-quality employment, job creation and preservation, social cohesion, social innovation, local and regional development, environmental protection, and people's well-being. Spain has been a leading country promoting social enterprise, and has national and more localized laws to define and develop the "social and solidarity economy," as will be discussed in Part II.[32]

The oldest form of corporation, the cooperative, can be considered a social enterprise. The cooperative business form provides the flexibility to pursue broader social and environmental purposes beyond profits. Cooperatives will be addressed more fully in Chapter 3.[33]

This is a time when a variety of social enterprise forms have been recognized in legal reforms. In the past decade newer corporate forms have been created all around the world that allow businesses to pursue both profit and social and environmental goals, and to stabilize their mission even when ownership or leadership changes. Considered "hybrids," these businesses take a variety of forms depending on the country of incorporation.

Social enterprise laws tend to fall into two categories: those that impact incorporation and those that confer a certification or status. Carol Liao discusses the community interest company (United Kingdom), the community contribution company, or C3(Canada), the low-profit limited liability company, or L3C (United States), the private governance B Corp certification (global), and the benefit corporation (United States). In addition, Florian Möslein examines certifications for business hybrids, such as B Corp, Social Enterprise Mark in the United Kingdom, Social Enterprises Certification in South Korea, and PHINEO Wirkt!-Siegel in Germany.[34]

In summary, the most profound transformation of the investor-owned corporation would target shareholder primacy and require all corporations to be responsible to a broader set of stakeholders. To address environmental sustainability, the government would require all corporations to have a triple bottom line purpose. Short of this, a variety of social enterprise legal forms already exist around the world to provide businesses the ability to broadly define their purpose to benefit stakeholders, such as benefit corporations and cooperatives. B Corp is the most widely used certification in the United States and globally, existing in sixty countries, along with the benefit corporation legal status in a majority of U.S. states, and so warrants a closer treatment. The next chapter provides an explanation and analysis of social enterprise design that shapes B Corp certification, benefit corporations, and cooperatives.[35]

3

Social Enterprise Design

Emerging and Enduring Values-Based Business Forms

Chapter 3 takes the corporate reform proposal to develop social enterprises further and deeply explores emerging and enduring forms. It explains the third-party B Corp certification and designing a business as a benefit corporation. The benefit corporation model legislation directly challenges shareholder primacy and provides an alternative enterprise design for multi-stakeholder businesses. Building on scholarship from a variety of disciplines, this chapter brings out the strengths and weaknesses of this alternative to investor-owned corporations. After discussing this emerging values-based business form, it shifts to the most enduring values-based business form: cooperatives. People have been organizing cooperative businesses for over 200 years to serve members' needs in a wide variety of economic sectors, from insurance and financial institutions to agriculture, electricity, health, education, and housing. The chapter provides a broad introduction to the history, theories, economic impact, and values and principles of cooperatives as a form of enterprise that could be utilized to address the most pressing sustainable development issues of the twenty-first century.

B Lab's Role and B Corp Certification

B Lab is a nonprofit based in the United States that says it "serves a global movement of people using business as a force for good." Since 2006, B Lab has been acting as a third-party certifier of companies that earn the B Corp label. It uses its B Impact Assessment to determine certification. In their formation documents, a prospective B Corp establishes that the corporate fiduciaries must consider the impact of their decisions on a variety of stakeholders that are not shareholders, such as the environment and the local, state, and national economies. In short, to become a certified B Corp, companies must receive a minimum score on the B Impact Assessment, include B Lab–specified commitments to stakeholders in corporate governing documents (amend the articles of incorporation), and pay an annual fee to B Lab. Carol Liao asserts that the mandate to influence corporate-governance documents makes the B Corp certification unique among corporate social responsibility certifications.[1]

Certified B Corps include a range of types of businesses at various stages of development, from new companies that are just launching to corporations that are well known for having a strong social and environmental purpose, such as Patagonia, Bigelow Tea, and Ben and Jerry's.

Certified B Corps may use the certification to market themselves to investors, consumers, and others. B Corp is a certification and not a legal entity; however, the nonprofit B Lab also promotes legal formation of businesses as benefit corporations with tools for jurisdictions interested in passing legislation. B Lab created and disseminates model legislation for this new form of social enterprise. Companies may want to incorporate as benefit corporations to build a distinct brand; provide legal guidance; protect the board of directors when they pursue social benefits; and protect the board from a hostile takeover.[2]

Möslein observes that some social enterprise laws require a majority of profits to be reinvested in social and environmental goals. For instance, the South Korean social enterprise law requires 66.6 percent of annual profits to be reinvested for social benefit. By contrast, Möslein says, the B Corp certification only requires 20 percent of annual profits or 2 percent of sales to be used this way; and the model benefit corporation legislation does not have any requirement for profit distributions.[3]

While somewhat optimistic about the certification, Liao cautions that B Lab is not equipped to closely monitor numerous companies effectively and has lower standards than some other certifications. Further, Dana

Brakman Reiser points out that nonshareholder stakeholders have no legal claims related to the B Corp certification.[4]

Yet, a combination of being B Corp–certified and employee-owned appears to produce strong positive results for stakeholders despite their lack of legal voice. The nonprofit Democracy Collaborative produced a report, *Mission-Led Employee-Owned Firms: The Best of the Best* (2019), that shows employee-owned businesses (including worker-owned cooperatives) that are B Corp–certified have the highest environmental and social impact scores among all the B-certified businesses. They studied employee-owned B Corps "because these companies have a commitment to social and ecological benefit embedded in their governing documents." Their study compared three categories with twenty firms in each, composed of businesses of similar industry and size. The first category was of B Corp, employee-owned firms; the second was of B Corp firms that are not employee-owned; and the third was of employee-owned firms that are not B Corps. B Lab provided the data, which resulted in showing the B Corp, employee-owned firms scored the highest of the three categories of firms in average B score, average worker score, and average environmental score. Thus, in terms of generating broad benefits for stakeholders, the optimal legal design appears to be an employee-owned certified B Corp.[5]

Benefit Corporation Legislation

Separate, but intertwined with the third-party B Corp certification, is the movement to create new statutes that allow businesses to incorporate as benefit corporations. Maryland was the first state in the United States to enact the model benefit corporation legislation in 2010. Now this model is the most widely adopted benefit corporation statute. By 2018, a majority of states (thirty-four of fifty) had enacted benefit corporation statutes. Corporations can utilize this corporate form at initial formation or in a later conversion, by amending the corporate charter. In 2015, Italy became the first country outside the United States to enact benefit corporation legislation. In 2020 there were legislative efforts ongoing in Australia, Argentina, Chile, Colombia, and Canada, but no other countries beyond the United States and Italy had enacted the legislation.[6]

Standard of Conduct

Benefit corporation model legislation is a direct reform of shareholder primacy. The fundamental distinction in benefit corporation statutes is to require such businesses to pursue multiple public benefit purposes beyond profit making: the model legislation requires a "general public benefit" but makes a "specific public benefit" optional. According to the model legislation, a required general public benefit means a "material positive impact on society and the environment, taken as a whole, from the business and operations of a benefit corporation assessed taking into account the impacts of the benefit corporation as reported against a third-party standard." The model legislation defines the optional "specific public benefit" as covering seven categories, including "protecting or restoring the environment."[7]

The model legislation creates a standard of conduct for corporate directors to consider multiple stakeholders when considering the "best interests of the benefit corporation," including employees, the community, and the environment. While this may already be optional in states with "constituency statutes," the model legislation makes this a mandatory duty. The legislation does not prioritize these considerations, but the corporation may prioritize them in its articles of incorporation. The model legislation includes explanatory comments. Regarding the conduct of corporate directors, the comments describe this section as the "heart of what it means to be a benefit corporation." The commentary says considering the interests of broad constituencies shows this corporate form "rejects the holdings in *Dodge v. Ford*, 170 N.W. 668 (Mich. 1919), and *eBay Domestic Holdings, Inc. v. Newmark*, 16 A.3d 1 (Del. Ch. 2010), that directors must maximize the financial value of a corporation." Thus, the benefit corporation model legislation directly challenges shareholder primacy and provides an alternative enterprise design for multistakeholder businesses.[8]

Transparency and Enforcement Requirements

Transparency and enforceability are important factors for social enterprise law. Third-party certifications are a form of private governance and lack legislative basis and regulatory oversight by a public entity. The benefit corporation statutes make this system more robust because they add a regulatory piece; but if the benefit corporation legislation lacks mechanisms to enforce the corporate purpose of a dual mission (profit plus social/environmental), it could ultimately undermine the social/environmental goals of the designation. This will dampen the social enterprise's ability to expand funding streams and create a strong brand as an environmentally sustainable business. B Lab misses an opportunity here to strengthen the law. In the explanatory comments on its model legislation, B Lab eschews governmental oversight to enforce the benefit corporation's purpose, and instead emphasizes enforcement through the statutory provisions on transparency and accountability.[9]

Section 401 of the model legislation requires the benefit corporation to prepare an annual report and specifies the contents, including the ways the corporation pursued the general public benefit. The comments explain a purpose of this reporting requirement is to prevent entities from abusing the benefit corporation status. If a corporation has a benefit director (in Section 302) or benefit officer (in Section 304), it has a mandatory duty to provide this annual report to shareholders.[10]

Despite the comments about an assessment against a third-party standard, the model legislation states that "[n]either the benefit report nor the assessment of the performance of the benefit corporation in the benefit report required by subsection (a)(2) needs to be audited or certified by a third party." This is another section where the model could be strengthened by clearly requiring a third-party audit or certification, not simply a self-assessment based on a third-party standard.[11]

It bears noting that with the emphasis on third-party standards, benefit corporation statutes that follow the model leave it to third parties instead of the government to develop standards for defining, reporting, and assessing the social and environmental benefits of these corporations. In this way, the model legislation places a lot of responsibility on private governance. The model does not provide even minimal guidance as it delegates this authority for standard setting to unidentified third parties. If this were a U.S. congressional delegation of standard-setting to an administrative agency, it might not stand up to the very loose nondelegation doctrine's "intelligible principle." That principle helps courts determine whether a provision violates separation of powers by delegating legislative power away from the legislative branch of government. This part of the law could certainly be strengthened so state legislatures are performing their proper lawmaking roles and not delegating such standard-setting to unelected bodies without any parameters for their work product.[12]

Further, the model benefit legislation lacks rigorous enforcement provisions. The model legislation's comments explain that a Section 305 "benefit enforcement proceeding" is the exclusive means to bring a claim to enforce the chapter. They further explain that a benefit enforcement proceeding may be used to enforce the obligations of a benefit corporation under Section 402(b) to post its benefit reports on its website and to supply copies of its benefit report if it does not have a website. However, only people who are listed in Section 305 can enforce the obligation; in other words, only the corporation directly or derivatively by a subset of specified shareholders can enforce this. Section 301(d) denies standing to a general member of the public to enforce obligations. So despite benefit corporations' recognition of stakeholders who are not shareholders, those broader stakeholders cannot enforce any of the corporations' obligations. A promise of public benefits is less meaningful if third-party beneficiary stakeholders lack a legal remedy for failure to deliver the benefit.[13]

In practice, according to Nancy Kurland's research, smaller companies, which make up the vast majority of benefit corporations, engage in self-reporting instead of hiring a third-party certifier. Further, benefit corporations frequently fail to post on their public websites the required annual reports. Kurland concurs that, under the model law, only shareholders can hold a benefit corporation accountable for these failings.[14]

These analyses by Kurland may inform how others view the trustworthiness of the benefit corporation status. A lack of adherence to standards, such as basic transparency in reporting, undermines the ability to provide a market advantage to these enterprises.

Impact investors in 2017 had $100 billion under management in social enterprises, such as benefit corporations, and many global consumers are willing to pay more to support companies with a social and environmental purpose. On a much larger scale, in 2020, BlackRock announced that it would be screening its investments for sustainability. Benefit corporations that have already been engaged in sustainability measuring and reporting will be better positioned to respond. As discussed below, however, the usefulness of sustainability screening to mitigate investment risk rests on measurable, comparable, and reliable indicators of social and environmental corporate behavior.[15]

Some scholars, such as Möslein, assert that social enterprise certifications may correct the information asymmetries that contribute to market imperfections, but only if they are designed properly. To be properly designed, a certification should be trustworthy, so the trademark means something positive to consumers and investors. A certification is a special kind of signaling instrument by a third party, the acceptability of which relies on the trustworthiness of the certifying body, the certification criteria, and compliance. According to Möslein, certifications with approval requirements rather than self-assessments, and with shorter certification terms followed by regular audits rather than perpetual certifications, seem more trustworthy to the public.[16]

Delaware, the leading state in the United States for business incorporations, created a law for public benefit corporations that differs from the B Lab model legislation in several respects: 1) the model requires annual public reporting, while Delaware requires biennial reporting to shareholders; 2) the model provides options for stakeholder considerations, while Delaware requires specific benefits to be included in the statement of business or purpose; and 3) the model allows shareholders to bring a benefit enforcement proceeding, while Delaware does not mention this. It is worth understanding these differences, especially because the British Academy's Future of the Corporation project recommends that Delaware's public benefit corporation law should become the norm.[17]

Kurland conducted a study of the oldest corporation to incorporate under Delaware's public benefit corporation law to explore accountability. She studied EA Engineering, Science, and Technology, Inc., a multistate consultancy with 500 employees. EA Engineering adopted an Employee Stock Ownership Plan (ESOP) and incorporated in Delaware as a public benefit corporation in 2014. ESOP is a "tax-qualified retirement [plan] created as part of the Employee Retirement Income Security Act ('ERISA') in 1974 that enable[s] broad-based employee ownership." Essentially, it creates a pathway for employees to own stock in the company for their retirement plan, but does not involve additional governance roles for employees. Because of the voluntary nature of compliance with the benefit corporation statute, Kurland observed at EA Engineering the importance of having an "adaptive learning framework" to incorporate environmental sustainability over legal accountability. An adaptive learning environment supports organizational change and progress toward goals.[18]

In summary, a broadly disseminated model around the world is the B Corp certification and its related benefit corporation legal status. In these early stages of development, with just a decade of experience, this social enterprise form is still malleable. There are areas where the model legislation is stronger than Delaware's benefit corporation statute, and both could be strengthened. The legislation would be stronger and the brand more distinctive with enforceable standards by stakeholder beneficiaries, guidelines for standard-setting, and a requirement of an unbiased audit by a third-party certifier such as B Lab.

Another shortcoming is that these social enterprises are not necessarily member- or employee-owned or democratically governed. Thus, they lack the added benefits of sharing wealth and control more equitably and democratically among those who are most directly impacted by the company. By contrast, the U.S. Federation of Worker Cooperatives asserts that an employee-owned cooperative "must create, in policy and practice, mechanisms for workers to make the decisions that affect the functioning and governance of the business." The Democracy Collaborative research shows that the strongest B Corp assessment scores for the environment, workers, and overall are in firms that are both employee-owned *and* B Corp–certified. The Just Coffee case study in Part II of this book involves a business that combines these design elements.[19]

Further, cooperatives are distinct from other corporate forms because they are the only form of business centered around membership, putting member and community benefit at the core of the model. Unlike the newer forms of social enterprises, cooperatives have perhaps the longest history of any corporate form and are distributed globally. We turn now to understand the history, values, and principles that distinguish cooperatives in the twenty-first century.[20]

What Is a Cooperative? A Typology for Analysis

An enduring form of social enterprise is the cooperative. Globally, cooperatives exceed multinational corporations in scope by serving over one billion members and providing over 100 million jobs. Cooperatives exist in all sizes and types, in all sectors of the economy, and in most countries in the world, complicating the ability to generalize. With that caveat, this section explains what cooperatives are, their history and economic impact, and the shared international values and principles that are their hallmark.[21]

When you think of fast-food icons Burger King, Taco Bell, Subway, and Pizza Hut, the word *cooperative* does not readily come to mind, yet these franchises are members of purchasing cooperatives. For instance, Subway franchisees are owner-members of IPC cooperative, which operates behind the scenes to run their supply chain. Similarly, Blue Diamond Almonds, Florida's Natural, Ocean Spray, Sunsweet, and Sun-Maid are familiar U.S. brand names, but most do not realize that they are part of the over 2,100 U.S. producer cooperatives, which together provide 190,000 jobs. Not many think "cooperative" when they read a news story from the Associated Press, and yet it too is organized as a cooperative.[22]

The Sustainable Economies Law Center provides legal services to cooperatives and other social enterprises. They emphasize the importance of the legal architecture of organizations and enterprises because they see this providing the structure for the larger economy: "Legal structures dictate how wealth flows through our organizations and how decisions are made. Traditional enterprise models are designed to grow the wealth of people who already have wealth, giving all decision-making power to those same individuals. By contrast, cooperatives put wealth and decisions into the hands of workers and consumers, building community well-being and transforming local economies."[23]

Cooperatives are a form of business that is owned and run by and for its members. They use democratic control with the principle of one member, one vote. Unlike a shareholder in a corporation, cooperative members are, by definition, people who use or work in the enterprise. People engaged in cooperatives are called cooperators.

In his treatise on U.S. cooperative law, Israel Packel opens with: "Cooperation is a basic tenet of a civilized society." Following that logic, cooperators can be seen as creating the foundation of a civilized business model. Perhaps they can become an antidote to the ruthless model of hyper-capitalism that currently dominates the global economy. The cooperative business model is flexible enough that it allows cooperators to pursue social and environmental purposes along with profits or a completely charitable purpose. Packel defines a cooperative as "an association which furnishes an economic service without entrepreneur profit and which is owned and controlled on a substantially equal basis by those for whom the association is rendering service."[24]

While cooperatives can be organized as nonprofits, they are typically neither purely charitable nor religious, but instead serve the economic or welfare interests of their members. Some of the nonfinancial benefits for members include the quality of product or service, ownership, control, and self-help (solving a shared problem or need).[25]

Cooperatives exist in a wide array of business sectors throughout the world and are more pervasive than one may initially think. When it is more advantageous and efficient to share resources, one may find a cooperative is the optimal platform for participation. One way to create a typology of cooperatives is to group them by member class. This aids in thinking through when it makes sense for members to band together to form an enterprise that is mutually beneficial. My typology groups cooperatives into four types, with examples of economic sectors where they operate, as shown in Table 1.

However, it is important to keep in mind that while a typology is helpful in sorting complex varieties into a manageable framework, some jurisdictions allow cooperatives to be a hybrid of these types. For instance, in the case studies in Part II, Caixa Popular is a cooperative bank with two classes of members, workers and other cooperatives. Consum is a cooperative grocery store chain with two classes of members, workers and consumers. Some jurisdictions also recognize a cooperative that serves the general society, such as the Italian Social Cooperative. These are more like nonprofit public-interest or charitable-purpose organizations in the United States. Italian Law

Table 1. Typology of Cooperatives

Membership Class	Economic Sector	Examples
Consumer	Grocery, housing, electric or water utilities, health, insurance	Consum grocery store, Cobb EMC electric utility
Wholesale purchasing	Grocery, hardware, drug stores, hospitals, schools, farmers	Ace Hardware, Growmark
Workers	Grocery, finance/credit union/bank, services, manufacturing, home health care	Mondragón, Just Coffee, Evergreen, Caixa Popular
Producers/farmers	Orange growers making juice, grape growers making wine, dairy farmers marketing under a shared label	Organic Valley, Bodegas Pinoso

381/91 recognizes this subset of cooperatives. While cooperatives generally pursue the interests of their members, an Italian social cooperative pursues the general interest of the community. The Italian National Institute of Statistics reports that prior to passing this law in 1991, there were just over 2,000 social cooperatives, and by 2011 there were more than 11,000.[26]

Cooperators and their cooperatives are linked globally through the International Cooperative Alliance (Co-op Alliance), a nongovernmental entity that has served this purpose since 1895. The Co-op Alliance believes that a more diversified and pluralistic global economy will minimize global economic crises like that experienced in 2007–2008, and that the way to build this into the economy is to grow a diversity of cooperative enterprises. The Co-op Alliance sees cooperatives as a way to counter the growing chasm in wealth inequality globally. It emphasizes that a key distinguishing feature revolves around cooperatives' ability to create wealth for their many members, including those who interact with the businesses as "service users, producers, independent business owners, consumers, and workers."

The Co-op Alliance continues that cooperatives are "not solely for the few who are rich enough to invest capital in investor-owned enterprises." In the 2020s, the Co-op Alliance aims for cooperatives to be known as the leading business model for economic, social, and environmental sustainability.[27]

Cooperatives' Global, United States, and Spanish Economic Contribution

The global economy is so dominated by multinational corporations and a collective fixation on the health of stock markets that rarely does one think about the cooperative form of enterprise. However, cooperatives play an enduring, socially beneficial role in the global economy and political life.

The United Nations asserts that cooperative businesses serve an important purpose in economic and social development around the world. In recognition of this, the UN General Assembly designated 2012 as the International Year of Cooperatives.[28]

In 2017, House Resolution 561 commended cooperatives as integral to sustainable economic growth in the United States. The resolution provides an understanding of the scope of the U.S. cooperative sector. Drawing on 2009 research by the University of Wisconsin Center for Cooperatives, the resolution recognized that there are 30,000 cooperatives in the United States that employ two million people. Additionally, worker-owned schools with teachers as member-owners account for $1 billion in revenue, provide 15,000 jobs, and offer 1,000 childcare or early learning centers.[29]

Data on cooperatives has improved since the 2017 U.S. Economic Census added a question about whether the business is organized as a cooperative. According to the National Cooperative Bank, the latest data shows that the top 100 U.S. cooperatives alone generated $228.2 billion in revenues in 2019. Some of the top-grossing cooperatives include the CHS, Inc. (a Fortune 100 business), Dairy Farmers of America, ACE Hardware Corp., HealthPartners, Inc., and Oglethorpe Power Corporation.[30]

House Resolution 561 highlighted that in rural parts of the country, where investor-owned utilities are scarce, cooperatives provide vital utility services of electricity, water, and telecommunications; cooperatives provide power to an astounding 18 million U.S. homes, schools, and businesses.

Over 100 million people in the United States belong to cooperative credit unions.[31]

In worker-owned cooperatives, workers both own and govern the business. The United States has approximately 350 worker-owned cooperatives employing over 4,500 people and generating over $500 million in annual revenue. The U.S. Federation of Worker Cooperatives emphasizes that "[s]uccessful worker cooperatives tend to create long-term stable jobs, enact sustainable business practices, and develop linkages among different parts of the social economy." The Federation sees worker-owned businesses as having a "direct stake in the local environment" and the power to decide to do business in a way that is sustainable.[32]

Cooperatives have a large presence in the European Union. In 2020, the European Commission reported there were 250,000 cooperatives in Europe with 163 million members creating 5.4 million jobs. According to the Spanish Business Confederation of the Social Economy (Spanish acronym CEPES), in 2019 data on all types of Spanish cooperatives showed there were more than 18,000 cooperatives with approximately 6.8 million members providing over 300,000 direct jobs. By revenues, the largest cooperative sectors are farmer, worker, and consumer types. Spanish cooperatives are found in all parts of the economy, but there are more in manufacturing, services, business, health, welfare, and education.[33]

In Spain, the worker-owned cooperative sector is a much more significant part of the economy than in the United States. According to the Spanish Confederation of Cooperatives of Associated Work (Spanish acronym COCETA), as of 2014 there were 17,000 worker cooperatives in Spain, directly employing 210,000 people, with revenues of €22.1 million. In the Valencian region of Spain alone, in 2016, there were 2,037 worker cooperatives, more than in the entire United States. The sector is now larger than it was before the global financial crisis, and it keeps growing. The 2016 numbers were 4 percent higher than in 2015. Valencians created 147 new worker-owned cooperatives in the region in 2016 alone. Time will tell whether cooperatives will be similarly resilient to the COVID-19 crisis. In total, there are 21,378 people in the region who are members in a worker-owned cooperative. By provinces within the region, the distribution of people working in cooperatives is: 13,195 in València, 6,243 in Alacant, and 1,940 in Castelló.[34]

Social scientists have developed a wide variety of theories to explain why cooperatives succeed or fail. Each theory clearly grows out of the scholar's discipline, so it is most useful to view them together in the search for a holistic, multidisciplinary picture. Birchnall provides a summation of the theories, including the supportive environment theory (Attwood and Bhaviskar), which emphasizes the presence of promoters and good legal and fiscal policies; social history theory, which emphasizes nationalism or political party solidarity as supporting a preference for charitable forms of organization; economics theory, which emphasizes how market failures and low barriers to entry lead to strong coops; and sociology theory, which emphasizes high levels of social capital supporting the emergence of coops.[35]

In the case studies in Part II, aspects of these theories are present to varying degrees. None of them alone entirely explains the conditions for successful cooperative business. Noteworthy in the theories is the muted role of law and environmental sustainability in understanding the conditions necessary for cooperative development.

Shah's theory of cooperative design and evolution looks to the design within a cooperative's governance structure. He asserted that one condition for success is that the purpose of the cooperative is of central importance to its members. In several of the case studies, a shared passion to create an environmentally sustainable business was the central condition for cooperation. As one considers maximizing the environmental sustainability of cooperatives, a critical legal-design question is how a cooperative's environmental sustainability purpose is drafted into the governance documents and structure of the cooperative to articulate this purpose in a durable way. This can be seen in a cooperative's bylaws and articles of incorporation, as well as its subsequent policies and practices.[36]

Early Cooperatives

Cooperatives have been part of the American economy since before the 1776 Declaration of Independence from Great Britain. The iconic American, Benjamin Franklin, started the first cooperative business in Philadelphia in 1752, the Philadelphia Contributionship for the Insurance of Houses from Loss by Fire; this is a mutual fire insurance company, which still exists today as the oldest property insurance company in the United States.

Franklin spearheaded this cooperative almost 100 years before twenty-eight people created what would become the most well-known cooperative in the world for its foundational values and principles.[37]

Started in 1844 in Rochdale, England, a group of primarily highly skilled artisans (flannel weavers, shoemakers, cabinet makers, and so on) formed the Rochdale Equitable Pioneers Society. Others observed their success and then adopted the Pioneers' "Values and Principles of Cooperation" into new cooperative ventures.[38]

The Rochdale Values and Principles became the architecture for the Co-op Alliance's 1995 "Statement on the Co-operative Identity." However, the Rochdale Pioneers Museum declares that "the co-operative ideal is as old as human society." The museum challenges us to recognize that the new idea is that conflict and competition are a principle of economic progress. Cooperative Values and Principles attempt to remind society what has been "forgotten in the turmoil and disintegration of rapid economic progress."[39]

Remnants of this older Rochdale period can still be seen in rural parts of the United States where Granges persist. After the Civil War, the Order of the Patrons of Husbandry, known as the Grange, formed to improve farming conditions. In 1875 the Grange endorsed the Rochdale Values and Principles. It actively promoted cooperative development and aimed to eliminate the middleman by directly linking farmers with manufacturers, and producers with consumers. Farmers participated in cooperatives to access lower prices for their supplies, such as seed, fertilizer, insurance, and credit. The Rural Electrification Act of 1937 paved the way for farmers to access capital and bring electricity to rural parts of the country that investor-owned utilities had refused to serve. Farmers formed rural electrical cooperatives, many of which still exist today. This will be discussed more fully in Chapter 8. Cooperatives still play a very large role in American agriculture, and the U.S. Department of Agriculture is the federal governmental locus of cooperative activity.[40]

The first cooperative in Spain was La Proletaria, founded in València in 1856. It started as a consumer cooperative and became a textile manufacturing cooperative. An aspect of cooperative development in Spain that is distinct from the United States is the role of the Catholic Social Movement, which in Spain has been deeply integrated into cooperative development. Catholic activists initiated the first rural savings bank, Caja Rural de Castellón, in 1903. Then in 1956 a Catholic priest led the creation of Mondragón, which has grown to be the largest worker-owned cooperative network in the world. The Spanish cooperative ecosystem and Mondragón model will be discussed more fully in Part II.[41]

Cooperative Values and Principles

Cooperatives globally share common values and principles, originally derived from Rochdale. This aspect of cooperatives makes them a very distinct, values-laden form of business, in contrast to the dominant form of investor-owned corporations. The founders of the cooperative movement, like its leaders today, wanted more than financial success for their businesses. According to the Co-op Alliance, the animating concern then and now is to promote "social justice" through a "jointly-owned and democratically controlled enterprise."[42]

The Co-op Alliance's 1995 Statement on the Cooperative Identity is so fundamental that it is included unedited here. The statement has become a reference point internationally for defining features in cooperative legislation. It may also be understood as a form of global private governance.[43]

The international cooperative values are:

SELF-HELP
In co-operatives, people help each other whilst helping themselves by
    working together for mutual benefit.
SELF-RESPONSIBILITY
Individuals within co-operatives act responsibly and play a full part in
    the organisation.
DEMOCRACY
A Co-operative will be structured so that members have control over
    the organisation—one member, one vote.
EQUALITY
Each member will have equal rights and benefits (according to their
    contribution).
EQUITY
Members will be treated justly and fairly.

SOLIDARITY

Members will support each other and other co-operatives.[44]

The Co-op Alliance encourages cooperatives to put these values into practice through seven principles, which are envisioned as guiding their action:

PRINCIPLE 1: VOLUNTARY AND OPEN MEMBERSHIP

Co-operatives are voluntary organisations, open to all persons able
to use their services and willing to accept responsibilities of
membership, without gender, social, racial, political or religious
discrimination.

PRINCIPLE 2: DEMOCRATIC MEMBER CONTROL

Co-operatives are democratic organisations controlled by their members, who actively participate in setting their policies and making
decisions. Men and women serving as elected representatives are
accountable to the membership. In primary co-operatives members
have equal voting rights (one member, one vote), and co operatives
at other levels are also organised in a democratic manner.

PRINCIPLE 3: MEMBER ECONOMIC PARTICIPATION

Members contribute equitably to, and democratically control, the
capital of their co-operative. At least part of that capital is usually
the common property of the co-operative. Members usually receive
limited compensation, if any, on capital subscribed as a condition
of membership. Members allocate surpluses for any of the following purposes: developing their co-operative, possibly by setting
up reserves, part of which at least would be indivisible; benefiting
members in proportion to their transactions with the co-operative;
and supporting other activities approved by the membership.

PRINCIPLE 4: AUTONOMY AND INDEPENDENCE

Co-operatives are autonomous, self-help organisations controlled by
their members. If they enter into agreements with other organisations, including governments, or raise capital from external
sources, they do so on terms that ensure democratic control by
their members and maintain their co-operative autonomy.

PRINCIPLE 5: EDUCATION, TRAINING AND INFORMATION

Co-operatives provide education and training for their members,
elected representatives, managers and employees so they can contribute effectively to the development of their co-operatives. They
inform the general public—particularly young people and opinion
leaders—about the nature and benefits of co-operation.
PRINCIPLE 6: CO-OPERATION AMONG CO-OPERATIVES
Co-operatives serve their members most effectively and strengthen
the Co-operative Movement by working together through local,
national, regional and international structures.
PRINCIPLE 7: CONCERN FOR COMMUNITY
Co-operatives work for the sustainable development of their communities through policies approved by their members.[45]

Cooperative Values and Principles are the essence of the cooperative identity. They establish the foundation on which, in 2002, the International Labour Organization built its Recommendation 193, which over 100 countries
have used to update their cooperative legislation.[46]

The Co-op Alliance, as the global steward of the Statement on the Cooperative Identity, provides advice on the practical application of these Principles and Values to educate cooperative leaders about how to implement
them on a day-to-day basis. The "Guidance" came out of a three-year participatory process that engaged cooperators around the world. While the
Co-op Alliance describes the Principles as "universal," it clarifies that the
Guidance is not "prescriptive" in the sense of a regulation that cooperatives must follow. It also recognizes that given the diversity of cooperatives
around the world, the interpretation and application of the Guidance will
not be uniform and will be democratically informed by local cultures and
traditions, and by the size and stage of development of each cooperative.
The Co-op Alliance provides Guidance on each of the Principles, but the
one most salient to the topic of environmental sustainability is its Guidance
on Principle 7, which is focused on sustainable development. This will be
discussed more fully in Chapters 5 and 6.[47]

In summary, while much of our economic attention is riveted on the rise
and fall of stock markets and the corporations they value, cooperatives have
been quietly exceeding multinational corporations in scope by serving over

one billion members and providing over 100 million jobs globally. People have been organizing cooperative businesses for over 200 years to serve members' needs in a wide variety of economic sectors, from insurance and financial institutions to agriculture, electricity, health, education, and housing. A typology of cooperatives groups them into enterprises composed of members who are consumers, workers, producers, or a mix of the three. Fundamentally, a values-based business form, the Cooperative Values and Principles are internationally recognized as influencing the purpose, direction, and governance structures of these businesses, including commitments to democratic participation of members, equity, and sustainable development.[48]

4

The Cooperative Difference,
Private Governance, and the Law

Distinguishing Cooperatives from
Investor-Owned Corporations

After laying the foundation with the Co-op Alliance's globally applicable Cooperative Values and Principles, we turn now to cooperative private governance and public law to show how Values and Principles shapes legal design for these social enterprises. Cooperatives are a form of enterprise that is distinct from investor-owned corporations in many ways. The fundamental distinction is that the cooperative identity of Values and Principles is shared by cooperatives across the world in all business sectors.[1]

The corollary for investor-owned corporations is maximizing shareholder value. Harken back to the critique of corporations in Chapter 2 and a growing call by reformers that center around requiring corporations to have a defined social and environmental purpose. The extent to which the cooperative identity—based on shared Values and Principles—is understood as a system of private governance, incorporated into legislation, promoted by cooperative federations, and built into each enterprise through governance documents and educational materials, are critical factors in maximizing cooperatives' potential for multiple benefits.

In Europe we see a general distinction in legislation, without a specific mandate about Cooperative Values and Principles. The EU requires cooperatives to have an objective or purpose, but when it comes to the European Company, the EU is silent about the company's purpose beyond making a profit.[2]

Federations play a role in disseminating Cooperative Values and Principles and encouraging incorporation into legal structures at the enterprise level, especially in the absence of mandatory legislation. For instance, the U.S. Federation of Worker Cooperatives includes the international Cooperative Values and Principles in its explanation of worker cooperatives.[3]

Translating this at the enterprise level, the U.S.-based Hanover Consumer Cooperative Society's general manager, Ed Fox, explains: "What defines us is . . . what we stand for. Our co-op is driven by a firm belief in socioeconomic equality, a just food system, the cooperative values and principles, and a well-nourished community cultivated through cooperation."[4]

Scholars such as Hagen Henrÿ, Gemma Fajardo García, and Antonio Fici have articulated a variety of features that further differentiate business forms. The primary distinction is that investor-owned corporations are, by design, investor-driven and focused on investment capital, so they are seeking the highest possible returns for their investors. By contrast, cooperatives are member-owned, member-driven, and focused on people, so they seek to produce positive results for their members, and results are not defined solely in economic terms.

While in corporations voting power increases as stock ownership increases, in cooperatives each member has one vote. Investors or employees of corporations are not engaged in democratic decision making (one person, one vote), while members in cooperatives are practicing these skills, which are essential for participation in a functioning democracy. Fici argues that democratic control, as required by a cooperative's governance structure, prevents external control of a cooperative and stimulates member participation in the cooperative and the economic life of a country.[5]

While those who own stock in a corporation may use the services or products of the corporation, they may have no need or interest in such use. By contrast, members of a cooperative use the services or products of the cooperative and are those who are primarily served or employed by the cooperative.[6]

Further, most of the cooperative's share capital must be held by members, any return on capital is at a limited rate, and part of the cooperative's reserves cannot be divided among the members. While a cooperative may raise capital from external sources, the terms of those arrangements must ensure democratic control by the cooperative's members and maintain the cooperative's autonomy. While corporations grow by expanding or merging, cooperatives generally don't merge, but form unions and federations to co-operate with other cooperatives, maintaining the autonomy of each entity.[7]

While corporations focus primarily on financial results, cooperatives focus on multiple goals and the way they achieve results. While corporations are seeking to implement corporate social responsibility when it is profitable, cooperatives have greater flexibility to prioritize how they relate to the community in which they are located and the larger society.

Because cooperatives generally do not need to produce a profit for investors, they have more resources to dedicate to serving broader purposes such as environmental protection or community development projects. Further, Cooperative Principles 5, 6, and 7 envision expenditure of resources externally to support nonmember education, other cooperatives, the community, and the environment.[8]

This social function is even more prominent in the emergent form of "general-interest" cooperatives. These enterprises are recognized in Italy as social cooperatives, in France as collective-interest cooperatives, in Spain as social-initiative cooperatives, and in Portugal as social-solidarity cooperatives. They primarily pursue the general interests of the community instead of focusing primarily on the mutual purpose of their members.[9]

All of these distinctive features of cooperatives derive from private governance. Depending on the jurisdiction, the features have been included in public law to varying degrees. The more these features are placed in public law, the greater the distinction between investor-owned corporations and cooperatives.

Private Environmental Governance

Some scholars argue that private environmental governance is essential to fill the environmental protection gaps left by a federal government that is actively deregulating. For instance, in the immediate aftermath of American president Donald Trump's announcement that the United States would leave the Paris Agreement on climate change, an outpouring of corporations declared they were still committed to reducing GHGs. For example, many major corporations (such as eBay, Citi, and Estée Lauder) and cooperatives (such as CROPP Cooperative/Organic Valley) formally committed to using 100 percent renewable energy on the RE100 platform (discussed below), although not legally required to do so.[10]

Even in stricter regulatory jurisdictions, private environmental governance can be used to encourage more aggressive changes within the private sector on quick time lines. Viewed as a parallel or complement to public law, Light and Michael Vandenbergh assert that private environmental governance "encompasses actions that private institutions . . . take that reduce negative externalities, manage common pool resources, and affect the distribution of environmental amenities." The government may be somewhere on the periphery, while nongovernmental actors take the lead in creating and enforcing these standards. For instance, if private governance is put into the form of a contract regarding a supply-chain standard, the government could play a judicial role in enforcing the terms. Other forms of enforcement include third-party certification or audits, consumer publicity, and boycott campaigns.[11]

The concept includes private standards an individual business sets for itself, third-party certifications, and standards and codes created by business associations. The standards can be sorted by instrument choice, similar to public law. They may be prescriptive and enforceable (such as a third-party certification), create property rights (such as an easement that maintains an agricultural land use), leverage markets (such as internal carbon trading and carbon offsets), or disclose information (such as sustainability measuring and reporting).[12]

Regarding prescriptive standards, a business may voluntarily impose an environmental performance standard or target on itself or be part of an association that creates standards for all its members. Either way, this will appear in the form of a goal or requirement. Sometimes these are seen in an entity's bylaws, internal policies, or annual sustainability report. Examples of such standards or targets are: to power the business's operations with renewable energy; to measure and report its carbon footprint; to eliminate plastic packaging; to adopt a zero-waste policy; and to sell only organic food. If the private prescriptive standard comes from an association, the third party should have the power to monitor compliance. The case studies in Part II of this book include a variety of examples of prescriptive standards for environmental sustainability that cooperatives have committed to implementing. These take sustainability measures beyond anything required by public law.[13]

Further, the instrument choice may be in the form of an ethical code that influences the fundamental purpose and norms of the business. B Corp certification and the Co-op Alliance's Cooperative Principles and Values fall into this category of private governance. As a third-party certification, B Corp has a system of oversight and enforcement because B Lab can remove the certification. The Cooperative Values and Principles, along with associated Guidance documents, can be seen as international private governance for cooperatives. Although the Co-op Alliance does not certify cooperatives or sanction them for failure to follow the Guidance, members of the cooperative can hold the entities to account through organizing the membership.

B Corp certification requires certain aspects of the multistakeholder values to be incorporated into governance documents of an enterprise. While the Co-op Alliance does not similarly require this, it is an option for cooperatives. The cooperative sustainability exemplars in the case studies show how the pathbreakers incorporate their social and environmental purposes into their bylaws or articles of incorporation. This legal design, while another form of private governance, gives greater legal durability. The enforcement mechanism lies in members of the cooperative exercising their democratic governance role and advocating for compliance. Ultimately, however, if there are no legal remedies in the jurisdiction where the cooperative is operating, the penalty to the cooperative will be loss of members who exit in protest.

Private governance exists as a parallel or supplement to legislation that incorporates the Co-op Alliance's Values and Principles and environmental legislation more broadly. It is a gap-filler when the law is absent. For example, some of the cooperative exemplars in Part II set up internal systems of carbon pricing and offsets, set zero-waste goals, and obtain all their electricity from renewables. Significantly, in cooperative bylaws and articles of incorporation, this private governance can be the result of the members' democratic participation, so it has a level of involvement and buy-in that is harder to achieve at larger scales of governance. A typology of private governance for cooperatives may assist in sorting the standard of conduct, the source of authority influencing the conduct, the private legal mechanism that is a best practice, and how the private governance is enforced. I show these factors in Table 2.

Table 2. Typology of Private Governance of Cooperatives

Standard of Conduct	Source: Enterprise, Alliance Principles, or B Lab	Governance Mechanism: Articles of Incorporation/ Association, Bylaws, Internal Policies, Sustainability Report	Enforcement: Certification, Member Advocacy, Exit of Members
SDGs/decarbon- ization: e.g., renewable-energy goal, internal car- bon market, zero waste, organic and local, Fair Trade	Principle 7; enterprise; B Lab	Articles or bylaws, internal policies, sustainability report	Member advocacy and exit; loss of certification
Concern for mul- tiple stakeholders in community where operating, including the environment	Principle 7; enterprise; B Lab	Articles or bylaws, internal policies, sustainability report	Member advocacy and exit; loss of certification
1 member, 1 vote	Principle 2; enterprise	Articles or bylaws	Member advocacy and exit
Open membership to those who use the coop's services	Principle 3; enterprise	Articles or bylaws	Member advocacy and exit
Limited return on member investment	Principle 3; enterprise	Articles or bylaws	Member advocacy and exit
Members benefit in proportion to their transactions with cooperative	Principle 3; enterprise	Articles or bylaws	Member advocacy and exit

(continued)

Table 2. (*continued*)

Standard of Conduct	Source: Enterprise, Alliance Principles, or B Lab	Governance Mechanism: Articles of Incorporation/ Association, Bylaws, Internal Policies, Sustainability Report	Enforcement: Certification, Member Advocacy, Exit of Members
Financial "surplus" partly used for reserve fund that cannot be divided among members	Principle 3; enterprise	Articles or bylaws	Member advocacy and exit
Maintain coop autonomy, governed by members	Principle 4; enterprise	Articles or bylaws	Member advocacy and exit
Educate members and others to contribute to coop development	Principle 5; enterprise	Articles or bylaws	Member advocacy and exit
Work with other coops	Principle 6; enterprise	Articles or bylaws	Member advocacy and exit

International Law

Cooperative law consists of the laws that define cooperatives; allow for their formation and legal identity; describe their organizational, operational, and financial structure; allocate surplus; define relations among constituencies, members, nonmembers, and other cooperatives; and determine the handling of dissolution, merger, demerger, and conversion. It further includes corporate, labor, tax, and antitrust laws that specifically address cooperatives.[14]

The segment of cooperative law to be examined most closely in this book are laws related to a cooperative's commitment and actualization of environmental sustainability. In other words, how cooperatives carry out Principle 7 and work for sustainable development. While the treatment of cooperative law compared to corporate or company law is sparse, the topic of environmental sustainability in cooperative law has been almost completely overlooked in the scholarly literature. None of the extensive treatises on cooperative law specifically address environmental sustainability in a separate section or chapter.[15]

Primarily, cooperative law will be found at the national and subnational levels; however, there are sources of international and regional (European Union) law as well. In a search for a body of public international cooperative law, Henrÿ identifies three key international instruments for cooperatives. These are the Co-op Alliance's 1995 Statement on the Cooperative Identity (Values and Principles), the 2001 UN Guidelines to support cooperative development, and the 2002 International Labour Organization Recommendation 193 promoting cooperatives. Each recognizes that legislation is necessary for cooperative development.[16]

Henrÿ argues that because Recommendation 193 explicitly refers to the Co-op Alliance's 1995 Statement on the Cooperative Identity, the Cooperative Values and Principles are now part of public international law. While this is contested, Fici points out that regardless, it demonstrates the Cooperative Principles "are a global 'persuasive' source of cooperative law." At a minimum, they can be seen as private governance internationally applicable, as described above.[17]

Moreover, although Henrÿ sees a growing respect for public international cooperative law, he worries that legislators have blurred the distinction between cooperatives and the dominant investor-owned corporation. He argues that these entities are quite different and advocates for legislation that seeks to distinguish them. Fici has observed a different, yet intertwined, type of problem in cooperative law. He sees a trend that many jurisdictions have replaced mandatory rules with options, "as a result of the relaxation or reinterpretation of some cooperative principles, including the democratic principle 'one member, one vote.'"[18]

Since cooperative law is mainly at the national or subnational level, Henrÿ highlights several features for cooperative law with advice for legislators that he draws from the International Labour Organization Recommendation 193:

1. Legislation should define cooperatives as "autonomous association[s] of persons united voluntarily to meet their common economic, social and cultural needs and aspirations through a jointly owned and democratically controlled enterprise." Care should be paid to giving economic, social and cultural aspects equal weight in law. Legislators should especially avoid giving too much weight to the enterprise element of the definition because this tends to blur the distinction between cooperatives and stock companies.[19]

2. Legislation should treat cooperatives equally compared to other enterprises. They should be "on terms no less favourable than those accorded to other forms of enterprise [. . .]. Governments should introduce support measures [. . .] for the activities of cooperatives that meet specific social and public policy outcomes . . ."[20]

3. Legislation should encourage cooperation among cooperatives, such as forming unions and federations to achieve economies of scale and cooperative value chains that link the producer to the consumer, while maintaining the autonomy of each cooperative. In other words, laws should "facilitate the membership of cooperatives in cooperative structures responding to the needs of cooperative members [. . .]."[21]

4. Legislation should encourage a cooperative audit to aid the members.[22]

5. Legislation should incorporate cooperative values and principles from the Alliance's statement on the cooperative identity.[23]

Of these recommendations, the advice to incorporate Cooperative Values and Principles into legislation is the one most tied to environmental sustainability. The Guidance on interpreting Cooperative Principle 7 (concern for community) goes into detail about sustainable development, drawing on concepts from the Brundtland Commission's Report, as discussed in Chapter 5.[24]

European Law

The European Union commissioned the Study Group on European Cooperative Law to assess national cooperative laws and the possibilities of cooperative law convergence in Europe. The Study Group includes in "cooperative law" the national acts or codes regarding "the institutional purposes and organizational structure" of cooperatives, including any provisions of labor, tax, competition, public procurement, or insolvency laws that are specifically dedicated to cooperatives.[25]

The Study Group has developed Principles of European Cooperative Law (PECOL). It drew these from an analysis of law in six European jurisdictions (Finland, France, Germany, Italy, Spain, and the United Kingdom). The Study Group describes the principles as the "'ideal' legal identity of cooperatives" and aims to articulate the common core of European cooperative law, based on existing cooperative law in Europe and the European Union regulation on the *societas cooperativa europaea*.[26]

The PECOL definition of cooperatives is "legal persons . . . that carry on any economic activity without profit as the ultimate purpose." The definition then branches into two categories of cooperatives, mutual and general-interest. Mutual cooperatives are what has been traditionally understood as cooperatives and are organized for "the interest of their members, as consumers, providers or workers of the cooperative . . . ." A general-interest cooperative is organized for "the general interest of the community . . . ."[27]

In the comments following the definition, the Study Group explains that this distinction has emerged in several countries to differentiate cooperatives that primarily serve a broader social purpose from those that primarily serve members of the cooperative. As examples of general-interest cooperatives, the Study Group cites those that provide social-health, educational, welfare, or cultural services to the community; or those that pursue economic activity with the purpose of integrating people who suffer social exclusion from the general labor market.[28]

The comments note further that while mutual cooperatives may have a complementary social function, general-interest cooperatives make this— for instance, environmental sustainability—their primary function. They highlight that this distinction impacts how membership is structured and how transactions with nonmembers are conducted. In a mutual cooperative,

interacting with members is an essential element of the cooperative's identity, so transactions with nonmembers create a special regulatory issue.[29]

Further, the distinction impacts cooperative governance. "Cooperatives
are directed and controlled by or on behalf of their members, who have ultimate democratic control through their governance system." Governance
structures must ensure that members democratically control and "can actively participate in policy making and major decisions, in principle on a
one member one vote basis." For governance of a general-interest cooperative, the structures should involve multiple stakeholders served by the
cooperative.[30]

One of the PECOLs is an obligatory nonfinancial audit and financial audit report by "qualified and independent auditors," at regular intervals, as
defined by law. The aim of the nonfinancial audit is to "verify that cooperatives pursue their objectives as defined by the law" and their bylaws "and
that their structure and activity are consistent with their identity as cooperatives." This is separate from a financial audit, which may be required by law.
The nonfinancial audit includes, but is not limited to, member participation
in cooperative governance; member democratic control; practices of cooperation among cooperatives; practices of cooperative social responsibility; the
manner in which the general interest has been pursued; and stakeholder
involvement in general-interest cooperatives.[31]

A nonfinancial audit measures success by the cooperative's contribution to social, regional, and sustainable development, beyond what is envisioned by corporate social responsibility. While European countries differ
in their legal approach to the audit, PECOL advocates for legislation that
distinguishes cooperatives from investor-owned corporations and insists
on a special audit to measure "member promotion and sustainable development." Nonfinancial reporting for cooperatives and corporations is discussed in detail in Chapter 6. Cooperative law of Spain and the United
States is discussed in Part II, leading into the case studies.[32]

Table 3 offers a checklist for those working on cooperative legislation at
the regional, national, and subnational levels. It is a distillation of principles
or best practices in European cooperative laws, international law, and private governance. Advocates and legislators can use Table 3 to compare their
existing or proposed legislation to these principles or best practices and
identify alignments and areas for improvement.[33]

Table 3. Checklist to Design or Assess Cooperative Legislation

Yes/No   Does the legislation . . .

- define cooperatives as autonomous associations of persons united voluntarily to meet their common economic, social and cultural needs and aspirations through a jointly owned and democratically controlled enterprise?
- give equal weight in its definition to each aspect of the cooperative, so profit is not the ultimate predominant purpose?
- include an option for a general-interest cooperative organized for the general interest of the community instead of the members?
- include an option for hybrid memberships, with worker and consumer members in the same cooperative?
- require a commitment to sustainable development goals?
- incorporate cooperative values and principles from the Co-op Alliance's Statement on the Cooperative Identity?
- require these to be part of the articles of incorporation or bylaws?
- specify democratic governance for policy making and major decisions on the basis of one member, one vote?
- encourage cooperation among cooperatives, such as forming unions and federations, and facilitate joining as members in such entities to achieve economies of scale and cooperative value chains that link the producer to the consumer, while maintaining the autonomy of each cooperative?
- require an annual financial audit?
- require an annual nonfinancial audit to aid the members and promote public transparency? This would include environmental sustainability; measures of member participation in democratic governance; member democratic control to maintain autonomy from investors, unions, or federations; practices of cooperation among cooperatives; practices of cooperative social responsibility; and the manner in which the general interest has been pursued and the stakeholders involved.
- include an exemption for small cooperatives under a certain size?
- treat cooperatives on terms no less favorable than those accorded to other forms of enterprise?

5

Sustainable by Design

What Is Sustainability?

People hold very different understandings when they use the term *sustainable*. In traditional business circles it means running a business profitably, and likely has no relation to environmental outcomes. When others use it, they may mean operating a business whose products restore natural resources and produce renewable energy. At its worst, the ambiguity of the term has allowed corporations pursuing business as usual to "greenwash" their behavior to gain market share.[1]

As used in this book, *sustainability* is often shorthand for the concept of "sustainable development." The 1992 United Nations Earth Summit defined sustainable development as development that creates economic, environmental, and social benefits that "improve the living standards of current and future generations, contribute to peaceful co-existence, social cohesion, social justice and social progress, and do so in a way that protects and does not degrade the natural environment."[2]

The 1987 Brundtland Report delivered at the Earth Summit explained that sustainable development "meets the needs of the present without compromising the ability of future generations to meet their own needs." It contains within it two key concepts: that an overriding priority is for the "essential needs of the world's poor" and that the ability to meet present and future needs may be limited by the state of technology and social organization.[3]

Importantly, the concept recognizes that ecological and other crises will persist in a world with widespread poverty and inequity. Some principles associated with sustainability are:

- The "polluter pays" to prevent and remediate harms.
- The "precautionary principle" should guide decision making, so when faced with uncertainty about environmental risks, one should err on the side of protection.
- "Intergenerational equity" should be prioritized by remembering that the current generation should make decisions that protect future generations.

In *Sustainability Reporting for Co-operatives: A Guidebook* (2016), the Co-op Alliance understands sustainable development as a paradigm shift from business as usual. It sets a goal of improving the quality of life of people, while recognizing "physical limits of the ecosystem, as well as cultural and societal limits in a framework of development." In this sense, the economy is nested inside society and the limits of ecology.[4]

In 2015, 193 United Nations member states unanimously adopted a further refinement of these concepts. They approved seventeen Sustainable Development Goals, which establish a consensus development agenda with articulated targets to be met by 2030. The Sustainable Development Goals are: 1) the elimination of poverty, 2) the elimination of hunger, 3) good health and well-being, 4) quality education, 5) gender equality, 6) clean water and sanitation, 7) affordable and clean energy, 8) decent work and economic growth, 9) the development of local industry, innovation, and infrastructure, 10) a reduction in inequality, 11) sustainable cities and communities, 12) responsible consumption and production, 13) climate action, 14) the protection of life below water, 15) the protection of life on land, 16) peace, justice, and strong institutions, and 17) the creation of partnerships to achieve the goals. These Sustainable Development Goals recognize the breadth of the challenges.[5]

Climate action and meeting the rest of the Sustainable Development Goals are inextricably intertwined. On the one hand, if we do not limit global warming, it makes achieving the other goals much more difficult.

On the other, if we do not simultaneously pursue the other goals, we have very little chance of limiting warming.

In order to avoid the worst effects of climate disruption, the 2015 Paris Agreement calls for national goals for GHGs to peak as soon as possible, and to achieve global GHG neutrality by 2050, meaning emissions of GHGs are entirely offset by removals of GHGs by sinks, such as forests. Since the Paris Agreement, scientists have emphasized the increasing urgency of making the needed changes, based on emerging understandings of climate science and protracted delays in countries' implementation of climate mitigation strategies. If we continue business as usual, the path we're on leads to climate-related disruptions and suffering by impairing health, livelihoods, food security, water supply, personal security, and economic growth, according to the 2018 report from the UN's Intergovernmental Panel on Climate Change. Further, the IPCC warned that the large majority of climate models were unable to limit global warming to +1.5°C without simultaneously pursuing international cooperation and alleviating inequality and poverty. The IPCC articulated a need for broad participation in this orientation toward sustainable development. An essential aspect of the economic transformation off fossil fuels should be achieving the Sustainable Development Goals.[6]

Businesses need not simply reduce their environmental impact, but should ensure their products and services do not harm the ability of future generations to meet their environmental resource, educational, and financial needs. And given the scope of the pollution, overuse, and extinction problems, business activity is most beneficial when it goes further and restores the environment.[7]

The case studies in this book show that system-changing sustainability flows from designing and building sustainability into the overarching organizational purpose, as reflected in legal governance documents and implemented in day-to-day practices. Creating a single organic product line or an employee transportation policy, while a step in the right direction, is not the system-changing sustainability the world needs. Fundamentally structuring the business to aim for the triple bottom line in all operations has a greater ability to shift the business in transformative ways.

Richardson and Sjåfjell argue that it is not enough for businesses to be more efficient by using fewer natural resources or creating less pollution relative to their economic activity because this fails to ensure environmental sustainability when the population and the economy are growing. They urge a legal requirement for all businesses: "all economic activity must avoid depleting non-substitutable natural capital or creating environmental externalities. It must invest more in clean, low-carbon technologies, climate adaptation projects, ecosystem rehabilitation and improvement, and other ways to build sustainability." Millstone similarly contends that the benefits of incorporation should only be extended to businesses that are committed to a social and environmental purpose.[8]

Sustainability calls for a fundamental shift to create lifestyles that use fewer resources. Instead of a culture seeking *more,* we need a culture of *enough.* Since investor-owned corporations have been driving high-consumption lifestyles in order to drive growth in sales and profits, the challenge is for businesses to operate with a different model. To accomplish this, sustainability must be part of the DNA of the business enterprise through legal design. Such businesses can be profit-seeking, but not profit-maximizing (putting profits ahead of environmental sustainability). Without serious reforms to shareholder primacy norms, as discussed in Chapter 2, there will have to be a strong business case for an investor-owned corporation to use a business model designed around the Sustainable Development Goals. In the absence of these two conditions, social enterprises will have the most flexibility to be sustainability pathbreakers.[9]

Is There a Business Case for Sustainability?

The "business case" for sustainability is often about saving money by using fewer resources or improving productivity. The limits of relying on the business-case approach are apparent when there is no business case, it is weak, or it requires long-term calculations that do not fit short-term profit goals. In *Green to Gold: How Smart Companies Use Environmental Strategy to Innovate, Create Value, and Build Competitive Advantage* (2006), Esty and Winston argue that leading corporations do more than comply with the law;

they incorporate environmental considerations into all of their operations, as follows:

- Design innovative products to help customers with their environmental problems, or even create new eco-defined market spaces.
- Push their suppliers to be better environmental stewards or even select them on that basis.
- Collect data to track their performance and establish metrics to gauge their progress.
- Partner with NGOs and other stakeholders to learn about and find innovative solutions to environmental problems.
- Build an "Eco-Advantage" culture through ambitious goal setting, incentives, training, and tools to engage all employees in the vision.

Esty and Winston's message was a counter to earlier studies that indicated the business case for sustainability was weak. According to Mozaffar Khan and coauthors, a variety of studies initially showed that "sustainability investments disproportionately raise a firm's costs, creating a competitive disadvantage in a competitive market."[10]

Yet, newer empirical studies show that sustainability investments create financial value, under certain conditions and in certain time frames. Robert Eccles and coauthors use a broader time period than earlier studies, examining company performance from 1993 to 2009. Khan, George Serafeim, and Aaron Yoon assert, "Eccles et al. (2014) identify a set of firms that adopted corporate policies related to environmental and social issues before the adoption of such policies became widespread and find that these firms outperform their peers in the future in terms of stock market and accounting performance." Light emphasized that the important point of this study is that a long-time horizon "may be crucial" not only for environmental protection, but for showing market value.[11]

The mixed results in the earlier studies may be due to the time line and the evolving nature of the data. The number of companies measuring and reporting sustainability, especially using the Global Reporting Initiative framework, has rapidly grown in the past decade, so that such practices are now the norm among the largest corporations. Further, until recently, there had been no guidance on the materiality of sustainability issues. However, the Sustainability Accounting Standards Board recently created a classification, and as of early 2014, had developed this for forty-five industries. Khan, Serafeim, and Yoon used this 2014 classification to develop a novel data set to measure firm investment performance. They found that firms with "strong ratings on material sustainability issues have better future performance than firms with inferior ratings on the same issues." Their results are "consistent with materiality guidance being helpful in improving the informativeness of ESG [environmental, social, governance] data for investors."[12]

Since Khan's research concludes that firms with strong ratings on material sustainability have stronger financial performance, a reasonable investor should seek to understand this aspect of a corporation before purchasing or trading its securities. This research could be used to help persuade the U.S. Securities and Exchange Commission to use its existing authority to interpret "materiality" to include environmental disclosures to inform the "reasonable investor." To date, that has not happened. In the commission's 2020 proposed rulemaking to "modernize and enhance" Regulation S-K, one of the commissioners criticized the proposal for failing to address reporting on climate change and other environmental, social, and governance concerns. Across the Atlantic, the European Union has already deemed such information important and now requires disclosures via its Non-Financial Reporting Directive, as explained below. This could provide a blueprint for the United States and other jurisdictions to pursue a similar level of transparency for investors.[13]

These nonfinancial disclosures are fundamental for those involved in corporate social responsibility and socially responsible investment, and, if legally required, could ensure that sustainability considerations lead to measurable, comparable, and reliable data, and related improvements. These movements within the corporate world are important to integrate and promote sustainable development at the enterprise level. The World Business Council for Sustainable Development frames corporate social responsibility as an ongoing commitment of ethical behavior and contributions to "economic development while improving the quality of life of the workforce and their families as well as of the local community and society at large." These more recent analyses by Eccles, followed by Khan and their collaborators, will bolster the business case for corporate social responsibility and socially responsible investment. Of course, better data with more transparency would assist in this grand project.[14]

The prospect of making sustainability a top priority for corporate leaders is shifting dramatically and rapidly, especially with institutional investing. At the beginning of 2020, BlackRock sent CEOs of the companies in which it invests a letter entitled "A Fundamental Reshaping of Finance." In it, Larry Fink, the chair and CEO, declared that "[e]very government, company, and shareholder must confront climate change." BlackRock ranked climate change as a top issue that has become "a defining factor in companies' long-term prospects." The letter continued that "climate risk is investment risk," and outlined the major features of risk, including floods, droughts, food costs, productivity declines from extremes, and uncertainties about mortgages and urban infrastructure needs in order to adapt to climate disruptions. Fink asserted that there will be a "significant reallocation of capital" in the near future. In response to this risk assessment, BlackRock is placing "sustainability at the center" of its investment approach. This will involve moving money out of investments that present high risks, such as "thermal coal producers," and creating new products that screen to exclude fossil fuels.[15]

Sustainability disclosures will see a major boost from this. BlackRock argues, "Each company's prospects for growth are inextricable from its ability to operate sustainably and serve its full set of stakeholders." To that end, the firm announced that the companies in which it invests need to produce sustainability data within that year (2020) to show how they are managing climate, including how the Paris Agreement's +2°C goal impacts them, how they are serving their stakeholders (not just shareholders), and what they are doing to enhance workforce diversity, supply-chain sustainability, and protection of consumer data. Significantly, Fink underscored the purpose of the corporation and identified social purpose as "the engine of long-term profitability." In reflecting on his forty years in finance, he distinguished climate change as a "structural, long-term crisis" different from any the world has experienced since the 1970s.[16]

BlackRock is moving the dial on corporate sustainability reporting by requiring it for all their investments. The 2020 gathering of billionaires and top corporations at the World Economic Forum in Davos followed that thread with its theme "Stakeholders for a Cohesive and Sustainable World," and invited teen climate activist Greta Thunberg to address them. This shift in focus could signal a turning point in the global economy and rapid movement of capital toward businesses that are taking the lead in forging a sustainable future through deep decarbonization. How businesses measure and report sustainability is going to take on a much greater significance in this reshaping of finance. Public laws could enhance a rapid transformation, but it is noteworthy that it is being led by a private entity with outsized capital influence. Still, as the next chapter will show, measuring and reporting sustainability does not necessarily lead to actual improvements. For sustainability results, new business models are essential, and governments should set clear boundaries that protect ecosystems and promote sustainable development rather than relying on variable voluntary efforts.

6

Measuring and Reporting Sustainability

THIS CHAPTER EXPLAINS how businesses measure and report sustainability, and critiques the limits of such a reporting system. It starts by describing the gold standard in reporting that most major corporations use as voluntary initiatives to show their corporate social responsibility and appeal to investors. Then it discusses the new law in the European Union that mandates non-financial reporting for certain businesses, and what data shows from the first years of implementation. It then turns to sustainability reporting in cooperatives to discuss the international guidelines for this sector and data on whether cooperatives are measuring and reporting sustainability. Finally, it delves into concerns about reporting not resulting in actual sustainability and the need for changes at a more fundamental level of the corporation's purpose and business models animating the private sector.

The Global Reporting Initiative—the Gold Standard in Corporate Sustainability Reporting

Sustainability reporting has become the norm for the world's largest corporations. Driven by investors, regulators, and public pressure, the use of some form of sustainability reporting among the world's largest 250 companies increased from 12 percent in 1993 to 93 percent in 2013. When scaling out to include the 500 large companies listed on the S&P index, in 2019, 90 percent published sustainability reports. Most large corporations use the Global Reporting Initiative (GRI) for sustainability reporting. GRI's Sustainability Disclosure Database contains at least 24,000 reports, and twenty-seven countries and regions reference GRI in their policies. Similarly, in an analysis of 2018 nonfinancial data reporting by the European Union's large publicly traded and financial companies, most of them (59 percent) used GRI to comply with the European Union's Non-Financial Reporting Directive.[1]

In 2015, GRI's Corporate Leadership Group examined sustainability trends businesses face and did a deeper dive into climate change, human rights, wealth inequality, and data and technology. Climate change was at the top of their agenda, as it is a critical global risk and prominent among the seventeen Sustainable Development Goals. They see a climate focus being driven by governments' Intended Nationally Determined Contributions under the Paris Agreement, related new funding sources, and climate-related financial and risk disclosure requirements for certain corporations. Related to climate reporting, the GRI contains a variety of standards and disclosures, such as energy consumption, reducing energy requirements of products and services, and tracking GHG emissions.[2]

Another forum for publicly declaring voluntary commitments to mitigate climate change is RE100, which has a growing list of companies committed to using 100 percent renewable energy within very short time frames. Launched in 2014, RE100 encourages the private sector to accelerate the push for renewables. More and more keep joining the list: in December 2018 there were 154 companies committed to RE100; by April 2019 this had grown to 169, and by April 2020 there were 235. This may seem like a very small number given the global total of companies; however, many of the companies on this list are publicly traded multinational corporations with massive energy demands, such as Apple, Coca-Cola, Citi, Danone, Google, Facebook, and Unilever. These private sector commitments to power their operations on renewable energy are one of the drivers for electric utilities shifting away from fossil fuels, as will be discussed in the renewable energy case studies in Part II. With the fairly universal adoption of the GRI sustainability reporting framework and the RE100 for investor-owned corporations, voluntary corporate commitments to sustainability are now very much in the mainstream globally for the largest corporations.[3]

In December 2018, the only cooperative on the RE100 list was U.S.-based CROPP Cooperative/Organic Valley, which is featured in the case studies in Part II. The RE100 list underreports the scale of the transition. The lack of cooperatives showing up on the list does not demonstrate a weaker commitment to sustainability. Many in the cooperative sector have made this shift without adding their name to this list: the pathbreaking cooperatives discussed below are almost all powering their operations on renewables with very little fanfare. They are driven by something deeper in their core DNA.

European Union Non-Financial Reporting Directive

The European Union has moved beyond voluntary disclosures with its Non-Financial Reporting Directive, enacted in 2014 and effective in 2017. Nonfinancial reporting is a way to get at the externalities of business without directly prohibiting or regulating them. The social costs of pollution, resource depletion, worker health and safety issues, wealth inequalities, and human rights abuses are known as externalities. Economists and accountants typically do not account for them at the enterprise level, and society/stakeholders are left to shoulder the costs. Yet, a growing cadre of academic and business leaders have pushed to change this by legally requiring corporate accounting for these nonfinancial matters. While it is unclear that legally required disclosure leads to improved environmental outcomes, Ioannis Ioannou and George Serafeim argue that it is good for businesses. They conducted an instrumental variables analysis suggesting that "increases in sustainability disclosure driven by the regulation are associated with increases in firm valuations."[4]

China, Denmark, Malaysia, South Africa, and the European Union have regulations requiring disclosure of environmental, social, and governance information. The law firm Frank Bold asserts that the Non-Financial Reporting Directive requires "[m]ore than 6,000 large companies" to disclose information on their "business model, policies, risks and outcomes regarding environmental, social and employee matters; respect for human rights; anti-corruption and bribery issues; and board diversity." The firm says the law's objective is to "lay the foundation for an integrated model of corporate reporting" that supplements and complements the "financial transparency" already required of companies, in order to "understand a company's development, performance and position, as well as the impact of its activity on society."[5]

The Non-Financial Reporting Directive requires companies to provide, at a minimum, the following information:

- a brief description of the undertaking's business model;
- a description of the policies pursued by the undertaking in relation to those matters, including due diligence processes implemented;
- the outcome of those policies;
- the principal risks related to those matters linked to the undertaking's operations, including, where relevant and proportionate, its business relationships, products or services that are likely to cause adverse impacts in those areas, and how the undertaking manages those risks;
- nonfinancial key performance indicators relevant to the particular business.

By December 2016, each European Union member was supposed to enact these requirements into their national laws. While the EU directive requires reporting, it does not prescribe a uniform report to use or specific types of information that must be disclosed. This complicates comparisons across companies and may prove to undermine the efficacy of this legal requirement. The audience for reporting is the investor community, civil society, and governments that seek to assess corporate social responsibility.[6]

The EU directive applies "to publicly traded corporations, banks and insurers with more than 500 employees." However, individual countries can impose more stringent requirements. For instance, Sweden, Iceland, and Denmark apply the requirements to all companies with more than 250 employees. Cooperatives are generally not covered by this EU reporting directive since its target is large, publicly traded corporations and financial institutions, yet large cooperatives may benefit from a similar reporting requirement tailored to them, as will be discussed.[7]

Ultimately, these corporate reformers are trying to use one of the largest economies in the world (the European Union) to change the global economy to align its activities with achieving a shared social and environmental purpose that increases prosperity for all. The aim of the EU directive is "to reorient capital flows towards sustainable investments and manage risks stemming from climate change, environmental degradation and social issues." From a purely pragmatic perspective, advocates for this law highlight that "mismanaged" human rights and environmental risks are bad for a company's economic performance due to costly accidents, litigation, supply-chain disruptions, damaged reputation, and failed investments.[8]

The Alliance for Corporate Transparency, a cohort of leading civil society organizations, has analyzed how the largest companies operating in the European Union complied with the law in the two years after it went into effect. The Alliance reviewed 105 companies' 2018 disclosures and 1,000 companies' 2019 disclosures of environment, social, employee, human rights, and anticorruption performance. Both years showed similar deficiencies. Overall, the Alliance found that half the companies failed to provide clear environmental information useful to assess issues, targets, and principal risks. Spanish companies fared worse than the European Union average: only 35 percent provided this kind of specificity about the environment. Nordic companies, at 92 percent, provided the most clarity. However, when focused on key issues and targets for improvement related to climate change, Spanish companies, at 63 percent, performed better than the European Union average of 50 percent.[9]

While the Alliance found that 90 percent of the companies reported on climate change, only 47 percent actually described the goal of their corporate climate policies and how they would be achieved. Reporting on GHG emissions is widespread, at 82 percent overall, and ranging from nearly universal reporting in the energy sector to 70 percent in the health care sector. However, only 26 percent of energy and resource extraction companies reported on how they were achieving a transition to below +2°C. Based on these findings, the Alliance recommends an update to the legislation to "clarify the requirement for the disclosure of companies' long-term transition plans to a zero-carbon economy and their economic implications."[10]

The researchers found a similar absence of meaningful information in reporting about water and biodiversity. For instance, only 24 percent of energy and health care companies operating in water-scarce areas reported on water use. Strikingly, the research showed that less than 10 percent dis-

closed lobbying expenditures and the public positions on which the corporations lobbied. The researchers found a correlation between company size
and report quality, with the largest companies (more than 50,001 employees) "presenting better results" than the smallest companies (fewer than
1,500 employees).[11]

The Non-Financial Reporting Directive is a prescriptive public-law strategy to increase access to sustainability information. Currently it does this
without setting any particular goals or purposes, or even a standardized
reporting format. How effective the reporting requirement will be in bringing about a change in corporate purposes and goals is unknown. How investors will respond to the nonfinancial disclosures remains to be seen. The
EU directive holds potential, but the analysis of compliance with the law
points out areas where it needs to be amended to make it more meaningful.

Further, it doesn't appear that reporting on externalities, without more,
leads to curbing them. In the analysis of the first year's data, researchers
say "the assessment of companies' reporting on their business model suggests a common disconnection between the non-financial statement and
the rest of the annual report." Leaving corporate transformation solely to
these voluntary means, even when coupled with mandatory nonfinancial
disclosures, will likely fall short of mobilizing the private sector at a scale
needed to move the global economy off fossil fuels in a manner that supports achieving broader Sustainable Development Goals.[12]

The short time lines for financial returns that investors demand pushes
against sustainability, despite new research showing positive long-term
financial data for companies with a strong commitment to corporate social responsibility. The short time lines and elevation of short-term profit
above all other goals may ultimately undermine the ability of sustainability reporting to produce improvements in sustainability outcomes, as discussed below.

If investor-owned corporations are required to pay for additional sustainability rather than externalizing those costs, as is the current status quo,
they can do so in several ways: by reducing returns to shareholders, reducing compensation to executives, reducing wages paid to labor, or increasing prices to consumers. Without reforming shareholder primacy and
excessive executive compensation (as measured by the gap between the

highest- and lowest-paid employees in an enterprise), workers will pay in stagnant wages and consumers will pay in higher prices. In other words, if addressing inequality is a primary goal (as it is in the Sustainable Development Goals) while deeply decarbonizing economic activity, legally binding reforms to shareholder primacy and executive pay in investor-owned corporations are essential.

Cooperative Reporting Guidelines

Similar to corporate sustainability reporting, cooperative reporting involves measuring, disclosing, and demonstrating accountability for organizational performance leading toward the goal of sustainable development. Ideally, such reporting would involve an integrated assessment of economic, environmental, and social impacts.

The tools cooperatives have been using for sustainability reporting are not uniform and lack metrics. Since cooperatives vary in size, structure, and sector, there is a strong argument for tools that are adaptive instead of totally uniform. To aid cooperatives in sustainability reporting, the Co-op Alliance released *Sustainability Reporting for Co-operatives: A Guidebook* in March 2016. The Co-op Alliance's Sustainability Solutions Group created this reporting framework, and they softly encourage all cooperatives to "consider" the framework.[13]

As discussed, the Global Reporting Initiative is the most commonly used sustainability reporting framework for corporations. Table 4 shows how the Co-op Alliance's *Guidebook* interprets the principles guiding the Global Reporting Initiative as applied to cooperatives.[14]

The Co-op Alliance's *Guidebook* does not advocate a uniform approach, but rather a review of existing sustainability frameworks and their relevance to a particular industry, relevance to a cooperative, cost, and requirements for verification. The sustainability reporting frameworks that cooperatives have used in addition to the Global Reporting Initiative (most commonly used by large cooperatives), are AccountAbility, B-Corporation, Carbon Disclosure Project, Global Alliance for Banking on Values (used by financial cooperatives), and International Integrated Reporting Framework.[15]

In addition to selecting a reporting framework, the Co-op Alliance's *Guidebook* recommends adopting a series of indicators that address Cooperative

Table 4. Principles Guiding Global Reporting Initiative Sustainability Reporting

Principle	Outcome
Stakeholder inclusiveness	The cooperative should identify its stakeholders and explain how it has responded to their reasonable expectations and interests.
Sustainability context	The report should present the cooperative's performance in the wider context of sustainability.
Materiality	The report should reflect the organization's significant economic, environmental and social impacts that can substantively influence the decisions of stakeholders.
Completeness	The report should include coverage of material aspects sufficient to reflect significant economic, social, and environmental impacts and to enable stakeholders to assess the organization's performance.
Balance	The report should reflect positive and negative aspects of the cooperative's performance to enable a reasoned assessment of its overall performance.
Comparability	The cooperative should select, compile, and report on information consistently that enables an analysis of performance over time.
Accuracy	The reported information should be sufficiently accurate and detailed for shareholders to assess the organization's performance.
Timeliness	The cooperative should report on a regular schedule so that information is available in time for stakeholders to make relevant decisions.
Reliability	The cooperative should gather, record, compile, analyze, and disclose information and processes used in the preparation of a report in such a way that they can be subject to examination, and that the quality and materiality of the information can be established.

Data from International Cooperative Alliance, *Sustainability Reporting for Cooperatives: A Guidebook*

Principles and Sustainable Development Goals: "A good indicator captures a broader trend, is unambiguous, is based on available data, and can be readily tracked over time."[16]

The *Guidebook* acknowledges that cooperatives could benefit from a new reporting framework that is built for them, but leaves the creation

Table 5. Sample Sustainability Indicators for the Cooperative Principles

Cooperative Principle	Sample Indicators
1. Voluntary and open membership	Number of members Diversity (age, sex, race, ethnicity, educational background, etc.) of members Diversity of the board
2. Democratic member control	Percent of members who voted in the board election Number of resolutions put forward by members
3. Member economic participation	Percent of capital provided by members Percent of total capital that is indivisible Indicator of member loyalty
4. Autonomy and independence	Percent of assets owned by outside investors
5. Education, training, and information	Percent of revenues expended on education, training, and information Number of participants by category (members, public, youth, directors) in programs run by the cooperative
6. Cooperation among cooperatives	Percent of revenues and expenses associated with other cooperatives
7. Concern for community	GHG emissions per member Percent of revenue allocated for community-based organizations Weight of food donated to community food banks Number of community organizations supported

Data from International Cooperative Alliance, *Sustainability Reporting for Cooperatives: A Guidebook*

of such a tool open for future development. Table 5 shows how the Co-op Alliance ties Cooperative Principles to specific sustainability reporting metrics.[17]

Finally, cooperatives, like corporations, need to decide the best way to present their sustainability reporting to the public. Several questions to be determined are whether the sustainability report stands alone or is part of an overall annual report; whether to use storytelling, cartoons, data visual-ization, and so on; and modes of dissemination to members and the public.[18]

Fiona Duguid researched the availability of nonfinancial measurement tools for cooperatives. Her 2017 findings were that while a variety of non-financial impact measurement tools are available, they are not standard-ized, very few are specific to cooperatives, very few use measurable indica-tors, and most emphasize the social over the environmental component of sustainability.[19]

Henrÿ argues for cooperative legislation that requires a nonfinancial au-dit, covering, among other things, ecological metrics. This requirement is not part of international law, so its applicability will vary from jurisdiction to jurisdiction. Italian law, for instance, requires regular audits of finan-cial and nonfinancial factors, and thus provides an example for legislation mandating such disclosures; however, it does not specifically require envi-ronmental sustainability as part of this. An Italian cooperative must have at least one registered auditor to examine the accounts. It is then subject to au-dit, typically every two years, covering a variety of aspects of the enterprise, including the "mutual nature" of the cooperative, effectiveness of member-ship, member participation in management, mutual relationships, absence of profit distribution purposes, and eligibility for tax and other benefits. An improvement on the Italian law would be to explicitly require measuring and reporting environmental, social, and governance metrics.[20]

Research on Cooperatives' Sustainability Reporting

If the hallmark of cooperatives is that they adhere to the seven Coopera-tive Principles, does adherence to these principles result in observable and predictable outcomes that differ from other corporate forms? In particular, do cooperatives have stronger environmental sustainability outcomes? How useful is sustainability reporting in answering these queries?

In a world that demands corporations absorb the true costs of doing busi-ness, some cooperative experts argue that cooperatives would be poised to outcompete with better prices and wages. Cooperatives have no share-holders other than members who have equal shares, typically have tight compensation ratios between the highest and lowest paid, and Cooperative

Principles and Values that encourage members to spread social costs across the membership more equitably. Designed to serve their members, cooperatives appear to be in a better position to provide lower prices than investor-owned corporations that are forced to stop externalizing costs. According to Williams, "it is now known that cooperative social bonds can lower the cost of almost any economic activity."[21]

Despite the structural advantages of cooperatives vis-à-vis investor-owned corporations regarding minimizing nonfinancial costs, data is lacking to firmly demonstrate this. Research on cooperatives' nonfinancial reporting of sustainability metrics shows that documenting sustainability is not a well-developed area for cooperatives. The Co-op Alliance's 2013 *Blueprint for a Cooperative Decade* makes sustainability a priority, but recognizes that it is not universally associated with cooperatives. The *Blueprint* urges change in this area to strategically position cooperatives as sustainability builders.[22]

The Co-op Alliance describes "sustainability reporting" as a mechanism that is common to corporations, but not as common in cooperatives; it cites research showing the largest 300 cooperatives are *not* engaged in sustainability reporting as much as Fortune 500 companies. For instance, in 2010, only twenty-two of the 300 largest cooperatives issued a sustainability report, and those twenty-two rarely commissioned an external entity to validate the results. By comparison, in that same year, 82 percent of the top 250 companies in the Fortune 500 used the GRI sustainability reporting framework. By 2013, sustainability reporting was almost universal among the top 250 corporations.[23]

In the report *Co-operatives and Sustainability: An Investigation into the Relationship* (2013), the Co-op Alliance analyzed connections between sustainability and Cooperative Principles and the degree to which cooperatives referred to sustainability in their annual reports and websites. The Co-op Alliance's research used a method of reviewing website and annual report content to assess whether cooperatives were communicating about sustainability and how. This approach scans for terms instead of specific outcomes (for example, reducing GHGs). The content analysis showed cooperatives were not communicating about the environmental aspect of sustainability as much as they were about the "community." Cooperatives described themselves in terms that emphasized the social rather than environmental and economic dimensions of sustainability.[24]

Following the Co-op Alliance's 2013 study, Fiona Duguid and Donna Balkan's 2016 study analyzed Canadian cooperatives' online sustainability communications. They applied a content analysis approach to the websites of 118 Canadian cooperatives, mutuals, and their federations. The research explored whether they report on sustainability-related activities and issues and/or communicate concepts related to sustainability through their websites and publicly posted reports.[25]

Duguid and Balkan posit that in contrast to investor-owned businesses, cooperatives are generally values-driven and so are more likely to take a long-term view of business decisions. Yet, they recognize that prior research has found globally cooperatives are not using any sort of sustainability reporting tool, and those that are, do not use one that is designed specifically for cooperatives. As such, cooperatives lag behind corporations on sustainability measurement and reporting.[26]

Of the 118 entities Duguid and Balkan studied, 77 percent had online sources that talked about sustainability. Only about half of the entities had online annual reports, and only 63 percent of the reports discussed sustainability. The entities produced reports with a variety of titles, such as "Corporate Social Responsibility Report," "Community Report," "Governance Report," and "Accountability Report"; only three had dedicated "Sustainability Reports."[27]

Further, the environmental aspects of sustainability were not dominant in these communications. The Canadian analysis showed the key terms cooperatives used most frequently were: 1) "community," 2) "economic," and 3) "social." The environmental aspect of sustainability ranked fourth in frequency among the terms used: cooperatives used it to encompass a commitment to reducing GHGs, carbon footprints, and the use of water, paper, and energy. The research found that in Canada, "the financial co-operatives—banking and insurance—are leagues ahead of the non-financial co-operatives" on sustainability reporting.[28]

In addition, Duguid and Balkan found the Canadian federations and umbrella organizations entirely lacked sustainability communications, which the researchers found "most surprising" given that these entities should be

leaders in this regard. Hence, they concluded there is a great need and opportunity for cooperative federations, umbrella organizations, and research
centers to expand their role in promoting environmental sustainability.
Last, the research pointed to the need for a follow-up study of what Canadian cooperatives are doing in terms of sustainability impact and what tools
they are using to measure it.[29]

In a 2014 study, H. Bollas-Araya and colleagues produced a comparative analysis of sustainability reporting by European cooperative banks and
investor-owned banks, based on GRI Sustainability Disclosure Data from
2000 to 2013. After the financial crisis, some scholars found banks using
sustainability reporting as a way to generate and maintain trust among their
stakeholders. They assert that cooperative banks have placed sustainability
at the core of their identity. The researchers' meaning becomes clear with
examples they provide, none of which involve environmental protection. By
"sustainability," they describe supporting the livelihoods of local communities by financing farms, fishing, and other small enterprises, supporting
"developing countries," and taking part in the life of their communities. In
their content analysis of banks' sustainability reports, the researchers found
a primary focus on social issues, such as workplace stability and supporting
local sports and arts groups; but what was missing was reporting on environmental sustainability. Also, they found investor-owned banks were more
likely than cooperative banks to produce formal sustainability reports.[30]

In summary, research on cooperatives in a variety of jurisdictions, including Canadian cooperatives and European bank cooperatives, shows that
cooperatives do not typically measure and report on environmental sustainability. Prior research shows that the *environment* is often not part of what is
meant when *sustainability* is discussed on websites or in reports. Research
similarly shows that Canadian federations and umbrella organizations are
not actively promoting environmental sustainability. Finally, research has
not demonstrated that environmental sustainability is part of the intrinsic
design of cooperatives. Instead, it shows that cooperatives are not communicating about their work to advance environmental sustainability.[31]

Lack of reporting and communication by cooperatives does not necessarily mean cooperatives are less environmentally sustainable. It could indicate an absence of a structured approach to measuring and communicating

about these goals because they have not been driven to do so, as corporations have, by regulations or an investor community.

Since the European Union is now requiring nonfinancial reporting for publicly traded corporations and financial institutions, the results may be instructive for a legal requirement for large cooperatives. However, the findings in the case studies in Part II of this book caution against reading much more into the absence of sustainability reporting. The case studies show that there are strong sustainability pathbreakers among cooperatives that do not formally measure and report sustainability nor have well-developed communications about it on their websites. Nonetheless, these pathbreakers had environmental sustainability at the core of their business model and purpose, and were showing actual improvements in their environmental performance. The approach to environmental sustainability reporting varied among the sustainability pathbreakers, based on the size of the entity. The smallest entity studied, with two employees and only €100,000 in annual revenues, built environmental sustainability into every aspect of the business, but did not do any official sustainability reporting. A midrange cooperative of 300 farmers excelled in incorporating sustainability into all aspects of their farming and business practices. Their actual implementation exceeded what they promoted in their marketing materials, and they did not produce a sustainability report. The first and largest consumer renewable energy cooperative in Spain is entirely designed around environmental sustainability, and yet does not produce a Global Reporting Initiative type of report.[32]

Although many of the sustainability pathbreakers did not report on sustainability, one that did do so found it very helpful. A worker-owned coffee roaster found that measuring and reporting its carbon footprint to peer businesses, as well as undergoing the reporting necessary to become a certified B Corp, has helped it improve sustainability outcomes.

Cooperatives could use sustainability reporting to aid them in assessing, understanding, and communicating their environmental sustainability value to the world. Yet, one should not equate lack of formal reporting with lack of environmentally sustainable outcomes. The case studies show that a fundamental driver for environmental outcomes is the business model and purpose of the organization. When legal design reflects a strong sustainability purpose, and this is coupled with measuring and reporting appropriate to the size and type of the business, cooperative pathbreakers are poised to accomplish more than required by environmental law.

Limits of Sustainability Reporting and the Need for New Business Models

Sustainability reporting is now the norm across the world's largest corporations, and is required by law for certain large corporations in the European Union. Yet, according to the Global Reporting Initiative's self-assessment, so far sustainability measuring and reporting has not generated improved environmental and social outcomes on the scale needed. In fact, in 2015, GRI's Corporate Leadership Group observed that measuring and reporting sustainability has not influenced "the decision-making processes that underpin the transition of governance structures, business models and resource consumption patterns to a sustainable level." This conclusion is based on seeing increases exactly where sustainability calls for decreases.[33]

Significantly, the GRI self-assessment highlighted the enduring global economic model of linearity instead of circularity as a particular problem. GRI's leadership observed that companies "harvest and extract materials, use them to manufacture a product, and sell the product to a consumer— who then discards it when it no longer serves its purpose." It reported an increase in the volume of raw materials entering the economy: 65 billion tons in 2010, expected to grow to about 82 billion tons in 2020.[34]

By contrast, a circular-economy business model designs products and services for durability and long life, repair and reuse. At the end of a product's useful life, it becomes raw materials for another product. A circular-economy paradigm is defined by the concept that nothing is waste: materials are by design circling through primary use, reuse, repair, repurpose, remanufacture, and recycle.[35]

The GRI Corporate Leadership Group provides suggestions for improvements. For instance, it urges companies to identify an economic model, such as the circular economy, and report on how it is making a transition to this model. Further, company reports could show how they are meeting specific Sustainable Development Goals. Another option is to place a monetary value on externalities and report on this in a much more transparent way in language understood by investors.[36]

If companies implement the GRI's suggested reforms, it could accelerate moving the world economy to be more sustainable. The push for publicly traded corporations to embrace corporate social responsibility is a positive trend that includes environmental sustainability. This often bumps up against the norm of maximizing profits for shareholders. True reform in the sense of a new business model, suggested by GRI, faces practical and ideological limitations. Some scholars worry that corporate social responsibility is just another marketing tool for businesses to maximize profits instead of securing enduring reductions in environmental harms and lowering resource use.[37]

Of course, there are shining stars in the investor-owned world, such as Interface, which has doggedly been pursuing environmental sustainability since 1994. There are alignments between environmental protection and restoration and profit that make the business case in certain situations. Interface's passionate founder and chair was able to lead the company to reframe its purpose and sustainability practices while growing into the world's largest producer of modular carpet. Similarly, Esty and Winston's case studies identify a variety of corporate exemplars that predate certified B Corps. They need to be encouraged to flourish. Yet, the reality is that these leaders are too few and are not significantly modifying the important trend lines in GHG emissions, extinction, water quality, resource use, and other measures of global environmental outcomes, within the necessary time frame.[38]

Legal Structure, Purpose, and Sustainability in Cooperatives

We are at a time when all businesses should be engaged in the undertaking at hand to promote the Sustainable Development Goals while deeply decarbonizing the economy. We need corporations within the existing system to make sustainability advances while we pursue reforms to transform the corporation. Chapter 2 discussed the purpose of investor-owned corporations and the push for corporate reform that many view as being at the heart of the problem: the shareholder primacy norm that prioritizes shareholder profit over environmental sustainability.

Harkening back to the call for reforming the purpose of investor-owned corporations, Richardson and Sjåfjell outline a corporate law reform that "embeds environmental standards in the governance of economic institutions in order to minimize the tensions their managers face between reconciling expectations that they act in the public interest while serving their private constituencies." They see the voluntary approaches of corporate social responsibility and socially responsible investing as complementary but insufficient to alter the purpose of economic activity and steer the economy toward long-term sustainability. While Richardson and Sjåfjell and other corporate law reformers are addressing the broader range of corporations, they recognize the current impediments to advancing their proposals.[39]

By contrast, GRI's suggested reforms to business models align well with cooperatives' legal structure, principles, and values. Cooperatives are positioned to immediately respond without any needed changes in law. Apart from larger cooperatives that have the human resources to undertake measuring and reporting, cooperatives are not widely using formal sustainability reporting. More fundamental than sustainability reporting, as recognized by GRI's Corporate Leadership Group, is the extent to which businesses are using models that incorporate sustainability into the core DNA of the enterprise. On this measure, cooperatives are in a good position to align more with the system-changing business models GRI suggests and ultimately could innovate more for sustainable outcomes.[40]

The case studies in Part II of this book explain how this is already occurring in key economic sectors that must be deeply decarbonized: energy, food and agriculture, water, finance, and trade. Although specific legal reforms could move this sector to be more effective and robust, as discussed in Part II, the sector is fully functional already and as a matter of private governance has an international focus on operationalizing the Sustainable Development Goals at the level of the cooperative enterprise.

At the Co-op Alliance's General Assembly in 2007, the members affirmed their commitment to address climate change and reduce its impact. The Co-op Alliance asserts that "co-operatives have been active in promoting sustainable development for over 150 years." It ties this track record to the structural difference of cooperatives being a democratically controlled business, and the Values and Principles difference of cooperatives caring for their communities, serving members not solely in economic terms, but with an eye on broader social, cultural, and environmental concerns. Not dominated by an economic growth goal, cooperatives have focused on building community wealth and prosperity, broadly recognizing the triple bottom line before that term was used in traditional business jargon. People who understand this sector can design governance documents to incorporate Cooperative Principle 7. The Co-op Alliance's Guidance on how to implement Principle 7 directly references the United Nations Sustainable Development Goals and explains how to translate them into enterprise-level activities.[41]

Moreover, sustainability is one of the priorities in the Co-op Alliance's *Blueprint for a Co-operative Decade*. The Co-op Alliance sees implementing the *Blueprint*'s recommendations as foundational to achieving its vision for cooperatives "to become the acknowledged leader in economic, social and environmental sustainability, the model preferred by people, and the fastest growing form of enterprise." In order to further advance this vision, the *Cooperative Growth for the Twenty-first Century* identified areas of emphasis. It recommended, among other things, addressing climate, energy, and other critical environmental challenges, and partnering with climate and food-sovereignty social movements. The sustainability exemplars in the case studies in Part II are tackling these challenges and more head on as they build their businesses.[42]

The Co-op Alliance's detailed Guidance document on sustainability clearly states that Cooperative Principle 7, "concern for community," is the locus of environmental sustainability for cooperatives. The Guidance uses the term "sustainable development" in interpreting Principle 7. It intended to mirror the term in the Brundtland Report, and its Guidance is informed by these UN-defined meanings. The Guidance insists that "[a]ll co-operatives have a responsibility and duty to consider and reduce their co-operative's environmental impact and promote environmental sustainability within their business operations and in the communities in which they operate." The Guidance observes that when "co-operatives work for the sustainable development of their communities through policies approved by their members," there is a strong emphasis on the community in which a cooperative is based. The Guidance also indicates *who* determines the proper balance between self-interest and community interest: the members should decide democratically. This challenges cooperatives to successfully run a democratically controlled and owned business that is environmentally sustainable and deeply committed to the community.[43]

The Cooperative Principle 7's "concern for the community" clearly defines a community as local, where the cooperative operates. If the understanding stopped there, cooperatives could be isolated, parochial enterprises. Yet, community is not understood solely as local. According to the Guidance, actualizing Principle 7 includes:

- using "ethical" supply-chain contracts;
- Fair Trade;
- promptly paying suppliers;
- trading with and supporting other cooperatives;
- using green consumerism;
- organic agriculture;
- renewable energy; and
- advocating for governments to implement environmental policies and initiatives.

This melding of an intense commitment to the local community with a concern for connecting and trading globally in a fair and ethical manner is the balance that is missing from some current internationally important debates. Cooperators' commitment to local development is not at the expense of, and in isolation from, global development. In this sense, the cooperators embrace multiple purposes, goals, scales of community, and solidarity, and reject the dualistic us-versus-them thinking that animates nationalist movements.[44]

Finally, with regard to the question of how sustainability is implemented and measured, the Guidance urges all cooperatives to audit their carbon emissions, seek to reduce dependence on fossil fuels, and ensure all the timber they use is certified as being sourced from sustainably managed forests, among other things. The kind of assessing and reporting the Guidance promotes was discussed above. Suffice it to say here that corporate social responsibility is an attempt to translate the UN Sustainable Development Goals into something that can be implemented at the enterprise level. Many large investor-owned corporations have corporate social responsibility departments and are quite adept at measuring and reporting nonfinancial metrics, but slick reports can exist in the absence of actual progress. Measuring and reporting sustainability is a newer area of focus for cooperatives, and research shows not many are doing it. Yet, Henrÿ has coined the term *cooperative social responsibility* (CoopSR) and argues that it can be stronger than corporate social responsibility because of the legal structure of cooperatives.[45]

Whether cooperative law makes it easier for cooperatives to implement Sustainable Development Goals (removes barriers and provides incentives) or legally requires them to do so (provides mandates) is a question to be precisely answered at the level of legal jurisdiction in which the cooperative is based. However, the overarching international private governance of Cooperative Principles can be generally seen as providing incentives for a more robust cooperative sustainability. First, cooperatives do not operate under a law or norm of shareholder primacy, so they do not have to weigh tradeoffs between protecting the environment and risking short-term shareholder profits. Instead, they need to weigh whether environmental protection is aligned with their members' interests.

Along these lines, Henrÿ asserts that because cooperatives are member-centered, operational decisions are more comprehensive than in corporations, which leads cooperatives toward "economy and ecology" solutions. By the same token, since members drive the focus of the cooperative, they are likely to include their concerns for a healthy environment. Further, in Henrÿ's ideal cooperative, the role of capital is neutralized because financial return on investment is not the primary goal of a cooperative. Cooperatives establish intergenerational solidarity by establishing and maintaining an indivisible reserve fund, with a manager who is obliged to protect the assets for future members and use a nonfinancial audit, which includes ecological metrics, that measures and shows the impact of the cooperative. In reality, all cooperatives do not fit Henrÿ's ideal, but the exemplars in the case studies provide proof that it is possible. Many of the features Henrÿ highlights are present in the sustainability pathbreakers' case studies.[46]

The Co-op Alliance similarly emphasizes Henrÿ's points about the member-centeredness of cooperatives. It argues that engaging in pursuits that promote economic, environmental, and social sustainability provides a "mutual advantage" for cooperatives because it will encourage membership renewal and expansion. This is the corollary to the publicly traded corporation's business case for sustainability. "Mutual advantage" urges the question of what the cooperative should prioritize when environmental sustainability does not result in a mutual advantage. Perhaps environmental protection costs the members money or is not valued by those who actively participate. This points to the importance of placing environmental sustainability in the governance documents of the cooperative to guide priority-setting from the legal foundation of the entity. The case studies revealed the importance of establishing such priorities in governance documents to anticipate and remove such conflicts before they develop.[47]

Similarly, the Co-op Alliance's Guidance points to the global financial crisis of 2007–2008 as proof that cooperatives are resilient and sustain local communities when put to the test. The Co-op Alliance observes that because cooperatives' economic activities focus on meeting their members' needs, they are less inclined to engage in financial speculation. The case study of Caixa Popular in Part II demonstrates this at a regional cooperative bank in Spain.

In summary, the Values and Principles that can shape a cooperative's purpose and day-to-day practices are more closely aligned with environmental sustainability than a publicly traded corporation's purpose to prioritize shareholder's interests. Cooperatives should be freer to pursue environmental sustainability than corporations seeking to maximize profits. Henrÿ and the Co-op Alliance make persuasive arguments, but they are theoretical. The case studies in this book provide an empirical analysis of these cooperative features to better understand how they encourage or require environmental sustainability. The case studies tangle with a variety of questions, ultimately building toward findings and lessons that span borders and legal jurisdictions. The core questions are:

- What laws and modes of private governance act as incentives, mandates, or barriers to environmentally sustainable cooperatives?
- How do sustainability exemplars build their purposes into the legal design of their businesses?
- To what extent are cooperative exemplars including the Co-op Alliance's Values and Principles, and an explicit commitment to environmental sustainability, in their governance documents?
- How does a purpose of environmental sustainability impact day-to-day practices?
- Are the cooperative exemplars using a circular-economy or a sustainable development business model?
- Is there a cooperative difference when it comes to environmental protection?
- Can positive exemplars be replicated and multiplied to shift the economy as we meet the climate imperative and deeply decarbonize?

These and other questions will be discussed throughout the following case studies, and ultimately distilled into the lessons presented in the final chapter.

PART TWO

CASE STUDIES: COOPERATIVE PATHBREAKERS TAKE CLIMATE

ACTION AND ADVANCE SUSTAINABLE DEVELOPMENT

TALK IS EMPTY WITHOUT ACTION. In this part of the book I bring you the case studies of cooperative pathbreakers who are showing how to implement the theories and ideas at the enterprise level. Part II frames the case studies against the backdrop of the United Nations' Intergovernmental Panel on Climate Change's 2018 report, with the aim of offering an accessible and strategic interpretation of what the report means for transforming the businesses that shape the global economy. The scientists' wake-up call for the world showed that the planet is already experiencing significant changes, and we have not even reached 1.5°C warming. Many land regions have warmer temperatures than global annual averages, and the rate of warming is two to three times higher in the Arctic. Sea levels are rising and already harming coastal communities with damaging flood waters, salt-water intrusion into drinking water, and destructive coastal erosion. People are experiencing weather extremes beyond floods: hurricanes, droughts, excessive cold and heat, raging fires, and blizzards. More people arc on the move, migrating in search of better conditions. All these experiences are slated to worsen and persist for future generations.[1]

However, the IPCC has a high degree of confidence that with international cooperation around key actions, we can collectively lessen the disruptions and suffering. The scientists reported that our collective future depends on our choices about whether and how quickly we reach net-zero global anthropogenic CO_2, and they urge action as soon as possible.[2]

If we continue business as usual, our economic system will continue to be a resource- and energy-intensive system, dependent on fossil fuels, that tends toward exacerbating economic inequalities. Economic growth, as measured by GDP, and globalization are tied to widespread adoption of greenhouse-gas-producing lifestyles around the world. These are lifestyles with high demands for transportation fuels (for individual cars and shipping products), abundant freshwater, and livestock products. If the world stays on this path, we are gambling on unknown technological breakthroughs to remove carbon dioxide to stave off the worst of the climate emergency scenarios. In a 2020 *New Yorker* article, Bill McKibben calls this +4°C world "impossible," while the +1.5°C world is "merely miserable."

If instead of settling for miserable we set our sights on building a new economy characterized by broad and robust international cooperation to reach sustainability and a commitment to social enterprises forming the backbone of our business activity, we could prosper while we mitigate and adapt. Prosperity goes beyond material or financial concerns to encompass quality of life, health, happiness, strength of relationships, trust in community, shared meaning and purpose, and satisfaction at work. In that sense, the prosperity-focused aim is to reduce the risks of the worst climate-change disruptions and achieve the Sustainable Development Goals. Such a move includes improving energy efficiency to reduce energy demands, switching to renewable energy to meet remaining needs, using sustainable and regenerative organic agricultural practices, human development along the lines of the Sustainable Development Goals, lower and healthier consumption patterns, such as fewer animal products, and well-managed and socially acceptable land systems that remove carbon from the atmosphere.[3] It also means tending to the dominant business structures, and using legal design to lock in a wholistic purpose, expand the class of owners, and sound a greater democratic voice in decisions.

The timing and scope of the challenges can be overwhelming when viewed as a whole. Clearly, governments are needed to set clear legal boundaries to protect human health and ecosystems, and set transparent goals and targets to deeply decarbonize. Yet, there are inspiring examples of businesses that are already making the necessary transitions and using the private sector as a force for positive change in advance of legal reforms. These deep-decarbonization pathways implicate key economic sectors, most of which are featured in this book: energy, food and agriculture, water, finance, and trade. The case studies are organized into these sectors to tell the stories of the cooperative sustainability pathbreakers' best practices. Readers can see how to operationalize these urgent and unprecedented goals by the examples of those who are not waiting for government action. Other important sectors, beyond the scope of this book but that warrant further attention by others, include how sustainability pathbreakers are deeply decarbonizing and meeting the Sustainable Development Goals with new approaches to forests, urban design and infrastructure, transportation, and buildings.

Why These Economic Sectors?

Energy, Food, and Agriculture

As I explained in *Law and Policy for a New Economy: Sustainable, Just, and Democratic,* a strategic approach is to focus on the largest sectors of emissions and target actions to rapidly transform those systems to reduce their GHG contributions. In the United States, the two largest sectors for GHG emissions are energy from fossil fuel combustion and agriculture. Based on 2018 data, the Environmental Protection Agency reports that energy to heat and electrify buildings and industry accounted for 61 percent and agriculture accounted for 10 percent of the total U.S. emissions. The U.S. numbers are consistent with a global assessment of where to focus attention for greatest reductions. The United Nations' 2014 Emissions Gap report identifies sectors and rates their overall emission-reduction potential as well as the barriers to implementation. Switching to renewable energy and sustainable agriculture are two sectors with the highest emission-reduction potentials globally. That assessment is underscored by a first-of-its-kind 2012 study that indicated food and agriculture may account for one-third of global greenhouse gases. The authors assessed *all* the components of the global food system, including growing, storing, and transporting the food. This food system study showed agricultural production caused the majority of the food-related emissions, followed by fertilizer manufacturing, and then refrigeration. Thus, the highest priority should be targeting deep decarbonization efforts on transitioning to renewable energy and sustainable agriculture.[4]

Water

Water is essential for life and economic development, so the availability of clean, reliable, and affordable water is fundamental to meeting many of the Sustainable Development Goals. Clean water and sanitation is a pressing need, which 2.2 billion people currently lack. Further, water is a priority for deep decarbonization because pumping, cleaning, and distributing water, and then later treating the wastewater, can be quite energy-intensive. It is not unusual for fossil-fuel–derived electricity and energy to be the highest costs related to operating a water treatment facility. As I have previously calculated, a municipal wastewater treatment facility with an average flow of ten million gallons (37,854 liters) per day uses as much energy as almost 3,000 average U.S. homes.[5]

Finance

Banking is included because of the critical role finance plays in determining whether we take the business as usual or the sustainable development path. Banks determine access to capital, with significant implications for the growth and development of institutions, communities, and cultures. Banks directly impact environmental sustainability when they provide capital for businesses and homeowners. In data from 2016–19, Rainforest Action Network documents that the top ten investor-owned banks have funded about $1 trillion in fossil fuels in the four-year period. The biggest financiers of fossil fuels are U.S. banks: JPMorgan Chase, Wells Fargo, Citi, and Bank of America. Whether banks invest in fossil fuels or renewables, provide low-interest loans for energy efficiency improvements, or green bonds for climate mitigation, banks strongly influence environmental outcomes globally.[6]

Trade

Trade is included in the sectors addressed in this book due to a combination of the climate impact from transportation-related emissions and achieving the broader Sustainable Development Goals of alleviating poverty and inequality, among others. Global emissions accounting under the Paris Agreement does not attribute any emissions from the international shipment of goods to any country, so in essence these are phantom emissions for which no state or party is accountable. Further, emissions related to international transportation (aviation and marine) have been expanding every year since 1990, and the International Maritime Organization projects substantial emissions increases—50 to 250 percent by 2050—in international shipping. So these trajectories are headed in exactly the wrong direction if the goal is deep decarbonization.[7]

Transition Vision

The biggest source of anthropogenic CO_2 results from our choices about energy investments and infrastructure. To deeply decarbonize, the IPCC recommends global energy demands need to shrink through conservation and energy efficiency, even while providing energy to those who currently lack it. The remaining demands should be met by renewable energy. This will entail more electrification of activities such as heating, cooking, and transportation, and innovations and deployment of dispersed energy storage capacity.[8]

For the food and agriculture transition, the IPCC indicates a need to restore ecosystems and promote soil carbon sequestration practices, and adopt less "resource-intensive diets" (that is, less meat). They also show less climate risk if some existing marginal agricultural lands are converted to forests (afforestation and reforestation) and the production of bioenergy instead of food crops, but highlight the societal conflicts this raises. Reducing the transportation miles between farmer and consumer reduces the carbon footprint of food. This all has implications for markets and grocery stores that typically provide the interface between producers and consumers.[9]

Given the scale and timing of such transitions, some argue it is best to keep a singular focus on moving off fossil fuels without regard to broader issues of the Sustainable Development Goals, such as wealth inequality, business ownership structures, poverty, access to water, and democracy. A singular-focus approach is inappropriate for the reality of the world today. These issues are inextricably connected. From a moral and ethical perspective, Pope Francis, in his 2016 encyclical *Laudato Si'*, called on all people to see that poverty, inequality, and climate change are a connected result of the current economic system and to demand ethical leadership that moves away from the profit motive as the singular goal of businesses. From a scientific perspective, the IPCC warned that the large majority of climate models were unable to limit global warming to 1.5°C without simultaneously pursuing international cooperation, and alleviating inequality and poverty. They also articulated a need for broad participation. Government at all levels, civil society, the private sector, indigenous peoples, and local communities are all needed, and the IPCC emphasized the importance of strengthening each sector's capacity for climate action. Whether from an ethical or scientific/pragmatic view, reversing the trends in wealth inequality for more broadly shared prosperity should be part of the new operating system for a fossil-free economy.[10]

The book's case studies respond to this multipronged charge, as the stories reveal lessons about increasing the capacity of the private sector to create platforms of participation for civil society and individuals to take climate action that advances the Sustainable Development Goals. In a variety of ways, the sustainability pathbreakers featured here are building climate-resilient development pathways at multiple scales in a variety of legal jurisdictions. We face daunting challenges of transforming the economy to move off fossil fuels while sustainably developing across the world in this decade. The IPCC tells us we cannot have a livable planet if we continue with business as usual. The time is here to restructure business to be far more democratic in ownership and governance. If we meet the challenge by cooperating as the sustainability pathbreakers have, the world we choose to build will be one that is far more prosperous and enjoyable.

7

Supportive Cooperative Ecosystems in Spain and the United States

ATTWOOD AND BHAVISKAR'S "supporting environment theory" of cooperatives posits that success is spurred by the presence of promoters, a good
legal and fiscal environment, and government support without government
control. The chapter is organized with this theory in mind. I insert the term
ecosystem in place of *environment* to foreground the interconnected system
dynamics at work in supporting cooperative ecosystems. An ecosystem refers to a community, including the relationships between all organisms and
the environment. People launch and grow cooperatives within a particular
ecosystem that includes the laws, government programs, and cooperative
support organizations that influence them.[1]

In the absence of supportive ecosystems, cooperatives tend to degener-
ate. Mitch Diamantopoulos's social-movement approach to cooperatives
suggests that degeneration follows a deemphasis and marginalization of
building cohesive networks of apex (top-level) and sector federations, plans
and proposals for technical and financial assistance, and targeting support
for emerging needs and opportunities. Social movement theorists observe
that well-established cooperatives often abandon the mission of developing
a cooperative movement in favor of responding to market forces and the individual cooperative's financial goals. A variety of scholars argue that investing in new cooperative development is an important investment in cooperative movement regeneration, and without it, the investor-owned corporation
will appear to be the only viable option. Thus, the existence of networks and

programs to promote regeneration appears to be an important feature in building a supportive cooperative ecosystem. The Valencian government's social-economy work, combined with the support systems built between the regional worker cooperative federation and a regional cooperative bank, described below, are yielding notable regenerative results.[2]

Further, the concerns about degeneration focus on the tendency of cooperatives to become larger and more complex over time, and to consolidate power among specialist managers at the expense of democratic participation of members. The sustainability exemplars in the case studies are a mix of well-established and start-up cooperatives. A factor unexplored in the social movement literature is the role of a clear commitment to Cooperative Principles and an environmental mission in governance documents to mitigate degenerative forces. Related to the concern about expert leadership, while cutting against democratic participation, the social movement literature does not assess whether expert leadership can coexist with a commitment to pursue strong environmental goals, and the role of clear governance documents in shaping this direction. These elements will be interrogated in the case studies.[3]

While only one part of the supportive cooperative ecosystem involves the government's role, in the United States there is an elephant in the room that needs to be acknowledged. Moving to a more egalitarian, inclusive, and democratic model of workplace ownership and governance does not advocate for socialism in the sense of government ownership of industry. While some argue for a stronger governmental role in enterprises, the cooperative model tends toward the opposite, with a more dispersed, more autonomous, and less concentrated form of ownership and control.

Although the government has a role to play in regulating the economy and carrying out programs that encourage economic development, the dominant U.S. approach is not working well for equity and sustainability. For instance, state and local governments commonly promote economic development by competing with each other to subsidize global multinationals to bring new jobs, many of which are part-time and require a further federal government subsidy in the form of food assistance and health insurance for low-paid workers. This couples public expenditure on private enterprises with no related benefit of community-based democratic ownership and control. For example, according to Good Jobs First, a nonprofit that has tracked state tax breaks, since 2000, U.S. government entities have given Amazon $1.115 billion in subsidies, surpassing the prior corporate beneficiary, Walmart.[4]

The pivot point for the political economy we need to create is to reorient government economic development programs to support an economy and business entities that provide more social and environmental benefits. Institutions of the social and solidarity economy, such as cooperatives, are flourishing around the world. There are important lessons to be learned from Spain in this regard. As explained below, they have established a national and regional governmental approach to encourage the social and solidarity economy that is yielding impressive results in terms of a growing contingent of worker-owned enterprises.

Sustainability exemplars in the following pages have developed within specific historical, political, social, economic, and legal contexts. In each story, I have tried to provide enough context for understanding without burying the reader in detail. Nonetheless, there are some common denominators of all the case studies in a particular country. While it is difficult to generalize without sacrificing some accuracy, this chapter identifies defining features of the organizational ecosystems in Spain and the United States. Within each ecosystem, there are institutional structures and laws that enable or impede the development of sustainable cooperative businesses. These include federations of cooperatives, national and local laws related to cooperatives, and laws advancing the social and solidarity economy more broadly.

Defining Features of the Spanish Supportive Cooperative Ecosystem

Cooperatives contribute a significant amount to the economy, directly employing over 300,000 people in Spain in 2019. The Spanish supportive cooperative ecosystem can be seen as including the dominant Mondragón model, the tiers of support from federations, and the laws and government

institutions that regulate these enterprises. Since the global economic crisis, the European Union has embraced the social and solidarity economy,
which is composed of enterprises organized with the purpose of pursuing
a triple bottom line. Within the European Union, Spain has been a leader
in developing the social and solidarity economy, of which cooperatives are a
strong component. There is much to be learned from the Spanish cooperatives and the government programs that encourage them.[5]

The Mondragón Model

No understanding of worker-owned cooperatives would be complete, and
especially one focused on Spain, without reference to what is now known
as the Mondragón model. Most studies of Spain's cooperatives for audi-
ences beyond the Iberian Peninsula center on Mondragón. The Mondragón
cooperative has inspired a spillover of cooperative development in Spain,
the United States, and throughout the world—what I call the *next-generation
cooperatives.*

My case studies venture into new territory with the next-generation cooperatives in the autonomous communities of València and Catalunya, which
have a long and rich history of cooperatives. The Valencian case studies fea-
turing Consum supermarkets and Caixa Popular's banks showed a strong
inspiration emanating from Mondragón. Similarly, every cooperative business leader I interviewed in the United States had ideas shaped by this
exemplar in some way. Here I provide a brief history of Mondragón's development and describe the model that has inspired worker cooperatives for
generations, regardless of legal jurisdictions, political borders, or cultural
backgrounds.

Founded in 1956 by Catholic priest José María Arizmendiarrieta (Arizmendi for short) and five Polytechnic School graduates in the Basque Community of Spain, Mondragón has grown to be the largest cooperative network in the world. Among all businesses in Spain, it is the seventh-largest
corporation. In 2019, Mondragón Corporación Cooperativa had 264 coop
businesses in a diverse array of economic sectors, fifteen technology sectors, and over 80,000 employees in operations spread over forty-one countries. This is twenty more countries than in 2007, showing that it is expand-

ing as a global workforce. Its businesses range from producing household appliances and bicycles to running supermarkets and schools.[6]

As a point of comparison, in 2014, the largest employee-owned cooperative in the United States employed a small fraction of this (2,000 people), focused on one type of business (health-care workers), and operated on a local scale (the Bronx in New York City). There are larger-scale U.S. cooperatives if one looks to farmers who have formed cooperatives for joint marketing purposes, CROPP/Organic Valley for instance, but again these exhibit nowhere near the scale and diversification of Mondragón.[7]

The founders built Mondragón in several key phases, reflecting a keen pragmatism and understanding of community needs. First, they formed a cooperative technology school to educate people for jobs in a local steel mill. Second, when the mill was about to go out of business, Father Arizmendi convinced the workers to buy it and turn it into a cooperative. Third, when they faced capitalization problems, Father Arizmendi led them to create a cooperative bank in 1959, Caja Laboral Popular (Working People's Bank).[8]

Two additional significant developments occurred in the late 1960s and early 1970s. In 1969, consumers and distributors created Eroski, which is a combination consumer- and employee-owned cooperative supermarket. In 2005, the supermarket had 13,200 full-time worker-owners. By 2017, it had 30,500 employees and was the third largest supermarket retailer in Spain, with over €6 billion in sales.[9]

In 1974, Mondragón added another element by creating Ikerlan, a research and development group. Thirty years later, scholars identified this research and development piece as a key element in the cooperative's ability to innovate and build successful businesses to compete in global markets. With fourteen research and development centers, its mission is to create wealth within society through entrepreneurial development and job creation.[10]

Mondragón is associated with a variety of positive practices linked to more equitable distribution of wealth. To name a few: maintaining a small gap between the highest- and lowest-paid workers, prioritizing the development and training of worker-owners, cooperating between the linked businesses to weather economic downturns, and supporting the democratic participation of the worker-owners.

This is a global enterprise and yet the highest salary is capped at eight times the lowest. Tight gaps between highest- and lowest-paid employees are one of the differences between cooperatives and investor-owned corporations. Mondragón has found no empirical evidence that a CEO's performance correlates with size of compensation. Instead they see it as linked to the manager's commitment to the cooperative values. At Mondragón, based on research from 2005, while workers were paid at or slightly above market rate, upper-level management were paid below it. The manager of Fagor, a domestic appliance manufacturer, earned seven times the lowest-paid employee; the bank manager of Caja Laboral earned eight times the lowest-paid; and the rest of the upper managers earned about 4.5 times the lowest-paid.[11]

Under its corporate umbrella of linked businesses, Mondragón uses a system of loaning money between the cooperatives that are better performing and those that are struggling, insulating them from the difficulties of the broader Spanish economy, which in 2017 had 18 percent unemployment overall. Instead of being laid off, worker-owners are more often given the option to move into new positions in a different coop. For instance, after loaning money between the coops to keep Fagor Electrodomésticos open, when the manufacturer of consumer appliances suffered serious financial problems in 2013 and filed for bankruptcy, Mondragón promised to pay 80 percent of the workers' salaries for two years, and to relocate as many as possible to other cooperatives in the network. This was no small feat in a country reeling from the global economic crisis, with 28 percent unemployment at the time.[12]

In worker-owned cooperatives, employees have the opportunity to participate in the decision-making process. This model of workplace democracy has clear benefits for teaching democratic principles, ensuring wealth equity, and improving the surrounding society, as Mondragón coops are reputed to have turned a formerly depressed area of the Basque Country in Spain into a thriving community. George Cheney's *Values at Work: Employee Participation Meets Market Pressure at Mondragón* (1999) delves into employee participation and organizational management and effectiveness. His deeply informed case study is relevant to questions about how to implement workplace democracy amid the market pressures of globalization.[13]

Richard Williams's *The Cooperative Movement: Globalization from Below* (2007) builds on Cheney's knowledge base. His team of researchers conducted sixty-one interviews with management and worker-members in a variety of countries to produce comparative cooperative case studies. The researchers explored what made cooperatives work well or fail, concluding that key factors are proper capitalization and a high rate of member participation. Their case study of Mondragón highlights elements that have made Mondragón successful, such as its culture and organizational/governance structure, its ability to access capital through its bank, Caja Laboral, and its research and development groups that focus on innovation and future markets. However, the researchers assert that "more important than any other factor" are the early successes spawned by the organization's "constant application of their basic [cooperative] principles." This final finding is corroborated in the case studies in this book. As will be discussed in the following pages, a common denominator of all the sustainability pathbreakers is the clarity of their purpose and principles in their governance documents and daily operations.[14]

Some assume that this worker-owner model translates into better environmental outcomes. However, much less analysis exists on Mondragón's approach to environmental sustainability, and there are no scholarly works on the subject. This is certainly an area that warrants further examination. This book provides a methodology that could be used for future research on Mondragón to build on the existing scholarship.[15]

According to the Mondragón Team Academy, which teaches entrepreneurial cooperators, Mondragón's sustainability metric of choice has been ISO 14001. This is an internationally recognized environmental management system that businesses use to measure and reduce environmental impacts. In 2012, Mondragón reported sixty certifications. It focused on implementing environmental management systems and performance of high strategic value, such as eco-efficiency, ecodesign, and minimizing impacts.[16]

Mondragón is placing its environmental efforts in particular on specific industries: renewable energy, energy efficiency, waste, and water. Mondragón also created a green community forum in 2013 for intercooperation about the green economy sector. Mondragón has articulated the importance of the Cooperative Principles and how it incorporates them into its corporate management model. There is very little emphasis here on protecting the environment. However, Mondragón does include the environment in its understanding of "social engagement" as part of being an "excellent company." It is one of seven areas of "social engagement," and the company asserts that it will go beyond legal compliance with environmental laws. Instead, "respect for the environment will be a mainstay of all our operations, leading to the implementation of environmental management systems and including this concept in the design of both processes and products." Environmental management appears again as one of ten items listed as "Commitment and Cooperative Identity." It mentions using indicators such as carbon footprint and water consumption, so there is some self-reported evidence of Mondragón using private environmental governance to make progress on internal goals.[17]

Mondragón has grown to be a multinational presence. Its interconnected web of cooperatives under the Mondragón parent corporation has allowed it to weather financial storms with minimal disruptions to its workforce. The case studies are populated by next-generation cooperatives. Mondragón inspired the founders of the Valencian cooperative movement, explored below, who then took a different approach and created more autonomous cooperatives. It also inspired the U.S. cooperative pathbreakers featured here. With Mondragón as a backdrop, we now turn to the Spanish cooperative federations to understand their role in cooperative formation and development.

Spanish Federations

The Spanish supportive cooperative ecosystem includes cooperative unions and federations. These support organizations are like a set of Russian nesting dolls, moving from the international, to the national, to the subnational autonomous communities. There are sector- and geographically based Spanish unions and federations, and most are members of the Spanish Business Confederation of the Social Economy (Spanish acronym CEPES).[18]

WORKER COOPERATIVES

CECOP-CICOPA Europe is the European Confederation of Industrial and Service Cooperatives. They have members that represent worker cooperatives organized at the level of nations. The Spanish Confederation of Cooperatives of Associated Work (Spanish acronym COCETA) is a member. COCETA is at the top of the national hierarchy of support systems for worker cooperatives and has represented enterprises in Spain since 1986. COCETA is comprised of seventeen member entities that represent different autonomous communities of Spain.[19]

COCETA highlights that cooperatives provide an ethical business model. COCETA emphasizes that this model is contributing to the socioeconomic growth of the locality in which it is located, the creation of stable employment, and the fight against exclusion, and is working toward social cohesion, integration of hard to employ people, and equality. Notably, COCETA does not mention environmental sustainability in its overall description.[20]

At the autonomous community level, since 1988, COCETA's member representing the Valencian Autonomous Community has been the Valencian Federation of Associated Labor Cooperative Companies (Spanish acronym FEVECTA). In 1985, the first Law of Cooperatives of the Valencian Community (Law 11/1985) led to the creation of FEVECTA as a representative for cooperatives in the community. FEVECTA provides its growing membership of 600 with legal services, advice on designing bylaws, and other services.[21]

FEVECTA distinguishes worker cooperatives from other workplaces by emphasizing doing business based on the seven Coop Alliance Principles. It says this informs a philosophy geared toward stable employment, valuing people and their work above capital, making decisions democratically, and having motivated workers who are also owners and managers. It promotes the worker cooperative as a great option for entrepreneurs to launch new businesses.[22]

In 2003, FEVECTA participated in updating the Law of Cooperatives of the Valencian Community (Law 8/2003 of March 24) as a member of the legal commission of the Confederation of Cooperatives of the Valencian Community. This law gives cooperatives greater flexibility and power to self-regulate.[23]

Worker cooperatives in València span multiple business sectors. To cite a few examples, Sampedro y Torres Impulso Cooperativo provides financial, accounting, auditing, and legal advice on labor, tax, and cooperative law. Serlicoop has 167 professionals providing home health care, cleaning, and counseling services. Martí Sorolla Coop is an educational cooperative of teachers that runs six schools. A variety of the case studies below feature Valencian worker cooperatives in additional areas, such as banking, agricultural products, and grocery stores.[24]

Food and Agriculture Cooperatives

Food and agricultural cooperatives have industry-specific umbrella federations. There is a national-level cooperative federation for Spanish food and agricultural cooperatives called Cooperativas Agro-alimentarias, España. Its focus is on defending the social and economic interests of its cooperative members nationwide.[25]

At the subnational level, the autonomous communities have organized agricultural federations. For instance, the Valencian community has an overarching federation of food and agricultural cooperatives called Cooperativas Agro-alimentarias, Comunitat Valenciana. Formed as a nonprofit, it represents agrarian cooperativism in that autonomous community. As of 2016, the federation comprised 374 cooperatives. It is designed to provide development assistance to its cooperative members on a community level.[26]

CONSUMER COOPERATIVES

Hispacoop is the Spanish Confederation of Consumer and User Cooperatives, which, some international cooperative experts assert, has a great record of sustainability initiatives. In 2017 it represented 170 regional federations and cooperatives (such as grocery stores, electricity, and cohousing) with more than five million consumer-members. Hispacoop's function is to represent consumer cooperatives before national and international fora and to defend, inform, and educate consumers. As to the latter efforts, it offers educational materials for consumers concerning, for example, the difference between "organic" and "ecological" food, how it is regulated, and how to read the labels.[27]

ENVIRONMENTAL SUSTAINABILITY WITHIN THE SPANISH FEDERATIONS

While the worker cooperative federations are important for the initiation and development of cooperatives, they are not yet providing a robust set of resources or supports for environmental sustainability. In a January

2019 review of CECOP-CICOPA Europe's publications on the federation's website, there was only one report on cooperatives and sustainable development. Similarly, the national-level COCETA provides "publications and tools" on its website. It targets topics such as youth engagement, women in cooperatives, and business innovations. COCETA did not have a clearly labeled area to find environmental publications. In a January 2019 review of its publications related to the environment, it provided only two, both from 2013, and neither was generated by the federation. For instance, there was a presentation by Som Energia, a cooperative featured in this book, regarding renewable energy and a presentation by an environmental services consulting firm.[28]

In a January 2019 analysis of the Valencian worker cooperative federation FEVECTA's website, there were no publications related to environmental sustainability. FEVECTA states commitments to "equality and Corporate Social Responsibility." What it means by that is gender equity, which at the time of the review was the entire focus of its webpage content on the topic of corporate social responsibility.[29]

By contrast, the federation in the Catalunya Autonomous Community took a different approach. Like València, this community also has a long history of cooperatives, perhaps even longer. In 1899 they organized the first cooperative congresses. Although the national government suppressed cooperatives and confiscated property starting in 1939, when Spain established itself as a democracy in 1981, cooperativism resumed again.[30]

In 1984, activists formed the Confederation of Cooperatives of Catalunya. This is the apex organization representing six federations in the community that represent agriculture, consumers, education, housing, transportation, and workers. A February 2019 analysis of its website revealed over a decade of "Sustainability Reports" explaining the history and background of the confederation and its commitment to supporting environmental protections. The confederation includes "sustainability" as one of six business values and ethics to complement the Cooperative Principles. Further, in 2006, the General Assembly of the confederation adopted a policy of corporate social responsibility that it says forms the core of the confederation's operational strategy. The policy incorporates the seven Cooperative Principles and the confederation asserts that "all" its actions reflect a commitment to sustainable development. Additionally, it commits to manage waste, water, and energy. Finally, the confederation has a corporate social responsibility manager dedicated to implementing this policy.[31]

The European, Spanish, and Valencian worker cooperative federations could improve articulating environmental sustainability as part of the Co-operative Principles and identity and take an active role in promoting this. The publications sections of their websites could be better developed to pro-vide clarity for cooperative managers about how to carry out Principle 7 and implement environmental sustainability. Specifically, the federations could provide resources from the Co-op Alliance, such as Guidance on how to implement the seventh Principle, environmental metrics, and best prac-tices for regular reporting. Further, they could learn from the example of the Confederation of Cooperatives of Catalunya, which has prioritized this through its operational strategy and employment of a manager to carry out its sustainability policies. The federations and apex organizations for coop-eratives could play a much more impactful role on deep decarbonization and sustainable development if they were to develop expertise in tracking carbon footprints and other sustainability measures, and provide software and training to their members to implement these initiatives at scale across thousands of work sites.

Spanish Cooperative Law

The Spanish supportive cooperative ecosystem includes laws and regula-tory structures that make it easier to function as a cooperatively organized business. The Directorate General for Employment at the Employment and Social Security Ministry regulates Spanish cooperatives. Since 1931, Spain has had a national law applicable to all cooperatives. A significant distinc-tion in Spanish law is the 1978 Spanish Constitution's inclusion of coop-eratives and in particular worker ownership. Article 129.2 of the constitu-tion directs "public authorities" to "efficiently promote the various forms of participation within companies" and to "encourage cooperative societies by means of appropriate legislation." The constitutional provision further directs public authorities to "establish a means to facilitate access by the workers to ownership of the means of production."[32]

The constitutional emphasis on worker-owned cooperatives reflects the importance of cooperatives in Spain in broadening the base of ownership.

As such, it has aided in regenerating the cooperative economy generally. The specific focus on workers over other forms of cooperative ownership (for example, consumer-owners) is notable as well. Not only is Spain home to the largest worker-owned cooperative in the world, which predated the 1978 Constitution, but the worker-owner cooperative is the most common type of cooperative found throughout the country forty years after this constitution articulated the national goal.

The current national cooperative law has been in place since 1999, the Law of Cooperatives (Law 27/1999 of 16 July 1999). There are also national laws for specific types of cooperatives, such as banks and insurance cooperatives, to name a few, and a national tax law for cooperatives. The Constitution's support for creating legislation and giving the autonomous communities legislative powers has encouraged each community to enact cooperative laws. While the autonomous communities have created fifteen distinct regional cooperative laws, most have the same elements. According to the Law of Cooperatives, the law classifies cooperatives as worker, consumer and user, housing, farmer, services, fishermen, hauler and transport, insurance, health care, education, credit, and social initiative.[33]

Spanish cooperative laws cover many topics, from formation to dissolution, but the area of interest here is focused on defining cooperative purposes. According to Law 27/1999, a cooperative is a company or *"Sociedad"* that is "established by people who associate to conduct business activities directed at meeting their economic and social needs and aspirations. . . ." Members are free to join and leave voluntarily, and they have "a democratic structure and operation, in accordance with the principles stated by the International Co-operative Alliance, as provided for in this Law."[34]

Thus, the international Cooperative Principles are specifically included in Spanish cooperative law. Spanish legal opinions say the Cooperative Principles are characteristics of this business form. Fajardo interprets the Spanish law as positioning cooperatives in a group of enterprises based on "mutual aim" instead of based on profit as commercial enterprises.[35]

Spanish cooperatives are formed with articles of association. In Spain, a cooperative acquires legal personality by "public document" (*escritura pública*). The bylaws are contained in the cooperative's statutes (*estatutos*). The cooperative registers its name to reflect both that it is a cooperative and to specify its autonomous community. For instance, "Coop. V." signifies a business organized as a cooperative in the Valencian community. While Spanish cooperatives receive favorable tax treatment, they are subject to the same accounting obligations as other types of business. Outside of Basque- and Navarra-based cooperatives, the national Tax Regime of Cooperatives Law 20/1990 of 19 December 1990 applies.[36]

By Spanish law, members of cooperatives have the right to participate in governance by attending and voting at the general assembly. All coopera- tives are required to have decision-making bodies consisting of a general assembly and a director. Some types of cooperatives are also required to have internal auditors and a surveillance council.[37]

Spanish Laws and Institutions to Promote the Social Economy

Spanish cooperatives also should be understood within the broader context of a "social and solidarity economy," which gained strength after the global financial crisis that started in 2007. Fundamentally, the social and solidarity economy emphasizes local control, community power, and democratic participation. The institutions that make up such an economy include cooperatives, along with other types of social enterprises.

In 2011, still roiled in the wake of the global financial crisis, Spain passed the national Social Economy Law, Law 5/2011 of 29 March. The preamble to the law reflected that this was part of a long trajectory of promoting the social economy that started with the first cooperatives in Europe in the eigh- teenth and nineteenth centuries. The preamble also roots the action in sup- port of the social economy in the Spanish Constitution, Articles 1.1, 9.2, 40, 41, 47, and 129.2. Part of the motivation for passing the law, as expressed in the preamble, was to commit the country to a model of sustainable develop- ment, with its triple bottom line of progress. Among other things, the law articulates principles to guide social economy enterprises. The principles include:

- putting people and social purpose above capital;
- using transparent and democratic management;
- promoting solidarity and inclusion of people at risk of social exclusion;
- generating stable and quality employment;

- fostering environmental sustainability; and
- retaining independence from public authorities.[38]

The Spanish Social Economy Law then describes the enterprises that compose this economy. These include:

- cooperatives,
- employee-owned companies,
- mutual societies (nonprofits),
- special employment centers (at least 70 percent of staff have disabilities),
- social integration enterprises (30–60 percent of staff are people who have been difficult to integrate into employment; 80 percent of revenue is reinvested in the company),
- associations (entities that defend the rights of people with disabilities and at risk of social exclusion),
- fishermen's guilds,
- agrarian societies, and
- foundations.[39]

Spain's Social Economy Law establishes a supportive cooperative ecosystem. The law charges the government to promote the social economy through a variety of means, including training, disseminating information, removing obstacles to new enterprises, and introducing social economy concepts in school curricula. The social economy law is meant to supplement but not supplant existing laws related to cooperatives and other social economy enterprises. The law specifies that the national Council for the Promotion of the Social Economy will be governed by the provisions of this law and be the central organizing entity for the state.[40]

Spain has inspired other European countries, such as France and Portugal, to similarly enact national laws. European government institutions have also been analyzing the social economy, with more than "200 official documents that acknowledge the importance of Social Economy in job creation, social inclusion, the fostering of entrepreneurship, access to social services of general interest or social cohesion."[41]

The social economy legal framework and government implementation appear to be showing positive results. According to the Spanish Business Confederation of the Social Economy, in 2019, there were over two million jobs in the Spanish social economy enterprises. In the cooperative sector nationwide, from 2012 to 2019, the number of jobs grew from 286,912 to 314,119. The autonomous communities with the highest number of cooperative employees in 2018 were: Basque (60,000), Andalucia (59,000), València (49,000), and Catalunya (46,800). When measured by the number of cooperatives, the leading communities switch: Catalunya (4,000), Andalucia (3,780), València (2,400), and Basque (1,600). With this legal and statistical framework in mind, one can better understand the progress made by the government of València and the city of Barcelona (within Catalunya).[42]

VALENCIAN GOVERNMENT PROMOTES SOCIAL ECONOMY

València is an autonomous community in Spain, containing the provinces of València, Alacant, and Castelló. It has a thriving sector of worker cooperatives, with more enterprises than the Basque Community, which is home to Mondragón. How this community is building a supportive cooperative ecosystem provides some lessons for other jurisdictions interested in spurring economic development with cooperatives, especially those owned by workers.[43]

On February 11, 2019, I met with government officials in the Autonomous Community of València to understand the government's role in their successful development of cooperatives. The first thing to note is the name of the government agency, which immediately signals a different approach. Not the Department of Commerce or Small Business Administration, but the Department of Sustainable Economy, Productive Sectors, Commerce, and Work (Conselleria d'Economia Sostenible, Sectors Productius, Comerç i Treball). Within this agency are the office of Social Economy and Entrepreneurship and the office of Cooperative Services. I was able to interview the directors of both of these offices, María José Ortolá Sastre and José Ignacio Martínez, respectively.

Ortolá Sastre explained the Valencian government's efforts to transform the economic system since 2015, when the new government was elected. They have four years of experience promoting a new sustainable economic model focused on the triple bottom line. She explained that this initiative is in direct response to the global economic crisis of 2007–8 and the prior government's support for an economy based on speculation, which contributed to the construction bubble and subsequent crash.[44]

The Valencian government's new economic focus is tied to Spain's 2011 national law on the social economy. Mirroring the national law, the Valencian government calls the new approach "la economía social," which it describes as an economic model with a human face. Similar to the national law, the social economy institutions include cooperatives, labor unions, foundations, social action associations, fishing guilds, employment centers, and other businesses.[45]

The Valencian government's support for cooperatives is part of this larger overall strategy to change the economic model. The agency provides support for hiring employees who are hard to employ and for technical assistance; makes direct investments in existing businesses; and provides funds to stimulate the creation of new businesses. València supports all varieties of cooperatives, which are sorted into categories: farming, workers, consumers, transportation, industry, housing, teaching, and credit/banking cooperatives.[46]

The agency's annual financial support for cooperatives is significant and has grown each year:

2015: 2.6M €	($3M)	
2016: 2.9M €	($3.3M)	
2017: 4.3M €	($4.85M)	
2018: 5.4M €	($6M)	
2019: 6M €	($6.8M)[47]	

Some U.S. states and cities have recently passed laws to promote cooperative development too. Consider New York City, which in 2015 launched a $1.2 million annual fund to grow worker-owned cooperatives. Managed by the city's Department of Small Business Services, by 2017 the fund had grown to $2.2 million, which resulted in 185 people hired by worker cooperatives.[48]

While this city investment is more than other U.S. cities, on a per capita basis it is not even in the realm of the Valencian government's support for cooperatives. In 2018, the Valencian Autonomous Community had an estimated 4.9 million people, compared to New York City's 8.6 million.[49]

There is a long history of cooperatives being part of the Valencian economy, dating back to the nineteenth century, that helps nurture this kind of
ecosystem for business development. However, the existence of more established cooperatives alone does not reinforce and grow the cooperative sector.

Not to be overlooked is the government leadership to accelerate innovation and alter the economic model. This can be seen in the Spanish national
law; in the Valencian research and planning documents, which informs its
daily work; and in the level of autonomous-community government investment in supporting the emergence and development of these enterprises.
Two of the primary planning documents are:

- Valencian Community Plan to Support and Promote Cooperatives
  (2018–19), and
- Valencian Government Action Plan to Transform the Economic
  System.

Further support for the creation and development of Valencian cooperatives comes from the region's largest cooperative bank, Caixa Popular, which
is owned by its workers and its cooperative members. The bank provides
critical capital at favorable rates for its cooperative members, requires its
members to adhere to the seven Cooperative Principles, and pays membership dues for involvement in the worker cooperative federation FEVECTA,
among other things. (See more about Caixa Popular in the case study in
Chapter 11.) In combination, València's supportive cooperative ecosystem
includes clear government objectives and funding priorities for existing
and emerging cooperatives, a cooperative bank to capitalize cooperatives,
distinct adherence to the Cooperative Principles within the national legal
framework, and a regional worker federation to advocate for worker-owned
cooperatives and provide technical assistance. It is in that context that one
may appreciate the statistics that as of 2016, there were 2,464 cooperatives
employing 75,000 people in this autonomous community. Two out of five
Valencians are members of cooperatives.[50]

CITY OF BARCELONA PROMOTES SOCIAL ECONOMY

The city of Barcelona is promoting cooperatives as part of its program to
support the social and solidarity economy. The city commissioner's Office

for Cooperative, Social and Solidarity Economy directs this program, which was started in 2015. The office has recently started to focus on strengthening local economic opportunities (jobs and wealth), with an emphasis on responsible consumption. The consumption focus includes a city plan (2016–19), support for sustainable consumer cooperatives, promotion of time banks to exchange time in a noncommercial way, reducing food waste, and a ten-year "commitment toward sustainability 2012–2022." Barcelona's time bank provides a platform to connect people who want to share a service involving their time, such as caring for an elder, in exchange for receiving a service of someone else's time, such as cleaning an apartment. It is seen as a strategy to build community and increase well being without involving the use of a currency.[51]

Barcelona's budgetary commitment to the social economy is impressive. In 2019, the city budget allocated over €24 million to fund the social economy. In addition to funding, the city provides training and advice for people participating in the social economy.[52]

Barcelona describes the social and solidarity economy as a set of socio-economic measures guided by values such as equity, solidarity, sustainability, participation, inclusion, and a commitment to the community. The measures strongly emphasize democratic participation in the economy and at the enterprise level. The city also discusses the transformational focus of work to reorient the economy so it serves human needs. Last, their efforts include supporting enterprises that improve society and the financing of solidarity initiatives with poorer countries.[53]

Barcelona is promoting a change in the socio-economic model toward a concept of a plural economy, which involves:

- Democratising the infrastructure of common resources (water, land, energy, knowledge, etc.), which must be managed by society or, where this is not possible, by administrations.
- Subjecting the market to the demands of the common good, and therefore regulating it democratically, through local, regional and national governments and the general public.
- Promoting a demonetised economy based on the principles of reciprocity and self-consumption.[54]

In summary, Barcelona's efforts show the municipal government strongly promoting a coordinated effort with businesses to transform to a social and solidarity economy.

Defining Features of the U.S. Supportive Cooperative Ecosystem

Cooperatives contribute a significant amount to the U.S. economy, directly employing two million Americans and owning $3 trillion worth of assets. The United States' supportive cooperative ecosystem can be seen as including the tiers of support from federations, the laws and governmental institutions that regulate these enterprises, and the long history of cooperatives serving as essential platforms to connect farmers to build economies of scale and bring electricity throughout rural America.[55]

U.S. Federations

WORKER COOPERATIVES

The U.S. Federation of Worker Cooperatives represents 6,000 worker-owners in cooperatives and other democratic workplaces (consumer coops, nonprofits, and so on). Founded in 2004, the federation's mission is to "build a thriving cooperative movement of stable, empowering jobs through worker-ownership." It provides education and training about cooperative development, one-on-one consulting and referrals to experts, and some discounted health and vendor benefits. Its explanation of what a worker cooperative is includes the Co-op Alliance's Values and Principles, and it emphasizes having a "direct stake" in the local environment and doing business in a sustainable way. In contrast to the tiered federation structure in Spain, where the national federation is composed of members representing each of the autonomous communities, the U.S. national federation is not subdivided into state-based federations of worker cooperatives. While it is connected to regional, state, and local federations, the system is not so well established as to have a representative federation in each state. There is a Western Worker Cooperative Convening, an Eastern Conference for Workplace Democracy, and a Federation of Southern Cooperatives. A variety of cities have their own federations, such as Buffalo, New York; Madison, Wisconsin; and Jackson, Mississippi. Unlike in Spain, however, representation is not organized around all major political subdivisions in the United States.[56]

FARMER COOPERATIVES

The National Council of Farmer Cooperatives is the national federation that represents almost 2,500 U.S. farmer cooperatives. The council has been operating since 1929 and asserts that the "majority of America's two million farmers and ranchers belong to one or more farmer cooperatives." In the United States, farmer cooperatives have been integral to helping individual farmers achieve economies of scale. The National Council of Farmer Cooperatives diffuses power by having a membership that includes state and regional councils.

In addition to farmers owning and controlling the production and distribution of agriculture, the National Council of Farmer Cooperatives' "values" include "stewardship of natural resources." The stewardship commitment, however, does not necessarily mean promoting sustainable, organic, and regenerative agriculture. Nor does it mean supporting efforts to prevent agricultural pollution from getting into water, as the federation has publicly supported weakening the ability of the Environmental Protection Agency to regulate water pollution. The legislative priorities of the federation are geared toward conventional farming and fending off regulation that would add costs to farmers. Supporting organic farm cooperatives is not mentioned. On climate, the council's priority is to "ensure any climate change initiative provides benefits and opportunities for farmer cooperatives without adding burdensome costs and regulations."[57]

CONSUMER COOPERATIVES

What later became the National Cooperative Business Association—Cooperative League of the United States of America (NCBA CLUSA) started in 1916, and by 1919 held conventions across the United States to establish itself as the central federation of consumer cooperatives. Membership includes consumer cooperatives in a variety of sectors including food, insurance, and electricity. The association works to advance the interests of cooperatives, such as advocating for favorable laws, holding conferences and learning events, and providing educational resources such as webinars.

Starting in 1956 and continuing today, its Consumer Cooperative Management Association puts on an annual training program for coop managers, including for food coops.[58]

The Consumer Federation of America has been the national federation for consumer cooperatives since 1968. It engages in research, advocacy, and education to support consumer issues. Founded by leaders from organized labor, farmers, electric cooperatives, and consumer advocates, the federation works on a broad array of national- and state-level policy issues, from auto loans to fuel efficiency standards and food safety to food prices and labeling. A September 2019 analysis of the federation's website revealed no particular emphasis on environmental sustainability or the Co-op Alliance's Cooperative Principles; however, it provides reports on sustainability issues, such as energy efficiency, reducing antibiotics in the food supply, and fuel efficiency.[59]

Across the board, the U.S. cooperative federations and apex organizations have ample room for improvement in supporting their members' environmental sustainability. None of them offer a clear vision for environmental sustainability and the value that cooperatives can add to the global shift to rapidly decarbonize the economy and meet the Sustainable Development Goals. Improvement could start with a grounding in the Cooperative Values and Principles and lead to an active role in promoting these principles, in particular Principle 7, to better articulate the cooperative difference and value to members and society. They could be providing publications, training, and tools for their members to set and achieve sustainability goals. In particular, they would be more relevant to today's challenges if they offered services to audit nonfinancial reporting on environmental, social, and governance metrics; track carbon footprints; make industrywide comparisons; and share best practices for sustainability improvements.

U.S. Cooperative Law

FEDERAL LAW

Unlike Spain, the United States lacks a national constitutional provision related to cooperatives and comprehensive national laws. Most of the relevant U.S. laws are state laws, so there is a more fractured approach. However, there are several important federal laws that impact U.S. cooperative development, creating some uniform expectations about the sector. It is hard to imagine how rural America would have developed and obtained electricity in the 1900s without cooperatives, facilitated by federal laws and policies. New Deal–era laws in particular have strongly influenced cooperative development, as will be explained more fully in Chapter 8.

As an overarching federal law, the Internal Revenue Code applies special tax rules to cooperatives, found in Sections 1381 through 1388 (Subchapter T). In order to qualify for the favorable tax treatment, the business entity must comply with three requirements: 1) the members have democratic control, 2) all net profits must be vested in and allocated to the members, and 3) capital must be subordinated (that is, power and control are not allocated based on the amount of capital invested). If a cooperative does not qualify, the higher rates of regular corporate taxation apply. Additionally, some cooperatives that are not organized to make a profit can qualify for tax-exempt status under special provisions in Section 521 (for farmers) and Section 501 (for telecommunications, electric, banks, and crop-financing entities).[60]

Cooperatives faced legal obstacles under the federal Sherman Antitrust Act of 1890 and the 1914 Clayton Act. While those laws were aimed at preventing large corporations from controlling too much of a market, these antitrust concerns were used to prevent a cooperative from becoming multicompany or multisector to such an extent that it dominates the market. In 1922, Congress passed the Capper-Volstead Act to remedy issues with the two earlier laws as related to agricultural cooperatives. This law expanded the definition of what cooperative activities are considered legal. Section 1 of the Capper-Volstead Act provided an exemption from antitrust enforcement for farmer cooperatives with agricultural producer-members by specifically allowing "[p]ersons engaged in the production of agricultural products as farmers, planters, ranchmen, dairymen, nut or fruit growers" to act together in associations, "corporate or otherwise, with or without capital stock, in collectively processing, preparing for market, handling, and marketing in interstate and foreign commerce, such products of persons so engaged."[61]

In order to qualify for the Capper-Volstead exemption, an agricultural cooperative must be operated for the mutual benefit of the members and "conform to one or both of the following" attributes: 1) not grant a member more than one vote in governance matters due to stock ownership or capital contributions; 2) not pay dividends on stock in excess of 8 percent annually; and 3) not deal in the products of nonmembers in an amount in excess of the value handled for members.[62]

The Federal Farm Loan Act of 1916 established a network of rural financing cooperatives owned by the borrowers. This provided access to capital for farmers and rural homeowners. This law was a response to several studies commissioned by Presidents Theodore Roosevelt, Taft, and Wilson, all of which concluded that rural Americans lacked access to credit. The solution, inspired by examples from Europe, was to create a cooperative credit system designed to maximize service to its member-borrowers instead of maximizing profit for the banks.

In 1978, Congress chartered the National Consumer Cooperative Bank to serve cooperatives more broadly and support community revitalization. Shortly after its launch, however, during the Reagan administration, Congress privatized the bank, and it was renamed the National Cooperative Bank. Over time, the bank has specialized in lending to cooperatives in housing, grocery, hardware, and purchasing, growing its assets. By 1999, the National Cooperative Bank was providing financing to cooperatives and their members in every U.S. state. To encourage energy-efficiency upgrades and solar panels, the bank started a consumer loan program with the National Rural Utilities Cooperative Finance Corporation to finance these improvements.[63]

In 1935, President Franklin Roosevelt used an executive order establishing the Rural Electrification Administration, with the goal of bringing electricity to rural parts of the country. In that year, the vast majority of rural homes (nine out of ten) lacked electricity. Thanks to New Deal laws supporting the formation of electric cooperatives, by 1953 cooperatives were delivering electricity to 90 percent of U.S. farms.[64]

These federal laws have facilitated the ability of cooperatives, especially those with farmer-members, to be a dominant part of U.S. agriculture and electricity provision. Today, a majority of the country's two million farmers and ranchers belong to cooperatives. Given this focus on agriculture, it is not surprising that the U.S. Department of Agriculture has been key to

developing cooperatives. The Agricultural Act of 2014 established the Inter-
agency Working Group on Cooperative Development and authorized the
secretary of agriculture to coordinate and chair this group.[65]

Starting in 1990, federal law established and provided funding for the
Rural Cooperative Development Grant program. This is a fund of money to
assist in the startup or expansion of rural cooperatives. The amount of funding varies, but in 2019, the fund had $5.8 million available in maximum
grant amounts of $200,000. This is less than the total amount the Autono-
mous Community of València, Spain, spent on cooperative development
for its comparatively small area in 2019.[66]

To build a leadership base for future federal legislative efforts, in 2016
the Congressional Cooperative Business Caucus was formed, open to all
U.S. senators and representatives. The most recent action of the caucus was
in October 2017, when Rep. Jared Polis introduced House Resolution 561,
a resolution commending cooperatives as integral to sustainable economic
growth across many economic sectors. According to the caucus, the cooperative sector has grown by 20 percent since the 2007–2008 financial crisis. However, to date, the caucus has not moved forward any substantive
national legislation.[67]

Advocates for worker ownership see the upcoming baby-boomer retire-
ments of millions of U.S. business owners as an opportunity to scale up
worker ownership. This could be a time when retiring owners convert their
businesses to worker cooperatives to preserve their missions and make
their workplaces more democratic. Project Equity's 2017 study reported
that baby boomers own 2.3 million businesses employing almost 25 million Americans. There are federal laws to encourage converting these businesses to worker-owned cooperatives. For instance, the federal tax code,
Section 1042, gives "special tax treatment" to an employer "who sells his/
her/its business to the employees who form a worker owned cooperative
to acquire the stock in the employer's corporation. . . ." In order to qualify,
the cooperative needs to meet Subchapter T of the tax code (see above), a
majority of the members must be employees, a majority of the voting stock
must be owned by the members, a majority of the board must be elected by
the members (one person, one vote), and a majority of earnings and losses
must be allocated to the members.[68]

In 2018, the Main Street Employee Ownership Act was signed into law. It impacts how the Small Business Administration (SBA) administers loans to Employee Stock Ownership Plans (ESOPs) and worker cooperatives. Under this law, for the first time employee-owned businesses are eligible for SBA Section 7(a) loans. It also fast-tracks loans to help companies finance conversion to employee ownership. Since capitalizing cooperatives has been a hurdle in their establishment and growth, this may spur expansion of employee-owned cooperatives.[69]

In summary, while the U.S. legal approach to cooperatives places responsibility for establishing rules related to incorporation and governance at the state level, federal laws impact tax treatment for cooperatives, antitrust exemptions for farmer cooperatives, programs resulting in the electrification of much of rural America, the initial charter for a national cooperative bank, favorable tax and financing for businesses converting to employee-owned cooperatives, access to grant money, and eligibility for loans from the Small Business Administration.

STATE LAW

In the United States, state law primarily determines the design, structure, and governance of cooperatives. The National Cooperative Business Association CLUSA International, the national apex organization that has existed for over 100 years, advocates for a fifty-state approach. It anticipates that the U.S. Department of Agriculture's State Cooperative Statute Website will be key to identifying statutes that can serve as models and lead to more consistency nationwide. There are approximately eighty-five statutes scattered across the fifty states that relate to cooperative incorporation and governance.[70]

While there is variation across the states, a lawyer may assist the formation of an entity that follows cooperative principles, but is formed as a limited liability corporation (LLC), a nonprofit, or a C or S Corp, rather than a cooperative corporation. According to the *International Handbook of Cooperative Law*, three important principles distinguish cooperatives from business corporations in most state statutes: "member control, distribution of cooperative benefits in proportion to use rather than capital ownership, and limited returns on capital stock." Many states have codified this latter principle by following the Capper-Volstead Act's maximum dividend rate on capital stock of 8 percent. Cooperatives are owned by their members, who usually have one vote per person or a proportional vote based on the amount of business conducted with the cooperative, as established in the bylaws. These cooperative features echo the Co-op Alliance's shared international Cooperative Principles.[71]

Like directors of an investor-owned corporation, cooperative directors also have a duty of care and loyalty, are entitled to rely on the business judgment rule in making decisions, and are responsible for discharging their duties in good faith in the best interests of the cooperative. A distinction, however, is that member-owners govern the cooperative, following the democratic principle, more closely than investors in investor-owned corporations. Barbara Czachorska-Jones and her colleagues assert that cooperative boards are typically very attuned to the wishes of the members. This assumption may be strained, however, in states whose cooperative laws allow a greater role for nonmember investors.[72]

Several states experimented with "new" cooperatives among agricultural producers in the 1990s. These allowed greater equity from patron investors and were geared toward producing greater wealth for members, but did not necessarily involve nonpatron investors having voting, ownership, and management rights. Only a limited number of cooperatives developed ways to attract nonpatron investors through this period. With Wyoming leading the way, followed by Minnesota, Tennessee, Wisconsin, and a few other states, legislators passed statutes to facilitate this.[73]

Then in 2007, the National Conference of Commissioners on Uniform State Laws adopted the Uniform Limited Cooperative Association Act (Uniform Act), which it slightly amended in 2011 and 2013. This is a model law for states to consider enacting. It is a modern alternative to the Uniform Agricultural Cooperative Association Act and is applicable to all types of cooperatives. The Uniform Act facilitates incorporating as a limited cooperative association, which may be a multistakeholder entity with various membership classes (workers, consumers, investors, community stakeholders) coexisting.

The Uniform Act is not meant to supplant traditional cooperative statutes, but to provide an additional option. According to one of the drafters, it is "intended to provide an unincorporated cooperative structure with centralized management but democratic member control as an alternative to a limited liability company, which has been a form of business many have turned to when the traditional cooperative form of business entity has not been receptive to outside investments." Oklahoma (SB1708), Nebraska (LB848), Utah (SB69), Vermont (HB21), Colorado (SB191), and several other states have adopted the Uniform Act.[74]

The Uniform Act's 2007 prefatory note says its biggest defining feature is that it encourages equity investment in cooperatives by allowing voting investor members who are not patron members. Allowing for investors to participate in this way may make it easier for cooperatives to raise capital. However, a concern is that in states that allow nonmember investors to have greater authority through voting and board positions, the cooperative may not qualify for the antitrust exemption under the Capper-Volstead Act or for federal tax relief under Subchapter T of the Internal Revenue Code.[75]

The Uniform Act tries to mitigate the role of the investor by emphasizing the Cooperative Principles and the social nature of the cooperative in an attempt to establish the primacy of member-patron interests over investor interests. To this end, a majority of voting power must still be held by patron members according to Sections 514, 804, and 405. The 2007 prefatory note states that the act "may be correctly perceived as protecting cooperative principles within state law in ways not possible under more general organizational statutes." It proceeds to explain that the Co-op Alliance's Principles "undergird and animate" many of its provisions and it reflects Cooperative Principles "except to the extent necessary to accommodate investor members" for those cooperatives that choose to have nonpatron investors. The exception unfortunately could allow a cooperative to create governance documents that veer from Cooperative Principles in order to attract investors.[76]

Further, the note specifically emphasizes the sustainability principle "Concern for community," which is addressed by Section 820 of the act. The note explains that this section "varies the law generally applicable to corporate directors . . . to allow the directors of a limited cooperative association to consider cooperative principles as well as a number of community constituencies in making decisions." Section 820 of the act says that a director "may" consider "(1) the interest of employees, customers, and suppliers of the association; (2) the interest of the community in which the association operates; and (3) other cooperative principles and values that may be applied in the context of the decision."[77]

Thus, the act makes this a discretionary allowance, but does not mandate that directors consider these stakeholders. These provisions make drafting the articles of organization/incorporation and bylaws a very important tool for carefully securing a strong sustainability purpose and commitment to the Cooperative Principles. If cooperative founders are interested in creating a multistakeholder cooperative in a state that has passed the Uniform Act, they can incorporate as either a limited cooperative association under that act or as an LLC. Then in the bylaws (for the cooperative) or operating agreement (for the LLC), they can specify the different ownership classes and levels of authority for each. For instance, the bylaws could create a class of investor members who are not allowed to vote or could create positions on the board of directors for nonmember stakeholders. Likewise, the bylaws could create a mandatory duty for the directors to carry out Cooperative Principle 7, the sustainability principle. There are examples of these bylaw provisions below.[78]

Going into detail about each state's cooperative laws is beyond the purposes of this book, yet it is useful to highlight a recent development in California, AB 816, which went into effect in 2016. The new law establishes a worker cooperative corporate entity, where previously the state law only defined consumer cooperatives. The law offers a statutory definition of a "worker cooperative," eases barriers to raising investment capital from within the local community, requires a class of worker-members, and requires that worker-members control the cooperative. The statute defines a "worker cooperative" as having "the purpose of creating and maintaining sustainable jobs and generating wealth in order to improve the quality of life of its worker-members, dignify human work, allow workers' democratic self-management, and promote community and local development in this state." The statute contains no further elaboration on what is meant by "sustainable jobs" or promoting community. However, one could still read the statute as reflecting some of the Co-op Alliance's Cooperative Values and Principles (for example, democratic self-management and promoting community). A stronger legislative model, however, exists with the Spanish cooperative law that directly references the Co-op Alliance's Cooperative Principles. According to Spanish Law 27/1999, a cooperative has a "democratic structure and operation, in accordance with the principles stated by the International Co-operative Alliance."[79]

In summary, while some of the Cooperative Principles are reflected in state statutes, there has been a trend toward loosening these requirements in order to attract investors to capitalize the enterprises. The sustainability principle appears in the Uniform Act as a permission for directors to consider multiple stakeholders, but is not a mandatory requirement. There is great variation among state statutes regarding what Cooperative Principles they emphasize, reflecting a sensitivity to the diverse array of business sectors that use the cooperative form and their members' needs. While this may increase the number of cooperatives operating within a jurisdiction, it sacrifices a shared understanding of what it means to be a cooperative. Further, it blurs the distinction between cooperatives and other types of enterprises. Without statutory reforms, cooperative founders still have the ability to include a strong purpose and adherence to all of the Cooperative Values and Principles in their formation and governance documents. One common feature of the sustainability pathbreakers is their clarity of mission, purpose, and values in their governance documents.

LOCAL LAWS: MADISON, WISCONSIN, CASE STUDY

At the local level, some U.S. cities have established programs to support worker-owned cooperatives as part of their economic development efforts. The Democracy at Work Institute (an affiliate of the U.S. Federation of Worker Cooperatives) promotes scaling up worker ownership by facilitating conversions of existing businesses. Their program, Workers to Owners, is a national collaborative that is starting to see some legislative success. In addition to the federal Main Street Employee Ownership Act, some cities are taking up the challenge and have recently passed laws to promote cooperative development. For instance, in 2014, the New York City Council created a $1.2 million fund, managed by the city's Department of Small Business Services, to grow worker-owned cooperative businesses. In its third-fiscal-year report on the program, the department documented that the fund had grown to $2.2 million. Prior to the fund, there were only twelve worker cooperatives in New York City, but by 2018 there were eighty. These are primarily small employers: in fiscal year 2017, the fund resulted in 185 people being hired by worker cooperatives.[80]

The city of Madison, Wisconsin, similarly launched a Cooperative Enterprise Development Program. In 2019, I interviewed Madison's business development specialist, Ruth Rohlich, the city's lead on this program. She explained that Mayor Paul Soglin was inspired by what was happening with New York's cooperative program. Initially the mayor proposed allocating $5 million over five years from the city's general fund to support new worker-owned cooperatives as a way to address economic inequality. Madison's City Council approved funding the new Cooperative Enterprise Development Program at a lower level: $600,000 per year for five years, starting in 2016, which it later scaled back. The new program is housed in the Economic Development Division, which considers worker-owned cooperatives as a business entity that can address inequities and build assets and wealth beyond a charity-type program.[81]

Madison's Cooperative Enterprise Development Program has the following goals:

1. Build capacity for organizations offering technical assistance and lending for new and existing cooperative businesses.
2. Convert existing businesses to worker cooperatives.
3. Create employment for the people of Madison who have been excluded from the traditional business model, including the formerly incarcerated, low-skilled, veteran and new American populations.
4. Provide fundraising and technical support for creative problem solving such as creating corner stores and food stores in neighborhoods where none currently exist, gathering spaces, neighborhood amenities, and working with trade unions to create union co-ops.[82]

Madison has a supportive cooperative ecosystem and has long been home to cooperatives. Prior to this program, there were seventy-five Madison-based cooperatives, including credit unions, mutual insurance companies, health, food, and so on. Some notable examples include Union Cab (worker, started in 1979), Isthmus Engineering and Manufacturing (worker, started in 1980), Community Pharmacy (worker, coop since 1991), and Willy Street Coop (consumer, started in 1974). Madison's ecosystem also includes active labor unions, loan funds, city economic development staff familiar with cooperatives, the University of Wisconsin Center on Cooperatives, and attorneys who specialize in this area.[83]

Yet the launch of new worker-owned cooperatives in Madison has been slower than in New York City. In 2016, Madison selected the Madison Development Corporation to run the new cooperative loan program. That entity funds low-income housing and already managed a couple of loan funds for the city. The city selected a newly formed Madison Cooperative Development Coalition to provide technical assistance. The coalition says its charge is to "form worker cooperatives that address income inequality and racial disparities by creating living-wage and union jobs." The selection of the technical assistance provider aligned with the city's goals for the program. But working with a startup coalition has involved some time delay. In the second fiscal year of the program, the money was split evenly between these two entities, with $300,000 to each. However, in 2018, the City Council cut its budget further: for the remaining three years of the program, the funding will only be $300,000 per year, and all of it will go into technical assistance the coalition provides. The loan fund has $600,000 in it and only one loan has been completed, which was to an existing cooperative. At the close of 2020, the coalition had provided technical assistance that helped launch four new cooperatives in bookkeeping, home health care, doula and postnatal care, and racial-equity consulting. In 2020, the coalition employed a part-time administrator and a full-time cooperative developer.[84]

Still, the idea holds unrealized potential. The city has been able to orient its survey of local businesses to identify businesses interested in succession planning and converting to cooperatives. The city is making direct connections with manufacturing companies with 500 or fewer employees and discussing conversion options. In the summer of 2019, in partnership with the University of Wisconsin Cooperative Center, it planned to hold two classes for these businesses on succession planning. The city has attorneys, accountants, and financial planners who are volunteers with the Coalition and available to assist in business conversions. The entities can apply to the Cooperative Development Coalition for a $10,000 grant to be used for conversion costs, worker salaries, and a wide array of other expenses.[85]

Rohlich reflected on the reality of supporting new worker-owned businesses involving people who have been excluded from the traditional business model. Madison's initiative is aimed specifically at creating opportunities for people of color to start and own cooperatives. This is critical for promoting greater equity because most of Madison's existing cooperatives and organizational leaders are not people of color. Further, when the prospective worker-owners are low-income, it makes capitalizing the new business very difficult. The worker-owners need to provide personal guarantees for a loan and many may have existing personal credit issues, which disqualify them from obtaining bank loans or even peer-to-peer Kiva loans. Since Madison's program is geared toward addressing inequality, the city wanted to find a way to assist in capitalizing, despite the financial risks. The coalition pooled resources among interested parties and allocated $40,000 to launch Soaring Independent's home health care business.[86]

While Madison's emphasis on democratic ownership as an antidote to racial and economic inequities shares similarities with Spain's social economy focus, Madison's program does not prioritize environmental sustainability. By contrast, the programs in València and Barcelona are multibenefit: they use a sustainable economic model focused on the triple bottom line. While Madison's program has a strong equity/equality purpose, if it extends the program after the initial five years, it could accomplish multiple benefits with further refinement to orient the program toward forming businesses based on the triple bottom line. Madison would then be on a stronger footing to meet the sustainability challenges of the next economy we need to build.

Inclusion of Cooperative Principles in Legal Formation and Governance Documents

While statutory direction is optimal for distinguishing cooperatives from other enterprises, it is not essential. If a business is incorporated as a cooperative, lawyers and founders can draft the Cooperative Values and

Principles into the articles of organization/incorporation and bylaws of a cooperative or the operating agreement of an LLC. Both bylaws and operating agreements are documents that regulate the internal affairs and governance of a business and are malleable. Future worker-owners may change them by following the methods prescribed within them, so the method of change is also a key part of this drafting.[87]

Numerous resources for U.S. cooperatives discuss the seven Cooperative Principles. *Tackling the Law, Together: A Legal Guide to Worker Cooperatives Generally and in Massachusetts* (2015), for instance, says that "cooperatives around the world generally follow" these core principles. Yet, the reality is a little murky because state law does not require it and very few cooperative lawyers focus on designing the Cooperative Values and Principles into the governance documents of these enterprises.[88]

In 2018, I attended a regular conference call that the U.S. Department of Agriculture holds with attorneys for cooperatives, in which we discussed whether and how the attorneys incorporated the Cooperative Values and Principles into bylaws and internal policies. The attorneys acknowledged this as a possible approach, but not a uniform approach or one required by law. Of course, there are other ways to articulate and promote the Cooperative Principles beyond ingraining them in the legal structure of the entity. For example, one of the attorneys noted that a food cooperative he represents displays the Principles throughout their stores to educate the public and employees.

However, incorporating the Principles into the legal documents sets a firm foundation for the members that distinguishes their entity from other corporate forms. Cooperative members can then work from this legal foundation to infuse the Principles into their day-to-day operations and sustainability best practices. This approach institutionalizes the Cooperative Values and Principles in a more enduring way.[89]

Real Pickles cooperative in Massachusetts provides a replicable model for how to effectively integrate the Cooperative Values and Principles and draft a strong sustainability mission and purpose into the governance documents of an enterprise. The documents also include a liquidation section designed to discourage members from being pushed to liquidate the cooperative to "cash out" for themselves when the business is successful.

Drafting a strong sustainability purpose into the articles of organization provides clarity for the present and future members of the cooperative. This gives a durability to focusing on sustainability in a way that leaving it up to making the business case to justify actions does not. Real Pickles' purpose, drafted in Article 1, "is to operate a small, democratically-organized business helping to build a regionally-based, organic food system that supports ecological and human health."

Real Pickles' bylaws further articulate this purpose with a mission and guiding principles, including an explicit commitment to the Co-op Alliance's Principles, as follows:

4. Mission & Guiding Principles
   a) Mission. Real Pickles is committed to promoting human and ecological health by providing people with delicious, nourishing food and by working toward a regional, organic food system.
   b) Guiding Principles
      i) We are committed to producing the highest quality traditional pickled foods available, using natural fermentation.
      ii) We buy our vegetables only from Northeast family farms and sell our products only within the Northeast U.S.
      iii) Our ingredients are 100% organic.
      iv) The foods that we sell are minimally processed.
      v) We are committed to remaining a small business.
      vi) The jobs that we create support the dignity of all workers of the Co-operative.
      vii) We are committed to the Co-op Principles.
5. Co-operative Identity. The Co-operative shall operate at all times on a co-operative basis for the mutual benefit of its Members and the advancement of its mission in a manner consistent with the Co-operative Identity as defined by the International Cooperative Alliance:
   a) Definition. A co-operative is an autonomous association of persons united voluntarily to meet their common economic,

social, and cultural needs and aspirations through a jointly-owned and democratically-controlled enterprise.

b) Values. Co-operatives are based on the values of self-help, self-responsibility, democracy, equality, equity and solidarity. In the tradition of their founders, cooperative members believe in the ethical values of honesty, openness, social responsibility and caring for others.

c) Principles. The Co-operative Principles are guidelines by which co-operatives put their values into practice:

    i)   Voluntary and Open Membership. Co-operatives are voluntary organizations, open to all persons able to use their services and willing to accept the responsibilities of membership, without gender, social, racial, political or religious discrimination.

    ii) Democratic Member Control. Co-operatives are democratic organizations controlled by their members, who actively participate in setting their policies and making decisions. Men and women serving as elected representatives are accountable to the membership. In primary co-operatives members have equal voting rights (one member, one vote) and co-operatives at other levels are also organized in a democratic manner.

    iii) Member Economic Participation. Members contribute equitably to, and democratically control, the capital of their co-operative. At least part of that capital is usually the common property of the co-operative. Members usually receive limited compensation, if any, on capital subscribed as a condition of membership. Members allocate surpluses for any or all of the following purposes: developing their co-operative, possibly by setting up reserves, part of which at least would be indivisible; benefiting members in proportion to their transactions with the co-operative; and supporting other activities approved by the membership.

    iv) Autonomy and Independence. Co-operatives are autonomous, self-help organizations controlled by their members. If they enter into agreements with other

organizations, including governments, or raise capital
from external sources, they do so on terms that ensure
democratic control by their members and maintain their
co-operative autonomy.

v) Education, Training and Information. Co-operatives
provide education and training for their members, elected
representatives, managers, and employees so they can
contribute effectively to the development of their cooperatives. They inform the general public—particularly
young people and opinion leaders—about the nature and
benefits of co-operation.

vi) Cooperation among Co-operatives. Co-operatives serve
their members most effectively and strengthen the cooperative movement by working together through local,
national, regional and international structures.

vii) Concern for Community. Co-operatives work for the
sustainable development of their communities through
policies approved by their members.

6. Activities. The Co-operative shall engage in such activities that
advance its mission and serve the mutual benefit of its members, and to do all things necessary, appropriate and proper
for the accomplishment, furtherance of or incidental to these
activities.[90]

Another example of a bylaw provision that includes a strong sustainability
focus and creates a mandatory duty for management is from the Louisville
Community Grocery draft bylaws:

TRIPLE BOTTOM LINE PRINCIPLES. The Cooperative will be operated according to Triple Bottom Line principles. The duty of the managers to maximize financial benefit to the members is supplemented by a co-equal duty
to responsibly pursue optimal benefit to concerns of the company's community, including its employees and associates and the health and safety of the
public, and of the sustainability of the biological environment.[91]

Membership in the Real Pickles cooperative is limited to employees, each
of whom is entitled to one vote. Real Pickles raises capital from nonpatron

investors by issuing preferred shares, but these do not come with any voting rights. The bylaws secure the consistency of the articles of organization by requiring a consensus vote of the members for any changes, which they define as a unanimous vote with up to one-third of all members abstaining. Similarly, any changes to the cooperative's purpose, mission, and guiding principles must be by a consensus vote of the members.[92]

Real Pickles' bylaws establish a strict pay differential in the basic DNA of the enterprise, advancing a goal of narrowing the wealth gap and the stated values of equity and equality. The bylaws state: "Pay Ratio. The pay differential (the ratio of hourly compensation for the highest paid employee to that of the lowest) is not to exceed 3:1."[93]

If the Real Pickles cooperative is liquidated, the articles of organization state, "the funds and assets legally available to be distributed . . . to the Cooperative's Members and Preferred Holders" are to reimburse them for the original price of their shares. Then "any remaining Available Funds and Assets shall be distributed to one or more organizations, selected in accordance with the bylaws, whose missions are consistent with the purposes of the Cooperative." The founders included this provision so worker-owners would not be tempted to pursue their self interest in cashing out of a successful business by selling it and abandoning its cooperative form and purpose. This is a structural constraint that attempts to balance the self-interest of the worker-owners with the interests of the multiple stakeholders who benefit from the cooperative's continued existence and success (organic farmers, local community, customers, and future worker-owners, to name a few).[94]

A final note about this exemplar is that the governance documents were the result of a variety of cooperatives and supportive organizations and attorneys providing ideas about best legal drafting practices. Real Pickles' process of developing these documents is a demonstration of the Cooperative Principles of "cooperation among cooperatives" and "promoting education" in action. Real Pickles learned from Equal Exchange (the oldest and largest fair trade coffee cooperative in the United States) about the importance of a liquidation clause to prevent current members from feeling pressured to extract value from the cooperative, to the detriment of other stakeholders, including future members. Real Pickles was able to work with their lawyer to build on sample bylaws from the Valley Alliance of Worker Cooperatives, an entity that supports the ecosystem of worker-cooperatives in western Massachusetts. The result is a model Real Pickles is willing to freely share to help others who are endeavoring to lock in a strong sustainability purpose to the fundamentals of a cooperative business.

8

Renewable Energy

The monumental effort of deep decarbonization requires a strategic focus on the largest sectors of emissions and an implementable plan to rapidly transform those systems to reduce their GHG contributions. In the United States, the European Union, and globally, the single largest cause of GHG emissions is fossil fuel extraction, processing, and combustion for energy. Emissions from electricity and heat production, industry, transportation, and other energy are all essentially caused by extracting, processing, and burning fossil fuels. In 2010, globally, reliance on fossil fuels contributed 70 percent of all GHGs, as shown in Figure 1.[1]

One decade after the emissions snapshot in Figure 1, global electricity generation is increasing and remains dominated by coal, which is the dirtiest of all fossil fuels. Overall, energy production from all sources grew between 2000 and 2018, and although production from renewables increased, coal increased too, which resulted in rising levels of GHGs globally. Total energy production in 2000 was 15,427 TWh (terawatt hours) and by 2018 it had expanded to 26,603 TWh. Comparing coal and renewables (not including nuclear), in 2000 there were 5,994 TWh (coal) and 2,836 TWh (renewables), and by 2018 there were 10,123 TWh (coal) and 6,799 TWh (renewables). An early 2020 report from JP Morgan noted that around 60 percent of coal-fired utilities are less than twenty years old, with a design life of fifty years. Given this reality, the International Energy Agency asserts that design life would need to be limited to twenty-five years in order to meet

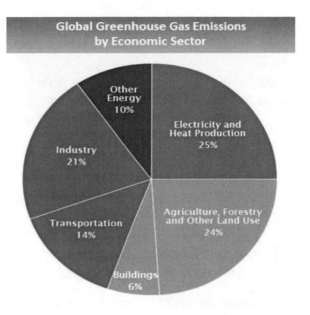

Figure 1. Global greenhouse gas emissions, 2010 (Credit: U.S. EPA, global emissions based on IPCC 2010 emissions, https://www.epa.gov/ghgemissions/global-greenhouse-gas-emissions-data#Sector)

the Paris Agreement's goal of limiting warming to +2°C. In other words, coal-powered utilities need to be retired early.

This is not as out of reach as it seemed even a few years ago. The U.S. Energy Information Administration showed that U.S. 2019 annual energy consumption from renewable sources exceeded coal consumption for the first time in over a century. Economic forces have pushed U.S. utilities to retire hundreds of aging coal plants since 2010, and coal is expected to fall by one-quarter in 2020. The cost of renewables has come down so much that it is now cheaper to produce electricity this way than from coal. The coronavirus pandemic, which reduced electricity demand, has hastened the decline of coal. The world's second-largest coal consumer, India, has prioritized lower-cost solar over coal in 2020.[2]

Creating electricity with renewable energy and moving more heating, transportation, and industry to renewable electricity is the great challenge and opportunity of this time on Earth. Countries are already moving in this direction, but need greater momentum. In 2017, 17 percent of the electricity in the United States and 32 percent in Spain came from renewable sources. In both countries, the largest share of renewable energy is produced by wind.[3]

This chapter sketches the pathways to sustainable development and deep decarbonization of electricity at the global level. Then it provides case studies of cooperative exemplars to demonstrate how different aspects of the transition are already under way in Spain and the United States. Each case study includes an overview of the cooperative's history, type, and economic sector; context based on its industry and country; best practices in environmental sustainability; and its governance and cooperative legal structure.

Pathways to Sustainable Development and Deep Decarbonization

To deeply decarbonize, the IPCC recommends global energy demands need to shrink through conservation and energy efficiency, even while providing energy to those who currently lack it. Renewable energy should meet the remaining demands. Around the world, 1.1 billion people lack access to any electricity. Those who have electricity obtain it primarily from fossil fuels: in 2015, fossil fuels powered 66 percent of global electricity generation. The world must now rapidly shift off fossil fuels in response to the climate emergency. The IPCC states that the faster we can accomplish this, the less suffering people will experience from climate change.[4]

The climate emergency calls for a disruption of the business model for energy, and that we undertake the following systemwide steps:

- Deploy comprehensive energy efficiency and conservation programs, including enforceable energy performance standards for buildings.
- Retire all fossil fuel electric generation plants that do not capture and store the GHGs they emit.
- Develop renewable energy generating capacity in the least environmentally destructive ways available (for example, on already built surfaces and brownfields, compatible with multiple land uses, and offshore).

- Plan for greater electricity demands due to electrification of transportation (mass transit and private).
- Incentivize the switch to electric vehicles (EVs), emphasizing fleets and platforms for sharing.
- Invest in storage to smooth out variability in renewable generation.
- Incorporate a diversity and abundance of customer-sited generation and storage.
- Emphasize democratic ownership and participation models that reduce energy poverty, ensure a just transition, and equitably share the benefits of the renewables revolution.

The European Union's "Clean Energy for All Europeans" legislative package, completed in 2019, promotes some of these steps, but not all and not aggressively enough. The package includes eight legislative acts, and EU countries have one to two years to incorporate these directives into national law. Some of the requirements of these new laws are: set binding targets of at least 32.5 percent energy efficiency by 2030 compared to "business as usual"; set binding targets of 32 percent renewable energy sources in the EU energy mix by 2030; require each nation to make ten-year climate and energy plans; redesign the energy market to better integrate renewables; and pay special attention to a just transition in coal regions.[5]

The renewable energy revolution poses a once-in-a-lifetime opportunity for co-benefits of democracy, bringing electricity to those who lack it, and equity. This transition has the potential for widely shared prosperity, not just a decrease in suffering. To unlock that potential, however, we also need to disrupt business as usual in the dominant business model. The co-benefits are more likely if the institutions that lead the way are democratically owned and managed with the explicit goal of bringing the benefits of ownership to those who have traditionally been marginalized by the current economic and energy system.

Unprecedented in scale and time frame, the shift to renewables offers a chance to build a new energy model that is not only environmentally cleaner but more resilient technically, and provides greater equity and ownership among a broader group of stakeholders. The modular nature of some renewables, such as rooftop and community solar, make this new form of

energy generation more readily available to shared ownership than the monopoly model of large fossil fuel generating stations. Even with utility-scale
renewable installations, there are possibilities for shared ownership and
more democratic governance. If the goal is a democratic economy that is
equitable and sustainable, the institutions that lead the renewable revolution should be built to reflect that.

The United Nations is wrestling with how to guide renewable energy development to align it with the Sustainable Development Goals. Meanwhile
the European Commission has put out a European Green Deal that calls
for decarbonizing energy on a tight time frame, but never mentions the
importance of structuring ownership and using new public investments to
simultaneously advance the social economy. Similarly, the Clean Energy for
All Europeans legislative package did not include anything about structuring ownership or prioritizing social economy energy cooperatives.[6]

Of course, many types of businesses are working on this transformation,
including the multinationals that have profited the most from fossil fuel
exploitation. Understanding how a cooperative renewable energy enterprise
fits into the transition is critical because it provides a type of enterprise that
lends itself readily to meeting multiple Sustainable Development Goals:
access to affordable, reliable, and sustainable energy (SDG 7), urgent action
to combat climate change (SDG 13), decent work and sustainable economic
growth (SDG 8), reducing inequalities (SDG 10), supporting sustainable
cities and communities (SDG 11), and responsible consumption and production (SDG 12).

Regulators have traditionally required electric utilities to deliver electricity cheaply and reliably, but now the world needs regulations requiring
utilities to deeply decarbonize electricity while keeping prices affordable.
Utilities are key to getting to zero carbon emissions. Some pathbreaking
electric utilities show how to rapidly transition in a way that balances cost,
reliability, and environmental impact, while promoting democratic ownership. Renewable energy cooperatives can engage at various points in the
energy system to facilitate these transitions and innovations. Unless one is
disconnected from the grid, there are four basic steps involved in accessing
electricity:

1) production (the facilities that generate electricity);
2) transportation (the high-voltage wires that move electricity from the point of production to transformer substations, "the grid");
3) distribution (moving electricity from substations to electricity consumers on medium- and low-voltage wires); and
4) commercialization (companies buying electricity on wholesale markets, then directing it to and billing consumers).

With the exception of transporting electricity on the grid, the cooperative pathbreakers featured here are playing key roles in the three other steps in the electricity system.

Energy Cooperative Pathbreakers

This chapter features Spanish renewable energy cooperatives involved in those three points in the electricity system. In the United States, it features an electricity distribution cooperative that is adding renewables and an entirely renewable power supply cooperative, both in a state that lacks any laws that incentivize such a shift. Spain's Som Energia is a nationwide, modern renewable energy cooperative that provides its consumer-members 100 percent renewable electricity and energy efficiency services. It is involved in production and commercialization of renewables, and at the close of 2019, it was annually producing 17 GWh (gigawatt hours) of renewable electricity. Cooperativa Eléctrica Benéfica San Francisco de Asís de Crevillent is an almost century-old electric cooperative that provides power to a Spanish municipality. It owns and operates the local distribution network and is now powering the community of 28,000 people solely with renewables. In 2010, this historic cooperative became the parent company of the Enercoop Group. By 2019, Enercoop had grown to be Spain's largest electricity cooperative purchasing group, annually supplying 50 GWh of renewable electricity to members throughout Spain.

In the United States, the adoption of renewable electricity at cooperatives lags behind the Spanish cooperatives, but there are some notable pathbreakers. Cobb EMC is a traditional electricity distribution cooperative, serving

200,000 members in the state of Georgia. It participates in Green Energy EMC, a cooperative that brokers contracts for renewables. In 2019, Cobb EMC had the most megawatts (MW) in solar of any U.S. cooperative (123.7 MW), had the largest percentage of solar of any eastern cooperative (40 percent on some days), and offered services to promote demand management and the use of electric vehicles to advance moving transportation off of fossil fuels. In 2020 other electric cooperatives announced plans to produce renewables in gigawatts instead of megawatts. For example, the Great River Energy cooperative plans to add 1.1 GW of new wind capacity by 2022, which shows Cobb EMC is a pathbreaker, but will soon not be an outlier.[7]

Som Energia (Electric, Consumer, Spanish National)

Overview of the Cooperative: History, Type, and Economic Sector

Founded in 2010 as a nonprofit cooperative, Som Energia is Spain's first renewable electricity consumer cooperative dedicated to 100 percent renewables. The idea for the cooperative came from a professor in Girona who knew about consumer renewable energy cooperatives in Germany and other parts of Europe, but when he came to Spain he saw there weren't any. So a group of professors, students, and friends formed Som Energia to pioneer this concept with an initial group of 150 members.[8]

Som Energia created a platform of participation that is highly effective. By 2019, they had built the largest Spanish consumer cooperative for renewables in terms of members and amount of renewable electricity. They are providing electricity marketing services nationwide in Spain, offering consumers the opportunity to join the cooperative. Members learn how to conserve energy and meet all their electricity needs with renewables. In 2019 alone the company experienced rapid growth: in March it had 55,766 members who were purchasing 13.56 GWh per year of renewable electricity, and by December it had 63,111 members who were purchasing 17 GWh per year of renewable electricity.[9]

Som Energia sees the current fossil fuel–based energy model as unsustainable and is committed to promoting an entirely renewable model. In addition to connecting people directly to renewables in a transparent way, it encourages members to "participate in a transformative social movement," grow the "social and solidarity economy," and "break the existing energy oligopoly." Members join by contributing €100 to a "social capital" fund.[10]

Other renewable energy consumer cooperatives have started to copy this model in different regions of the country, which Irene Machuca, an elected member of Som Energia's Governing Board who I interviewed for this book, sees as a positive development. There are no other nationwide consumer cooperatives for renewables in Spain, but there is a nationwide renewable power supply cooperative, featured next.[11]

Som Energia performs two principal functions: it produces renewable energy from sun, wind, biogas, and water (small hydro), financed by its members; and it buys additional renewable electricity to supply its members who have contracted for this service. Those contracts are transparently guaranteed as renewable through Spain's official Certificates of Guarantee of Origin, and don't require the consumers to make any technical changes, like installing solar panels on their home.[12]

Industry Context

To contextualize how Som Energia is attempting to transform the energy model, it helps to understand the scope of the energy transformation needed and how the electricity system works. As stated above, electric cooperatives can engage at various points in the energy system: 1) production, 2) transportation on high-voltage wires, 3) distribution to consumers, and 4) commercialization.[13]

The transportation and distribution points are monopoly-controlled. To clarify the use of "monopoly" here, the high-voltage transmission network is a monopoly in the strict sense because in Spain it belongs to a single private company, Red Eléctrica de España, S.A., which is the system operator. The medium- and low-voltage distribution networks are "natural monopolies" in that it makes no sense to duplicate the network of wires and substations. The distribution networks are owned by a variety of companies, but only one operates in and thus monopolizes each local distribution network. In addition to Spain's largest investor-owned electricity companies (Iberdrola, Endesa, Naturgy, EDP-Spain, and Viesgo), there are more than 300 small distribution companies, and presently only seventeen are cooperatively owned. Cooperativa Eléctrica Benéfica San Francisco de Asís de Crevillent, featured in the next case study, is one of these distribution companies. In Spain, consumers have a choice of companies for the production and commercialization points in the chain, and this is where Som Energia operates. It builds new renewable production facilities and contracts with other producers to guarantee renewable energy for its consumer-members.[14]

Som Energia is actively building projects to generate more renewable energy, despite a turbulent regulatory environment. When the cooperative started in 2010, Spain and other European countries offered generous subsidies for renewable energy production through feed-in tariffs (high rates for long fixed terms). Initially the cooperative developed projects during this time period, but as Spain realized the financial impact of the feed-in tariff policies, the government significantly scaled them back. Som Energia has adapted to these policy fluctuations to finance and develop new projects through its members' contributions.

Despite these problems with getting the prices right, Spain is producing a lot of electricity from renewables. In 2017, 32 percent of electricity generation came from renewables, according to the national Spanish Electricity Network. This was down from a total installed capacity of 46 percent due to a lack of hydroelectric power generation that year. There will be variations like this; however, what is notable is that Spain's renewable generation is practically unchanged since 2013, the year the government scaled back incentives for renewables. Since 2008, the most significant renewable energy source in Spain has been wind, which in 2017 accounted for more than half of all renewable generation and 18 percent of the nation's electricity.[15]

Spain's renewable production could expand greatly under its current government. Elected in 2018, Spain's prime minister, Pedro Sánchez, merged the environment and energy departments to create a new Ministry for Ecological Transition and Demographic Challenge (Ministerio Para La Tansición Ecológica y El Reto Demográfico), which Teresa Ribera Rodríguez leads.[16]

Encouraged by the Paris Agreement, Spain's new government initiated several changes in 2018. In October 2018, it passed a royal decree that abandoned the "sun tax," which will make it more affordable for people to put solar panels on their property for their personal use. The sun tax was part of the prior government's energy package, initiated between 2013 and 2015, to scale back subsidies for renewable energy due to a €28 billion tariff deficit. In April 2019, the Council of Ministers approved a royal decree establishing the regulatory framework for individuals producing renewable energy for their own consumption. This allows individuals and groups to produce renewable energy either entirely for self-consumption or to be fed into the grid for net metering on monthly electricity bills.[17]

In late 2018 Spain announced a draft Climate Change and Energy Transition law that set a goal of 100 percent renewable power sources for electricity by 2050, 74 percent renewables by 2030, and a phase-out of coal. As of 2020, the Sánchez government has not been able to finalize the plan, but the draft text of the law aligns it with economic recovery from the coronavirus pandemic. To meet the goals, some estimate the country would have to install at least 3,000 MW of new renewables every year for the next ten years, creating up to 350,000 new jobs every year. The scale of the development makes nimble and innovative cooperatives like Som Energia essential to the national transition.[18]

The International Institute of Law and the Environment (Instituto Internacional de Derecho y Medio Ambiente) advocates for this new law to include quantifiable emission reduction targets, including an interim target of a 40 percent reduction from 1990 levels by 2030, on the way to zero emissions by 2050. Applying the "polluter pays" principle, the institute supports establishing climate taxes on the most polluting companies. It further asserts that the country needs a roadmap for decarbonization that includes an orderly and progressive closure plan for coal-fired power plants. The end of coal could come as early as 2025. This would align Spain with other countries that have already committed to eliminating coal-burning power plants: Canada, France, the United Kingdom, Austria, Sweden, the Netherlands, and Italy.[19]

Concurrently, Spain plans to ban new licenses for fossil fuel drilling and fracking. The draft law sets a goal of improving energy efficiency by 35 percent in eleven years, with extra requirements for government building leases. While the law will also ban new cars with gasoline or diesel engines, implementation of that provision would be delayed until 2040.[20]

In April 2019, the Spanish government approved the National Strategy to Combat Energy Poverty, with a goal of reducing the number of people affected by at least 25 percent by 2025. Since improving energy efficiency reduces energy bills, this is one part of the new law to reduce energy poverty.[21]

Som Energia's Machuca was more circumspect about these legal changes. She says they will need some time to see if the new policies will really be implemented and not switched again with the next election. As of May 2020, the Climate Law and the Integrated National Energy and Climate Plan were still in draft form, as submitted in February 2019 to the European Commission for evaluation.[22]

As with other countries, investor-owned corporations also serve Spain's energy market, with five dominating the industry. They are primarily fossil fuel companies that have added renewables to their portfolios, much like a large agribusiness corporation that adds an organic product line. When discussing the cooperative difference for energy, I asked Machuca about the prices consumers pay for renewable electricity. She responded that it is not that the cooperative delivers the best price for renewable energy. In fact, the prices the consumer pays really are not very different if the provider is an investor-owned corporation or a nonprofit cooperative. Machuca asserted that "Som Energia's objective is not to be the cheapest company, but to provide a good balance for its employees and to change the model of energy production." She explained that the cooperative wants to increase access to renewables for everyone and engage people in a process of learning about energy use to focus on reducing consumption.[23]

Since rolling out renewable capacity quickly is essential for mitigating climate disruption, we discussed which type of enterprise can move quickly to mobilize funding and develop new projects. Machuca said that "both types of enterprise can get projects going about as fast as one another." The difference comes in the financing. She explained, "the larger investor-owned companies have more resources to get these facilities started. They can also influence the law to benefit them."[24]

Yet, Som Energia has access to a different type of financing, one that provides more equitable participation in this emerging area. Som Energia does not finance its projects through bank loans or traditional shareholder investments. Rather, it provides direct investment options to connect its members to renewable projects with very small barriers to entry. According to Machuca, "Som Energia gets capital from its members in amounts as low as €100 to join, and some opt to invest more than this. With these small amounts of money, the cooperative opens up opportunities for more people to invest in renewables. Ultimately, Som Energia is working for a social benefit, not a financial one."[25]

Som Energia has used two mechanisms to finance building new renewable energy-generation facilities: starting in 2012, it accepted voluntary member contributions to its "social capital" fund and reported paying 2,033 members a total of €133,000 from interest generated in 2018. Currently, the cooperative has almost €7.5 million in the fund, which has financed the majority of its projects. The social capital projects' current return on investment is 1.75 percent, so this is a slow-earning investment of members' money.[26]

However, Som Energia is not accepting new contributions to social capital. Instead they have accepted €3.7 million from 3,800 members in "kWh Generation." The company had to create this new investment vehicle because the government withdrew incentives for new renewable projects. This allows members to provide a "loan without interest" in order to finance generating renewable energy that they will use for twenty-five years. Som Energia uses these investments to fund new renewable generation and manages the projects without any economic margin. It recommends that consumer-members make an investment that equates to 70 percent of the kilowatt hours used per year, and then encourages members to continually reduce energy consumption through conservation. Members start recouping their investment in very small amounts twenty-four months after they have made the contribution, with the promise that it will be fully paid back at the end of twenty-five years. Members start receiving the renewable energy that is produced when the project is functioning (which takes from twelve to eighteen months). The members see this reflected on their electric bills from Som Energia, which the cooperative predicts but does not guarantee will be discounted.[27]

The cooperative has been able to use its membership base to finance new projects very quickly. For instance, in October 2017, during a seven-day period, it raised the needed social capital of €5 million to build three solar projects. In September 2015, during a brief two-hour window, it raised €800,000 to build a hydroelectric project.[28]

Environmental Sustainability

The entire purpose of Som Energia is focused on environmental sustainability. Its consumer-members are purchasing 100 percent renewable electricity and have the opportunity to learn how to reduce their energy consumption. Unlike a traditional electric utility, Som Energia has built a community among its cooperative members by establishing local groups where members can get together in person. It is here they can talk about how to conserve energy, learn from each other, and undertake projects collectively to reduce their energy usage.[29]

Additionally, Som Energia encourages members to produce renewables on their own homes and buildings. It publishes a guide and provides training to encourage movement in this direction, while also working to reform Spain's laws. Several of the local membership groups have organized buying clubs to purchase solar kits at discounted prices for members to install on their homes. On April 1, 2019, several local groups put out requests for proposals for entities to install solar on 100 roofs. The cooperative members will choose the winner, and in order to be considered the business must be local, a cooperative or nonprofit, and committed to the values of the social economy. The cooperative expected at least 1,000 home installations in 2019.[30]

Governance and the Cooperative Structure

Som Energia locks its purpose of renewable energy production into its internal bylaws or *estatutos*. Article 2 defines its "purpose" as the "marketing and production" of electrical energy from "renewable sources" and providing heat and power. The bylaws establish a required allocation of financial "surplus" of 20 percent to a reserve fund and 10 percent to an education and cooperative-promotion fund.[31]

As is typical of larger cooperatives, Som Energia uses a mix of direct and representative democracy in its governance. The cooperative has a General Assembly of all of the members (one member, one vote), a Governing Board, auditors, and a team of about seventy employees for day-to-day operations. The members elect the Governing Council, which is responsible for implementing the guidelines the members set in their General Assembly. By its bylaws, the General Assembly must be held once a year and adopt resolutions by a simple majority vote. Machuca commented that "about 2,000 people come to the annual General Assembly, but many others participate through the year in multiple other ways."[32]

Som Energia has a highly developed participatory democracy strategy that utilizes virtual participation online and locally in person among its members across Spain. In a website dedicated to member participation, the cooperative articulates a month-by-month sequence of events, from surveying members to making proposals, to discussing/debating, to finally defining the issues before a vote of the General Assembly. The organizational structure includes the utilization of local groups of members organized by geography, who meet in person to discuss and debate the issues. The cooperative uses this process to elect committee members, develop budgets, make strategic plans, and create internal regulations. It even provides a video tutorial on how to participate in the 2019 process.[33]

For 2019, Som Energia transparently includes all of the proposals that have been discussed, the number of votes, and the status of the proposals. For instance, proposals generated by the members that year and submitted to votes included whether hydroelectric dams should be part of renewable resources, whether a third of the Cooperative Education and Promotion Fund should be allocated to support local initiatives for ecological transition, and whether the cooperative should help people affected by natural disasters with an energy subsidy, among other topics.[34]

Additionally, the cooperative organizes another annual national gathering for its members called "Som Energia School" that is more relaxed than the General Assembly. It provides an in-person meeting point where people can attend training seminars, exchange ideas, debate issues, and get to know one another.[35]

Som Energia offers the best example I've seen for utilizing a mix of national and local gatherings, in-person and online, and making wide use of technology to connect members virtually. The cooperative members are practicing democracy, direct and representative, which is another important cooperative distinction, and one that should improve civic engagement.

Summary

Machuca proudly describes Som Energia as "unique" in Spain. "We are changing the energy model. We want others to copy us so we can counter the energy oligarchy that is dominated by fossil fuel companies and the government. When one joins the group, one is not only purchasing renewable power, but learning how to conserve energy and be more responsible consumers. We are teaching each other to change habits in our houses and this will move us forward to use less energy overall."[36]

The first ten years of Som Energia's development have been marked by fluctuating regulatory policies. Despite this, 150 members launched Som Energia and grew it into Spain's first and largest consumer cooperative for renewable electricity. At the start of 2020, the dream of a renewable energy cooperative was a reality for 64,000 members (up from 55,766 in 2019) supporting the production of 17 GWh per year (up from 13.56 GWh per year in 2019) of renewable electricity, and it continues to grow. If the new Spanish laws go into effect that set a goal of 74 percent renewable electricity by 2030, ban new cars with gas and diesel engines by 2040, and cut carbon emissions to net zero by 2050, Som Energia is poised to use its model as a base from which to expand membership and make these national climate protection goals a reality.[37]

Cooperativa Eléctrica Benéfica San Francisco de Asís de Crevillent (Electric Distribution, Consumer, Spanish Local)

Enercoop (Electric Purchasing, Consumer, Spanish National)

Overview of the Cooperative: History, Type, and Economic Sector

Founded in 1925 as the Cooperativa Eléctrica Benéfica San Francisco de Asís de Crevillent (or Cooperativa Eléctrica), this cooperative of consumer-members brought electricity to the small town of Crevillent, Spain, in the Valencian Community. As was typical of the time, the founders named the cooperative after the patron saint of the town, Saint Francis, who is known for his love of nature. Their choice of namesake foreshadowed their cooperative's ability to later become a powerful force in the fight against climate disruption by transitioning electricity from fossil fuels to renewables. They forged a path to sustainable energy by switching the entire community they serve to 100 percent renewables and became the parent company of the Enercoop Group, a renewable-energy purchasing group whose portfolio makes it the largest electricity cooperative in Spain.

The cooperative's initial and enduring objective has been to provide all its members with electricity of the best quality and price. They have been able to maintain this commitment even as an early adopter and now leader in renewable electricity and energy conservation.[38]

There has been a close relationship between the cooperative and the municipal government. At the time of Cooperativa Eléctrica's founding, Crevillent was a small town with big dreams of building textile factories. The town needed energy to power its development plans, but large electric companies were not serving the community. After the launch of Cooperativa Eléctrica, they were able to develop a base of industry to produce carpets and other textiles, while bringing electricity to all who lived in the town. The cooperative's first important action was to partner with the City Council of Crevillent, which acquired land where the cooperative built the first energy project to distribute power lines to its consumer-members.[39]

The Cooperativa Eléctrica owns and maintains the distribution network, the transformation centers, and substations in Crevillent. It has more than 400 kilometers of medium- and low-voltage power lines.[40]

In more recent decades, the cooperative's Governing Council and membership developed a heightened awareness about environmental protection as they grew more concerned about climate change. On its ninetieth anniversary, Cooperativa Eléctrica reexamined its role given the climate emergency and the need to stop using fossil fuels to produce energy.

Since the early 1990s it has added different forms of renewable energy to its mix of energy supply. In 1991, it acquired a hydroelectric facility in Calasparra (Murcia, Spain). In 2007, it joined a large solar facility, El Realengo, in Crevillent. As of 2019, everyone who lives in Crevillent (28,000 people) relied on 100 percent renewable electricity, which the cooperative says is at the lowest price.[41]

In 2010, this historic cooperative became the parent company of Enercoop, thus expanding its reach to the national level. This new energy cooperative works with a portfolio of entirely renewable energy. It generates, distributes, and markets renewable energy on the national Spanish energy market.

Industry Context

In 2009, Spain's new electricity market laws went into effect, which facilitated the emergence of new businesses to bring more renewable electricity into the market. Enercoop was able to emerge as a cooperative purchasing group, with three cooperatives. By 2019, there were twenty-one participating cooperatives and social economy companies, supplying power to more than 100,000 Spanish consumers. It is among the top three independent energy companies in the Valencian Community.[42]

In less than a decade, Enercoop grew to manage more than 50 GWh annually produced by renewables in Spain and Portugal. The company reports that its energy portfolio avoids more than 15,000 tons of CO_2 per year. It currently has revenues of more than €50 million per year.[43]

While Crevillent no longer has such a strong textile and carpet industry, the cooperative has adapted to the new market conditions and provides people and businesses both within the town and outside it with renewable energy. Other cooperatives featured in this book participate in Enercoop. The Cerveses Lluna beer cooperative, discussed in Chapter 9, is a member of Seneco cooperative, which obtains renewable energy from Enercoop. Som Energia, discussed above, is part of this purchasing group. Enercoop also supplies public entities. In 2019, it was selected to supply an additional 4.28 GWh per year of renewable electricity to three public university campuses in València.[44]

Som Energia and Enercoop launched around the same time and have both grown rapidly. This is in contrast to the trendline of Spanish electric cooperatives. Many of the historic Spanish electric cooperatives that were critical to electrifying rural parts of Spain have closed, perhaps in the face of competition from five big electricity companies. In 2019, only seventeen historic electric cooperatives, those that control local distribution, still existed, and fifteen of these were in the Valencian Community.[45]

By comparison, in the United States, in 2019 there were sixty-three power-supply (generation and transmission) cooperatives analogous to Enercoop Group. They supply wholesale power to local electric distribution cooperatives, which are spread across forty-seven states. The United States has 834 rural electric distribution cooperatives providing electricity to approximately 13 percent of the U.S. population (forty-two million people), as discussed next in the Cobb EMC case study.[46]

Enercoop is supplying more renewable electricity than any of the U.S. cooperatives. It is not entirely clear why Enercoop and Som Energia were able to emerge in 2010 as entirely renewable energy cooperatives and thrive in Spain. Part of the explanation is that legal and market dynamics in Spain, as part of the European Union, have been more favorable to the renewable energy transition and to competition in marketing electricity services.

The European Union has promoted renewable electricity for well over a decade. The European Union's latest renewable electricity directive is EU 2018/2001 of the European Parliament and Council of December 11, 2018. The European Union directive promotes "local energy communities." Enercoop, which grew out of the historic electricity distribution cooperative in the town of Crevillent, exemplifies this local character. Another aspect of the directive promotes a broader model of renewable energy communities. Enercoop also represents the broader model by serving a national group of cooperative and social enterprise members with entirely renewable energy.

This focus on renewables came after a late-1990s liberalization of Spain's electricity sector, also following European Union legislation. The Electricity Sector Act of 1997 opened generation, transmission, distribution, and marketing to free enterprises that the state regulated. One of Spain's leading environmental law professors, Germán Valencia Martin, explained that this law opened electricity generation and marketing to any entity and guaranteed access to transportation and distribution networks (the low- and high-voltage wires); it required large electric companies to separate their activities (production versus marketing) into different business entities; it facilitated a wholesale market pool to fix the price of electricity; the grid operator that had previously been part of the government became a publicly traded company; and Spain created a public regulatory body to supervise the entire system. Today, this government body is the National Commission on Markets and Competition (Comisión Nacional de los Mercados y la Competencia).

In 2007, in response to European Union legislation, Spain amended the law and ensured that electricity would be supplied on a competitive basis, opening consumer choice to any commercial company, starting on January 1, 2009. One year later, climate-change activists formed two of the cooperatives in the case studies, Som Energia and Enercoop, to serve as platforms that allowed greater participation and expansion in renewable electricity production. With this legally open market, today there are around 400 electricity marketing companies in Spain.[47]

The construction in 2007 of the Realengo solar plant occurred in response to Spain's laws encouraging the generation of electricity from renewable energy. Spain's renewable energy laws have changed over the years, but its origins are in Royal Decree 661/2007, of May 25, 2007. This law was aligned with Spain's national goal to have a minimum of 29.4 percent of total generation of electricity produced from renewables by 2010. The decree established a system popularly known as "premiums for renewables."[48]

Part of the 2007 legal reforms in favor of renewables involved establishing a national system of regulating "guarantees of the origins" of electricity produced by renewable sources. The guarantees of origin are certificates from Spain's National Commission on Markets and Competition that certify the electricity so it can then be marketed to consumers who want to purchase renewable electricity. This system is national and prohibits the dual use of a renewable guarantee to prevent fraud. The government issues guarantees of origin in MWh for monthly periods of time.[49]

Enercoop, Som Energia, and the other new marketing companies use these government guarantees of origin to market renewable electricity they purchase on the national wholesale market. Enercoop buys the electricity it supplies to its subscribers on the national wholesale market, and then provides this electricity to the historic service area of the people who live in Crevillent and to the variety of cooperatives and social enterprises that are members of the purchasing group.

Environmental Sustainability

Understanding Enercoop's business philosophy helps to distinguish it from other power-supply entities. Enercoop promises consumer-members that its business criterion is not "strictly economic" and that it prioritizes ensuring that the customer "enjoys the most affordable price and that all the energy distributed comes from 100% clean sources."[50]

The cooperative has a variety of renewable-generation facilities in which it participates, but is not necessarily the owner: minihydroelectric power stations in Calasparra, Rio Ferreira (Portugal), and four Aquaventus plants; El Realengo solar plant (this is the largest, producing 20 GWh per year) and Barrosa solar plant; and rooftop solar at their headquarters, on an industrial warehouse, and a funeral home.[51]

The cooperative is working to change energy consumption behavior and has interacted with students in all the local schools in Crevillent to teach about responsible energy consumption and renewable energy. In 2019, Enercoop announced a new initiative to help its consumer-members lower their electricity bills. It will engage in an "energy literacy process" by offering members a phone app with tools that show consumers how to optimize energy consumption and reduce their costs.[52]

The cooperative is also optimistic that new Spanish laws favoring individuals producing solar power on their homes and businesses will help their members reduce costs. It is currently studying how to assist members to engage in self-production so they can promote distributed solar.[53]

Last, Enercoop is interested in the European WiseGrid Project, which aims for a near future where electric vehicles will be tied into the grid as energy-storage systems. The cooperative sees these vehicles as sometimes providing energy to the network and sometimes taking it, at optimal times, to expand their collective battery storage.[54]

For both Enercoop and Cooperativa Eléctrica, sustainability has long meant a broad concern for the community in which they are operating. The budding industrialists who founded Cooperativa Eléctrica built into the entity's purpose a concern for the community. By allocating net financial gains to be used to fund socially oriented projects in the town, the cooperative continually reinvests in the local community. Using its Cooperative

Education and Promotion Fund, it supports social work such as financing two housing projects for the elderly and the disabled. Every year it contributes more than €400,000 to local social activities, sports, culture, and cooperative promotion.[55]

Summary

As has been true since its founding, today each member of Cooperativa Eléctrica and Enercoop, regardless of how much energy they purchase, has one vote. The historic electric cooperative for Crevillent and the modern purchasing group cooperative declare they have always put their consumer-members at the forefront of their priorities. They exist to serve their members and to provide energy at the lowest price. Due to regulatory and market dynamics in Spain, this lowest-price energy is renewable. So the cooperatives have been able to be early and aggressive adopters of clean energy. Enercoop has an energy portfolio that is 100 percent solar and hydropower. By 2019, Enercoop was ensuring renewable energy for twenty-one cooperatives and social economy companies, supplying 50 GWh annually to power more than 100,000 Spanish consumers.[56]

Cobb EMC (Electric Distribution, Consumer, State of Georgia)

Overview of the Cooperative: History, Type, and Economic Sector

Founded in 1938, Cobb EMC, a nonprofit, consumer-owned electric cooperative, distributes electricity to members in four counties in Georgia, in the southeastern United States. Fourteen commercial and 489 residential consumer-members founded Cobb EMC. Like other electric cooperatives of its era, Cobb EMC was formed in response to President Roosevelt's New Deal federal program that provided funding and other incentives for cooperatives to bring electricity to rural areas. Passed over by investor-owned utilities, it was the responsibility of this new cooperative to build the infrastructure to distribute electricity (setting poles and wires) across 432 square miles of rural Georgia. As Cooperativa Eléctrica had done in Spain almost a decade earlier, Cobb EMC made it possible for people in its service area to turn on electric lights for the very first time.[57]

By 2019, it had grown to serve about 180,000 residential and commercial consumer-members, placing it among the ten largest electric cooperatives in the country. It offers the third-lowest electricity rates of all of Georgia's electric cooperatives and ranks among the top three cooperatives in the nation for power reliability. Over the course of this cooperative's history, Georgia's cities have expanded and engulfed the rural areas, so now this "rural" electric cooperative is really more suburban/urban in terms of population density.[58]

In 2012, Cobb EMC put itself at the head of the pack by working with the Georgia Public Service Commission to get one of the first utility-scale solar facilities in Georgia established. Then Cobb EMC purchased the lion's share of the electricity generated by the new Sandhills project, a 146 MW solar installation that started producing electricity in 2016 and was the largest solar project east of the Mississippi River. In 2019, Cobb EMC had the most solar in its energy mix, in terms of percentage of energy generated by solar on a given day, of any electricity-distribution cooperative east of the Mississippi. Nationwide, it had the most solar of any distribution cooperative (123.7 MW), as reported by National Rural Electric Cooperative Association. In a state that lacks any legislation or regulation to promote renewable energy (no renewable energy targets, no zero-emission vehicle standards or goals, no net metering, no plan for climate change adaptation), this cooperative pathbreaker is an anomaly and invites a closer look. To learn more about the company's motivations and goals, I interviewed Tim Jarrell, Cobb EMC's vice president of energy supply, and Kristen Delaney, vice president of marketing and corporate communications. At the time of the interview, Jarrell had worked for the cooperative for twenty years, while Delaney had recently started there.[59]

Industry Context

The U.S. Energy Information Administration reported that in 2018, renewables generated 11 percent of total U.S. energy and 17 percent of electricity. Renewables include, in order of most power generated: biomass, hydropower, wind, solar, and geothermal. The National Renewable Energy Laboratory reported the United States had 47.5 GW of installed solar in 2017, mainly provided by utility-scale and rooftop projects. The Smart Electric

Power Alliance (SEPA) sees solar as holding great potential for growth in the United States. SEPA documents that in 2018, the nation added 20 percent more solar capacity than existed in 2017, so this case study features an electricity-distribution cooperative that is leading in solar specifically.[60]

SEPA identified utility-supplied solar as the main driver in this growth. When divided by type of ownership structure, investor-owned utilities' solar installations far outpace total capacity by cooperatives. However, cooperatives are starting to organize around the renewable energy revolution as well. In Georgia, for example, in 2001, the Georgia electric distribution cooperatives formed a new cooperative to serve their needs to obtain renewable electricity. Nationwide, the cumulative installed capacity of solar across all cooperatives was 1,112.9 MW in 2018. Yet, due to legal and market barriers, U.S. cooperatives lag behind other types of electric utilities and tend not to own the new renewable-energy generation, as described further below.[61]

The U.S. electric cooperatives serve one in eight Americans, spread over 50 percent of the U.S. land base. Since 40 percent of that electricity is powered by coal, the U.S. electric cooperatives hold tremendous untapped potential to make the kind of shift the pathbreaking Spanish cooperatives have already made. Some U.S. cooperatives are transitioning, despite the odds, because it makes sense financially and is consistent with Cooperative Principle 7 to care for the community. Several factors—scale; the historic role of electric cooperatives; the aging stock of coal-burning power plants, which opens the opportunity to replace them with something cleaner and cheaper; and the potentially greater benefits of distributed renewable energy to rural cooperatives (in terms of lower transmission and infrastructure costs, economic development, and resiliency benefits)—put cooperatives in a strong position to catalyze change across rural America.[62]

NEW DEAL COOPERATIVES ELECTRIFIED RURAL AMERICA

Without strong federal policies that established and promoted electric cooperatives almost a century ago, it is hard to imagine how rural America would have developed during the 1900s. As the United States faced the Great Depression, only 10 percent of U.S. farms had electricity. In the urban parts of the U.S. that had electricity, ownership and control were highly concentrated, with seven holding companies controlling 42 percent of all private power generation at the end of the 1920s. These investor-owned utilities rejected electrifying rural America, even when offered federal subsidies, because of the cost of serving poor and rural farming communities scattered across half the U.S. land mass.[63]

One of the great successes of President Roosevelt's New Deal was to spur the establishment of electric cooperatives that powered farming communities, providing necessary infrastructure for improved health, quality of life, and economic opportunities. The program was not simply a form of federally subsidized financing for electrification, but support for a specific type of governance and control. These smaller, locally controlled cooperatives were to present a counterbalance to the corporate consolidation that dominated urban electricity. Cooperatives offered a model for democratically controlled economic institutions.

In 1935–36, a combination of federal legislation and executive order established the Rural Electrification Administration (REA, today the Rural Utility Service or RUS), provided hundreds of millions of dollars in subsidized loans, gave preference to consumer cooperatives, and offered cooperatives engineering, management, legal, and financial assistance to get them established. This strong federal policy approach sought to bring the benefits of electricity into rural America, with the co-benefits of economic development and democratic ownership. Farmers were already familiar with organizing themselves within producer and marketing cooperatives (see Chapter 9 for the CROPP Cooperative/Organic Valley case study). Within two years, this combination of farmer involvement and strong New Deal federal government support led to the creation of 350 cooperatives, spread across forty-five states, electrifying 1.5 million families in farming communities.[64]

Not only did the federal policy transform rural electrification at lightning speed, but these electric cooperatives have had staying power. The National Rural Electric Cooperative Association reports that in 2019, there were 900 U.S. electric-distribution cooperatives providing vital services to forty-two million people, still primarily in rural areas. Cooperatives keep the lights on in rural America and, directly or indirectly, support almost 612,000 jobs each year.[65]

The federal policy aims of having democratic institutions leading rural electrification has been troubled in practice. The Rural Electrification

Administration's support of cooperative development echoed the international Cooperative Values and Principles: cooperative members should be actively involved in democratic practices on the basis of one member, one vote; and the member-consumers own the cooperative, keeping control local. However, a 2018 study by Debra Jeter and others asserts that self-dealing and other bylaw violations have been rampant at U.S. electric cooperatives, as the members have not actively engaged in oversight: "Too often they operated less as little republics than as small fiefdoms under the thumb of self-perpetuating, virtually self-appointed boards of directors."[66]

FEDERAL POLICY ENCOURAGED COOPERATIVES
TO RELY ON FOSSIL FUELS

Further, U.S. electric cooperatives supply or distribute energy that is heavily reliant on fossil fuels, such as coal. Cooperatives currently provide 11 percent of all the electricity in the U.S., and most of that is from sixty-three power-supply cooperatives that sell the wholesale electricity to distribution cooperatives and others. In response to a federal mandate in the 1978 Industrial Fuel Use Act, many cooperatives invested in coal-fired power plants, typically financed with thirty-year bond obligations. Congress repealed the mandate in 1987, but by then the damage was done: two-thirds of cooperatives' coal-fired generation had been built in response to the law. While a negative legacy for climate change, that federal law too shows the power of a strong federal policy approach to rapidly transform U.S. electricity generation.[67]

Forty years later, the dirty coal legacy continues. According to the National Rural Electric Cooperative Association, in 2017, coal fueled 40 percent of the total generation of the electric cooperatives. Further, of the top ten most carbon-intensive power plant emitters in the United States in 2017, six were cooperatives. This underscores the path-dependence created by investments in power generation, which have ripple effects for decades. Now, those coal-powered plants produce electricity that is generally more expensive than renewables. Many facilities, built during that 1978–87 period, should be at the end of their useful lives; we are now at another pivotal point when the path for cooperatives can be determined, with the potential to bring the co-benefits of cleaner and cheaper renewable electricity and democratic ownership throughout rural America.[68]

It is important to note that when viewing electricity generation across the United States, by no means are the power-supply cooperatives responsible for the most total tons of CO_2, since they produce relatively small amounts of electricity. The 2017 air emissions data showed that CO_2 emissions from U.S. power producers are highly concentrated, with nearly 25 percent of all emissions coming from five companies and 50 percent from only sixteen companies. Although power-supply cooperatives own a lot of fossil fuel generating facilities and produce some of the dirtiest power (based on CO_2 per unit of electricity), the largest overall power generators and sources of CO_2 emissions are investor-owned utilities. Of the sixteen companies emitting 50 percent of the total CO_2 emissions for power plants in the United States, all are investor-owned utilities. To reduce the worst impacts of climate change, all power producers that do not capture and store their GHGs will need to switch off fossil fuels by 2050. Federal policy that narrowly targets and prioritizes rapidly decarbonizing those sixteen power producers would be the most direct way to cut in half the GHGs from U.S. electricity. That, combined with a separate policy that incentivizes cooperatives, especially in ways that puts ownership of renewables in the hands of cooperatives and their members, would bring the benefits of clean renewable energy and democracy building to America.[69]

FEDERAL AND STATE POLICY COULD PROMOTE COOPERATIVE RENEWABLES

Cooperatives have fewer financial and public-law incentives than investor-owned utilities to encourage switching to renewables. Gabriel Pacyniak highlights that three drivers of clean energy are relatively ineffective with cooperatives: federal renewable energy tax credits (which are useless to a nonprofit), state renewable energy mandates (which often exempt cooperatives), and the ability to shift costs from investors to captive ratepayers (which doesn't apply because the cooperative member-owners are the ratepayers). Despite this, the National Rural Electric Cooperative Association's CEO, Jim Matheson, testified before Congress that the share of renewable energy that electric cooperatives provided to their members increased from 13 percent in 2009 to 17 percent in 2016. Cooperative-provided renewable energy is growing, but not nearly at the rate of other utilities and not fast enough to meet the climate imperative to quickly transition off fossil fuels. Further, cooperatives find it more financially beneficial to contract for renewable power than to generate it, since they cannot use the tax credits. The Cobb EMC case study reveals how a cooperative can add renewables to its energy supply mix in the face of these public-law obstacles.[70]

The investment in renewables and energy conservation that has to be made in the next decade to meet the climate emergency is unprecedented. Yet, it is analogous to bringing electricity to rural America almost a century ago, a feat that was accomplished on a swift time line with a commitment to democratic ownership and control. If the federal government fails to address the disincentives for electric cooperatives to own renewable-energy generation, it will result in the renewable energy transition further concentrating ownership and wealth in investor-owned corporations.

Moreover, some of the same multinational corporations that brought us climate disruption are positioning themselves to be the big financial winners in the most profound transition the world has experienced since the industrial revolution. For instance, Netherlands-based Shell Oil's newly acquired U.S. subsidiary, Silicon Ranch, is one of the largest solar power producers in the country. Shell has systematically acquired companies in many parts of the renewable-power supply chain, from power production to battery storage and car charging stations. As discussed below, Silicon Ranch is a key partner in expanding renewable generation with the Georgia cooperatives. Like the concentration of power in electricity holding companies the New Deal sought to counter in the 1930s, the United States will need another variation on the New Deal to disrupt this dynamic if there is to be a balance in who owns and controls renewable power in America.

If the growth of Spain's Som Energia and Enercoop could be replicated beyond Spain, cooperatives could be even more important for the co-benefits they can offer: decarbonization and democratic ownership and control. The U.S. government has played a system-defining role in rural electricity. The New Deal set in motion a system where the REA/RUS financed cooperatives' power-generation and -transmission projects and offered legal and technical support. Imagine what a similarly ambitious federal program could do today to orient electric cooperatives and other utilities to rapidly and deeply decarbonize electricity in America while building democratic ownership. A modern or "green" New Deal could incentivize and mandate reasonable

rates while retiring coal-burning power plants early and financing distributed renewable-generation capacity, storage, demand management, energy
efficiency, and electric vehicles. It could further enhance education and
oversight of democratic governance, and require programs to ensure that
low-income people benefit from the renewable energy transition.[71]

Environmental Sustainability

When states set strong climate and energy policies, they can create powerful drivers for switching to renewable energy. For instance, Colorado's
HB 1216 climate law authorizes GHG regulation to meet an economywide
80 percent reduction target, and New Mexico's SB 489 Energy Transition
Act sets a mandate of 100 percent zero-carbon electricity by 2045. Tri-State,
a generation and transmission cooperative, noted that these state laws were
factors in its historic January 2020 announcement that it was closing all of
its coal plants and mines and not developing any new ones, while boost-
ing its renewable portfolio to produce 50 percent of its electricity, up from
9 percent in 2019. As part of this announcement, Tri-State will be rapidly
adding 1 GW of new renewable generation by 2024, bringing their total
renewables to 2 GW. Tri-State provides electricity to forty-three distribution
cooperatives and others in Wyoming, Colorado, New Mexico, and Nebraska.
In recent years, it has provided some of the dirtiest energy in the country
in terms of GHGs per unit of energy. The 2020 announcement includes
abandoning a proposed coal-burning power plant in Kansas, the Holcomb
project, despite having won a permit battle in the Kansas Supreme Court to
allow the new facility. This incredible transformation results from a combination of drivers: strong state policies, distribution cooperatives that agi-
tated to pursue renewable goals, and the economics of renewables providing lower-cost electricity.[72]

Even without public-law reforms, however, we can learn from sustainability pathbreakers to use the cooperative sector to leverage needed changes,
albeit on a smaller scale. In 2019, SEPA identified Georgia as the state
where cooperatives have the most solar, having installed almost 24 percent
of the U.S. total for cooperatives. When viewed against all installed solar
in the United States, regardless of form of business entity, Georgia ranked
tenth in the nation, with 1.5 GW. As stated above, Georgia has zero statutes

or regulations that support renewable energy. Despite this, Georgians are benefiting from clean solar because the private sector is driving changes.

THE BUSINESS CASE FOR SOLAR IS STRONGER THAN FOR COAL

The business case for solar is strong, even when stacked against government policies that favor fossil fuels. During the COVID-19 pandemic, as people stayed home and demand for electricity declined, a variety of countries accelerated their shifts away from coal. Sweden and Austria closed their last coal-powered utilities in 2020. The United States Energy Information Agency predicts coal will be 10 percent of U.S. electricity by 2025 (down from 50 percent in 2010). More and more major corporations are powering their businesses with renewables. For example, in the largest deal yet between an electric-distribution cooperative and a private company customer, in 2018 Walton EMC agreed to build 200 MW of solar to power a single new customer, a Facebook facility in Georgia. The cooperative sees this development as a game changer. The move by Facebook is consistent with its internal goal to run all operations on 100 percent renewable energy by 2020. Many large corporations have set similar goals and are part of an urgent push from the private sector to decarbonize the electricity they use. If an electric cooperative is not poised to facilitate this, as Walton EMC was, the private company may simply opt to locate elsewhere or self-generate under a Direct-Access program, as Apple has done in some states.[73]

In a survey of U.S. cooperatives, SEPA found a growing interest in solar, even in states not associated with strong solar growth. Renewable energy in some areas means cost savings and self-sufficiency. For instance, in parts of North Dakota, it is far more cost-effective to install a solar-powered pump on a water well to provide for cattle than it is to string electrical lines and set poles to reach the area.[74]

Cobb EMC's sustainability emphasis calls for reducing environmental impacts while keeping customer costs as low as possible. Forward-thinking, Cobb EMC wanted to be one of the first cooperatives in Georgia to add utility-scale solar to its energy portfolio in 2012. Today, that 2012 decision is still the largest part of Cobb EMC's solar portfolio. It purchases 111 MW from the Sandhills Solar Facility. Cobb EMC purchases the solar through a long-term agreement with Southern Power, a subsidiary of the investor-owned Southern Company. Being an early adopter of solar in the state has not undermined its ability to offer a fair price to consumers, as it still charges less for electricity than most Georgia cooperatives.[75]

COBB EMC'S RENEWABLES, EFFICIENCY, AND ELECTRIC VEHICLES

Among electricity-distribution cooperatives east of the Mississippi River, Cobb EMC has the largest amount of solar in its overall mix of electricity generation. In 2019, Cobb EMC's energy mix consisted of 123 MW of renewable energy, mainly from solar. It will add another 25 MW, starting in 2021, by participating in a project with Green Power EMC. On an annual basis, renewables comprise about 9 percent of its energy supply. This varies by season. Based on 2019 figures, when the sun is shining and demand is low, such as in spring and fall when there is less need for air conditioning, as much as 40 percent of Cobb EMC's power on a given day can come from the sun. Still, nuclear is about 41 percent, natural gas is 43 percent, and coal is about 2 percent of its annual energy mix.[76]

Cobb EMC has encouraged its members to voluntarily participate in renewable energy by rooftop generation, community solar, and renewable energy credits (RECs). In recent years, however, the company has focused on adding more solar to its overall energy supply and passing on those benefits to all of its customers. Although operating in a state that lacks such a policy, Cobb EMC created a Distributed Energy Policy to facilitate consumer-members becoming energy generators. The policy encourages members to install renewables and connect to the Cobb EMC distribution system to take advantage of bi-directional metering (energy consumption and generation). This policy applies to solar, wind, and biomass. The cooperative offers a variety of easy-to-access web-based services to promote self-generation, such as a financing arrangement with a local credit union, a solar calculator, a list of approved contractors, and information about tax credits. The cooperative has structured this arrangement to allow homeowners to use the electricity they generate to offset the electricity they use, all at the normal retail rate. If the homeowner produces more electricity than they use, the cooperative pays a credit for this excess energy that enters the grid. However, Cobb EMC reduced the price it pays for excess renewable energy from an additional $0.0683 per kWh for systems installed before July 1, 2015, to $0.03828 per kWh for systems installed after that date, citing that this was the cost of the cooperative's avoided energy and capacity. This policy allows an annual recalculation of the avoided cost rate and normally doesn't change materially. Given that the retail cost of electricity in this area is lower than the average U.S. consumer pays, the payback period for consumers generating their own renewable electricity is longer than in states where electricity costs more. Not surprisingly, given the payback period, the customer participation rate in rooftop solar has been relatively low, with only eighty-four homes adding solar by 2019.[77]

Cobb EMC originally offered members the opportunity to participate in community renewable-energy projects, but ended the program as the company increased its overall renewable-energy portfolio, which benefits all members with the economics of the additional renewable energy. Still, it is worth explaining how community solar programs work, as they are a popular strategy for consumers and utilities to accelerate the transition to renewables. Community solar programs do not involve siting renewables on the members' property and do not necessarily require any member financing. Consumers can get the benefit of solar even if they live in an apartment or own a home that is not suitable for solar panels, and do not need to worry about maintenance or insurance for the installations. Utilities obtain a new stream of financing for solar, can meet new demands from customers for renewables, and keep energy production local.

There are three basic methods for structuring community solar projects: 1) customers pay a premium rate for solar on their utility bill based on monthly electrical usage; 2) customers pay upfront capital costs for X number of panels, and then all the energy the panels produce over their twenty or thirty years of life is credited to the customer-owner's utility bill; or 3) a third party develops a large solar installation and customers sign up for its power, receive a utility-bill credit for the portion of their electricity produced by solar, and then pay a portion of that credit to the developer. The last of the three is mainly used by investor-owned utilities. Cobb EMC used a variation on the first type for its community solar program, offering a simple premium billing option for members. Customers paid $25 for a set number of output from a community solar facility, but didn't know until the end of each month how much the facility would generate because it varied based on sunny days. At its height, close to 400 members participated in this program, according to Cobb EMC's Jarrell.[78]

While investor-owned utilities have installed more solar capacity, there are more electric cooperatives with community solar projects. SEPA, which maintains a comprehensive database of community solar installations, reports that in 2017 the United States had 734 MW of installed community solar capacity, more than half of it installed in 2017. Of the utilities with projects, 160 were cooperatives in thirty states, thirty-seven were public power utilities, and thirty-one were investor-owned utilities. Additionally, leading cooperatives have installed community solar for members in places not necessarily associated with being "green": North Carolina, Nevada, and Kentucky.[79]

Community solar is a small part of the total renewables picture. However, this is a growing area that could accelerate faster and more broadly with favorable policies; there are thirty-three states without a shared solar policy to facilitate this. The majority of community solar exists in three states that have policies with significant financial incentives for subscribers: Colorado, Minnesota, and Massachusetts. A customer survey across all types of utilities showed a high interest in community solar. Cooperatives and other utilities could use this strategy to encourage people to participate more broadly in the solar transition, but it needs to be structured in a way that makes sense financially for the participants.[80]

In response to commercial members' requests, Cobb EMC launched a new renewables program in 2020. It allows customers to purchase renewable energy credits. This is a simple opt-in process where customers pay one-half to one cent more per kWh for renewable credits to offset the electricity they use. As in the company's community solar program, consumers will be able to choose to participate on a monthly basis. The cooperative will first offer this to commercial customers and then to residential customers.[81]

To substantially or wholly switch a utility's energy supply to renewables requires the ability to store electricity for times when the renewable system is not able to produce (for example, at night for solar, or when it isn't windy for wind power). Jarrell says that Cobb EMC is exploring battery storage options with Green Power EMC and others. Cobb EMC is interested in utility-scale renewable projects with battery storage in the future.[82]

Electric vehicles provide additional storage capacity when tied into a utility's demand-management system and, when powered by renewables, are an important strategy for moving transportation off fossil fuels. Cobb EMC has an EV program for businesses and residences. For businesses, it offers grants of $500–$5,000 for the cost of purchasing and installing EV supply equipment that the businesses can use for their EV fleets. It encourages public access to the EV charging equipment as well, consistent with its goal of developing EV charging stations throughout its service territory. This is a new program for the cooperative, on which it has spent less than $50,000; fewer than ten chargers have been installed, but Cobb EMC expects that number to increase over time.[83]

A key part of sustainability is using energy more efficiently while reducing costs. Starting in 2016, Cobb EMC restructured its rates to offer a peak-demand–based rate to encourage residential members to shift usage to off-peak hours. As of 2019, 50,000 members had taken advantage of this policy, which can give members lower-cost electricity, and system peak demand was reduced by 6 MW. Connecting this new rate to its EV promotion, the cooperative advertises a nighttime rate that is totally free for 400 kWh and uses this to encourage people to switch to EVs and charge their cars overnight using the free energy. The peak daytime electric charge, by comparison, is $0.1350 per kWh. Cobb EMC is seeing consumers respond to price signals with these different time-of-use rates, resulting in greater energy affordability for all its members.[84]

RELATIONSHIP BETWEEN DISTRIBUTION AND GENERATION/TRANSMISSION COOPERATIVES AND THE CREATION OF GREEN POWER EMC

Distribution cooperatives like Cobb EMC typically contract with a generation and transmission cooperative to obtain their power supply. As described above with the Tri-State announcement, the generation and transmission cooperatives can significantly shape how much and how quickly renewable energy gets added to the overall supply. Cobb EMC gets its energy primarily from contracts with Oglethorpe Power, its generation and transmission cooperative. However, Cobb EMC can choose to obtain part of its power-supply requirements, above the contracted amount, from other entities. This, along with the creation of Green Power EMC, has made it possible for Cobb EMC and other Georgia cooperatives to expand their renewable-energy portfolios.

In some areas across the country, cooperatives are required to receive 100 percent of their energy supply from their generation and transmission cooperatives. Typically, these contracts do not allow the distribution cooperative to purchase from other suppliers. In Georgia, if a cooperative has needs above what it is contracted to receive from Oglethorpe Power, it can seek out other power purchase agreements to meet the need. Having the flexibility to obtain power from other sources provides some additional market opportunities to maintain stable power costs, says Jarrell.[85]

In 1997, the cooperatives who purchased from Oglethorpe Power forced a restructuring and broke it into three companies: power supply, transmission, and Georgia system operations. Oglethorpe generates electricity for thirty-eight electric-distribution cooperatives in the state. Then, in 2001, Green Power EMC was formed as a nonprofit renewable energy cooperative composed of these same thirty-eight electric-distribution cooperatives. Green Power EMC's electric cooperative members are providing electricity to almost half of Georgia's population. Oglethorpe partners with Green Power EMC, where Oglethorpe provides management services to this related cooperative, explains Jarrell. Because of the link between Oglethorpe Power and Green Power EMC, the distribution cooperatives can become members of this newer renewable energy cooperative and thereby add renewable power to their power-supply mix.[86]

Green Power EMC's function is to find, screen, analyze, and negotiate power-purchase agreements with renewable power providers. The participating electric distribution cooperatives contract with Green Power EMC for the renewable energy. These distribution cooperatives, such as Cobb EMC, either offer their members a choice to purchase green power for an additional cost or simply add the renewables to their wholesale power purchases and spread the costs and benefits evenly across all their members.[87]

By 2018, Green Power EMC had 238 MW of renewable energy projects, and that is going to almost triple by 2021. Its projects involve solar

(87 percent), wind, landfill gas, hydro, and biomass. It continues to add new renewable generation to the grid; in 2019, Green Power EMC announced two new deals that will rapidly expand its renewables, bringing its solar power capacity to 639 MW by 2021. The cooperative does not build or own these projects, as the financial incentives are not applicable to nonprofit cooperatives. Instead, Green Power EMC purchases the electricity produced over a thirty-year period on behalf of its electric cooperative members. The owner and developer of the solar installations for the newest deals is Silicon Ranch, which is Shell's U.S. solar company. It describes itself as "one of the largest independent solar power producers in the country."[88]

BROADER AREAS OF CORPORATE SOCIAL RESPONSIBILITY

Additional areas of corporate social responsibility include corporate recycling programs, air quality improvements from reducing the use of Cobb EMC trucks in favor of computerized meters to read electricity usage without a site visit, conservation education programs, and grants from their community foundation. The cooperative recycles retired meters, switches, computers, and other electronics, as well as metals, plastic, construction debris, wood poles, lamps, and ballasts. Further, it has run a recycling program for its members since 2016. The April 2019 Recycling Day event attracted about 1,200 members to drop off recyclables and take home a gift of LED lightbulbs. On that day the cooperative collected for recycling 15.89 tons of paper and 38.13 tons of metal.[89]

Cobb EMC's member education includes online resources offering tips on reducing energy demand while saving money in the process. It further provides free energy audits to identify the most cost-effective energy-saving improvements in customers' buildings. And it offers seminars on energy conservation throughout the year.[90]

Another way the cooperative carries out Cooperative Principle 7 to care for the community is through the Cobb EMC Community Foundation, which members fund through Operation Round Up. Through the foundation, they typically give $1 million per year in grants to the local community.[91]

In coming years Cobb EMC will be looking at defining future renewable energy goals. Currently, its strategic plan does not include a specific goal for renewable energy or energy conservation, but does include a focus on sustainability. The company is in the process of analyzing its options to determine what the goal can be while keeping rates low and reliability high. Adding renewables must make sense financially, and according to Jarrell, some solar contracts the cooperative has now are cheaper than coal and comparable to natural gas. If its board sets a goal consistent with the IPCC recommendations, it will not only be caring for its community, but setting the pace for the 900 other U.S. electric-distribution cooperatives powering rural and suburban America.[92]

Governance and the Cooperative Structure

As is typical of cooperatives, Cobb EMC follows a democratic governance model. Cobb EMC's bylaws establish that each member is entitled to cast one vote, attend an annual meeting, vote for representatives on the board of directors, and have access to records. Cobb EMC's member resources online invite members to an annual meeting to learn about progress and ways to save on electric bills. The more telling part of the company's governance is whether members participate.[93]

Of Cobb EMC's 180,000 members, 14,375 members voted in the 2018 board of director elections. Members are able to vote by mail and online, and 29,159 active customers are registered to vote electronically. The annual meeting is coupled with a member appreciation event. In 2019, 3,200 members attended the annual meeting. They brought family members with them, swelling attendance to about 10,000 people at the county fair grounds for a day organized around enjoyment. The business meeting part of the day is a more formal "state of the cooperative" meeting that provides updates to the members and allows anyone a chance to speak.[94]

The members are charged with electing nine members to the board of directors to serve three-year terms. The board is composed of members who represent a particular geographic district in the cooperative's service territory. The cooperative imposes term limits on directors of four terms (a total of twelve years) and has a conflict-of-interest provision in its bylaws.[95]

Board meetings are open to members as long as they register to attend, per the corporate policy of open meetings. According to Jarrell, members can access meeting minutes and the cooperative's financials on a members-only portal on the website, and minutes are available within ten days of the meetings.[96]

Although the bylaws do not contain a description of the purpose of the cooperative, Cobb EMC includes the seven international Cooperative Principles on its website and emphasizes the cooperative difference as a member-owned nonprofit. It asserts the company "strive[s] to be a good corporate citizen and neighbor, as well as a recognized leader in the utility industry." It describes itself as a nonprofit, member-owned electric cooperative with a vision to become "the energy industry benchmark through innovation that exceeds member and community expectations." It further asserts that Cobb EMC is "consistently recognized for competitive rates, reliable power, and a commitment to renewable energy and giving back" to the local community.[97]

Summary

The story of Cobb EMC gives an example of private-sector tenacity and progress, even in a state that has room for improvement in climate and renewable policy incentives. Starting in 2012, Cobb EMC led U.S. cooperatives east of the Mississippi River in terms of the percentage of solar energy as part of its overall supply. By 2019, solar powered as much as 40 percent of the cooperative's daily demand. Nationally, in 2020, it had the most MW in solar of any electric cooperative. While this is hopeful, it is not enough in the industry.

To take the renewables transition to scale within the time frame prescribed by the IPCC scientists, federal and state government action is necessary. While state laws can set transformative goals and mandates for 100 percent zero-carbon electricity or economywide GHG reductions, this is a piecemeal approach to something that requires a uniform national target and effort akin to the New Deal or the Apollo program that put the first human on the moon. Many of the dirty fossil fuel power-generating facilities the cooperatives use should be nearing the end of their useful lives, and the economics are on the side of investing in clean, renewable energy. We are at a crossroads where federal policy could strongly incentivize cooperatives to bring the co-benefits of renewable electricity and democratic ownership throughout rural and suburban America. Combining this with a federal policy that narrowly targets the sixteen investor-owned power producers responsible for 50 percent of the nation's GHGs from electricity would be the most direct way to reduce GHGs and bring the benefits of clean, renewable energy to America.

9

Food and Agriculture

AFTER ENERGY, FOOD AND agriculture is the most important economic sector to deeply decarbonize. The U.S. emissions inventory for 2018 attributes 10 percent of GHGs to agriculture. For many, the seemingly simple act of eating a meal relies on an increasingly complex and global food system. The food system involves manufacturing and distributing inputs such as fertilizer, pesticides, and seeds; producing agriculture; processing, distributing, and marketing products; and disposing of waste. Food systems are more than market transactions, but also involve a larger web of institutions, regulations, and relationships. In a first-of-its-kind 2012 study, Sonja Vermeulen and others assessed the GHGs associated with *all* the components of the global food system, including growing, storing, and transporting the food. The study indicated food and agriculture may account for a much greater share of global GHGs, 19–29 percent. In the system study, agricultural production yielded the vast majority (80–86 percent) of the food-related emissions, followed by fertilizer manufacturing, and then refrigeration. In this sense agriculture is a significant source of GHGs. Yet, agriculture is also becoming understood as a sink for GHGs. Scientists are identifying agricultural practices that remove carbon from the atmosphere and store carbon in soils. Understanding how to shrink the carbon footprint of food and increase agriculture's ability to be a carbon sink is the key to deep decarbonization.[1]

In addition, food and agriculture are essential to one's quality of life. Several of the Sustainable Development Goals relate to food and agriculture. The first two goals of ending poverty and hunger, the third goal of good health and well-being, the eighth goal of decent work, and the twelfth goal of responsible consumption and production all relate to sustainable food systems. Climate disruption challenges food security globally, making it more difficult to meet these Sustainable Development Goals. Extremes of temperature and water stress (floods and droughts), as well as new pests and diseases, all impact the growth of crops. Climate change impacts yield and quality. When there are disruptions in food production and distribution, low-income individuals and communities are the most vulnerable.[2]

This chapter sketches the pathways to sustainable development and deep decarbonization at the global level. Then it provides case studies of cooperative exemplars to demonstrate how different aspects of the transition are already under way. Each case study includes an overview of the cooperative's history, type, and economic sector; context based on its industry and country; best practices in environmental sustainability; and its governance and cooperative legal structure.

Pathways to Sustainable Development and Deep Decarbonization

For the food and agriculture transition, the IPCC indicates a need to restore ecosystems, promote soil carbon sequestration practices, and adopt less "resource-intensive diets" (that is, diets composed of less meat and calories obtained with fewer miles between farmer and consumer). GHG emissions from food production are created by fossil fuel–powered tractors, trucks needed to carry food to market, and appliances to store and cook foods. Methane and nitrous oxide are many times more potent GHGs than CO_2. These GHGs are emitted from enteric fermentation in ruminant livestock and released from tilled and fertilized soils, respectively.[3]

Peter Lehner and Nathan Rosenberg's 2017 analysis reviews the science and policy levers for U.S. agriculture, and identifies practices that are organic, utilize perennial crops, and no-till pastures as the most promising ways to not only reduce GHGs but to sequester carbon. Another piece of the puzzle is to reduce the transportation miles between farmer and consumer in order to shrink the carbon footprint of food and build locally diverse and resilient food systems. This all has implications for markets and grocery stores that typically provide the interface between producers and consumers.[4]

In short, the practices for deeply decarbonizing the food system involve growing food on land that is not replacing forests, without a lot of chemical and mechanical inputs (chemical fertilizers and pesticides and large machines), on diversified farms, with minimal processing and refrigeration, as close as possible to the consumer. Farming that is organic, perennial, and no-till relies more on labor and less on fossil fuel–powered machines. Farming that goes further and integrates forests is desirable for deeply decarbonizing food and agriculture. All these factors play a role in reducing the GHGs associated with food and need to be accomplished in ways that support affordability for consumers and stable incomes for farmers and grocery store employees.

Deeply decarbonizing food and agriculture will entail reforming laws, including those Lehner and Rosenberg identify, that impact farm subsidy and conservation programs, trade, tax, dietary guidelines, government procurement policy, financing for carbon farming, and putting a price on carbon. However, before and during these public-law reforms, there are social enterprises that are leading the way as sustainability pathbreakers. The legal reforms will take us further, but the private sector does not need to wait for them.[5]

Cooperatives can influence the food system in ways that promote the Sustainable Development Goals and deeply decarbonize. They are ideally suited to build platforms for producers, grocers, and consumers to participate in the sustainable food transformation. Cooperatives have long been a traditional mode for farmers to work together to find economies of scale in the United States, Spain, and many countries throughout the world. This makes them a familiar and accepted form of business organization for farmers. Cooperatives and their support organizations or federations can facilitate farmer-to-farmer knowledge sharing about organic, perennial, or other carbon-sequestration practices; build community links between farmers, groceries, and consumers; and facilitate political activity to promote policies for food systems that involve fewer GHGs and more reasonable financial terms for farmers and consumers.

Food and Agriculture Cooperative Pathbreakers

This chapter features case studies of two Spanish grocery store cooperatives, Consum Cooperativa and bioTrèmol; two organic beverage cooperatives, Bodegas Pinoso and Cerveses Lluna; and a U.S. farmer cooperative, CROPP Cooperative/Organic Valley. I start with the most common connection between people and their food—the grocery store—giving examples of both a large chain and an ultralocal, purely organic store. Then I discuss a traditional farmer, value-added cooperative that has weathered economic changes, was an early adopter of organic perennial-crop farming practices, and is currently the largest producer of organic wines in Spain. Next, I feature a modern worker cooperative crafting organic beer, also the largest producer in Spain. Last, I discuss an organic farmer cooperative that scaled up to be the first organic food company with annual revenues over $1 billion.

The case studies show food and agriculture cooperatives organized around a variety of member-owner classes, operating at vastly different scales, and filling important niches in the food system. All are committed to fair compensation for their members, good prices for consumers, and strong sustainability practices. Most are powered entirely by renewable electricity. While they diverge around ownership classes, all have well-defined values and principles that include environmental sustainability and have written these into the legal design of the businesses. This sustainable legal design sets the course for day-to-day practices, leading to overall improvements in environmental sustainability that go beyond anything required by environmental laws.

Consum Cooperativa (Grocery Store Chain, Worker and Consumer, Spanish National)

I was fielding audience questions from the Fulbright fellows in València, Spain, after a talk on cooperatives and was asked, "Are cooperatives just for the wealthy who can pay more for organic food?" A great question that went directly to wealth inequalities and the lack of access to affordable food, grown without pesticides. The fellow was from the United States, where her only experience of food cooperatives would have been of a certain ilk—specialty stores known for higher prices.

When I arrived in Spain, my apartment's property manager pointed me to the nearest grocery store, Consum. Perhaps it was because the sign did not include the "cooperative" designation, but when a person working at the cash register handed me a brochure to join, I assumed it was a "rewards" program like that found at other chain grocery stores. I shopped there for almost a month before learning from the Spanish consumer cooperative federation Hispacoop that Consum was a cooperative. Consum's appearance and prices make it look like any other large grocery store, but its cooperative features became clearer through my research. The United States has no comparable food cooperative with a chain of hundreds of stores operating under a single management system for the entire country, or even a single region. Consum competes with large investor-owned grocery stores, so its story is important to show that large-scale worker and consumer hybrid ownership, paired with strong environmental practices, is viable.[6]

Overview of the Cooperative: History, Type, and Economic Sector

This Spanish grocery chain is a hybrid, owned by its 14,000-plus workers (95 percent are worker-owners) and more than three million consumer-members. The company is designed and operated according to a strong set of values, and its commitment to environmental sustainability runs deep. In 2019, Consum Cooperativa powered 97.7 percent of its energy needs for its 707 supermarkets with renewable electricity.[7]

In 1975, a group of 600 founding members started the cooperative when each contributed 3,000 pesetas (about €18), and ten members applied for a bank loan to launch the social enterprise. They opened their first supermarket in Alaquàs (València), with six workers. Over forty years later, by 2019, Consum had grown to be the largest consumer cooperative on the Spanish Mediterranean.[8]

Consum's birth coincided with the Spanish dictator Francisco Franco's death. At that time, workers were a key force organizing to build a new democracy. Spain was in a deep recession related to the 1973 oil crisis, so it

was a time when people were searching for economic stability while struggling to build a new democracy. More people were moving from rural areas to major cities, such as València, which created greater demand for
housing and food in these areas. These factors all contributed to creating
fertile conditions for forming cooperatives. Consum's founders were very
active with trade unions, democratic political parties, and Catholic groups
committed to social justice. Inspired by the example of Mondragón in the
Basque Community, they were trying to establish a cooperative movement
in the Valencian Community in the 1970s. Their strategy was to create
housing and other worker cooperatives that responded to the community's
economic and political needs.[9]

Consum's founders wanted to influence the consumer sector by creating
a consumer cooperative grocery store that could compete with the large
European chain stores that were entering Spain. Given this aim, the fact
that, to the consumer, they look indistinguishable from other large grocery
stores is understood as part of their design. However, the cooperative difference emerges when one looks at Consum's mission, values, and practices.
Its main objective was to make local purchases as a group and eliminate
intermediaries between producers and consumers in order to offer the best
products at the best prices to its members. This was at a time when inflation was over 20 percent, so lowering the costs of basic necessities like food
was critical.[10]

In 2010, Consum revised its cooperative strategy and redefined its mission. The short version is: "Committed workers, satisfied customers." Carmen Picot, Consum's leader in corporate social responsibility, says that "cooperative values and principles are important because they give an identity
to a company and reflect its commitment to society and employees." Consum does not use the Co-op Alliance's Values and Principles, and instead
defines its values as follows:

- Listen to the client, workers, suppliers, and the environment.
- Contribute products, services, training, information, welfare, innovation, and development.
- Be responsible, meaning commitment, honesty, and sustainable
  action.[11]

Picot describes Consum as a "social economy" company that cares for and takes care of its stakeholders. Unlike an investor-owned corporation that is focused on profits for its shareholders, Consum is focused on serving its stakeholders. Consum defines stakeholders to include workers, consumer-members, suppliers, franchisees, social entities, and society in general. Its values include respecting the environment, not compromising future generations, increasing local wealth, and supporting local communities where the cooperative operates. Consum accounts for how it is implementing its values in its annual sustainability reports, which are publicly available from 2006 to the present, as discussed below.[12]

Industry Context

Consum is competing within the Spanish retail grocery market. As reported by the U.S. Department of Agriculture, in 2017 Spain's retail food sales were $118.7 billion. Spain's retail food market includes a diversity of options, from supermarkets and convenience stores to specialized eco stores and farmers markets. The greatest sales growth in 2017 was online. The retail food industry has been consolidating and the total number of retail outlets has decreased over the past decade.[13]

By contrast, Consum has been growing, adding employees, opening new stores, and selling online. In 2017, Consum was the seventh-largest retailer in sales volume in Spain, and the largest in the Spanish Mediterranean. It opened thirty-eight new supermarkets in 2017, with a total of 707 in the Mediterranean region. Consum ended 2017 with sales of €2,518.7 million, 7.45 percent more than the previous year. This provided a "surplus" of €51.7 million, an increase of 10.47 percent over the previous year. When Consum's sales and surplus grow, its members benefit too. In 2017, the cooperative distributed the surplus to its members: €31.4 million to the workers and €35.6 million to the consumers. Although not a leader in organic retail products, Consum has been able to stay competitive in Spain's retail market while pursuing other strong sustainability initiatives.[14]

Eroski, part of the Mondragón cooperative, is the largest cooperative grocery store in Spain, but Consum is not far behind. There is no comparable consumer cooperative grocery store on this scale in the United States. The closest is National Co-op Grocers, which the member stores organized to unify and strengthen the purchasing power of natural food cooperatives, while supporting their autonomy. The organization represents coops that collectively run 200 stores (compared to Consum's more than 700), with about one-third the consumer membership of Consum. National Co-op Grocers provides a common brand for marketing, stronger positions for purchasing as a group, a platform for exchanging information, and support for sustainability initiatives. Its more than $2 billion in annual aggregate sales makes its combined power a modest player in the U.S. food market. U.S. food coops have been leaders in promoting natural and organic foods. Consum carries organic food, and 99 percent of its suppliers are from Spain, but it was not organized to predominantly feature organic foods. The subject of the following case study, bioTrèmol, is more similar to U.S. food cooperatives in its scale and focus on organic/local foods.[15]

Environmental Sustainability

The European Community of Consumer Cooperatives, which represents cooperative retailers across Europe, distinguishes consumer cooperatives based on their values and principles. It positions "sustainability" at the "core of consumer co-operatives' actions." It emphasizes Cooperative Principle 7, concern for the community, as supporting "pioneering action in the domain of social and environmental sustainability." Some practices the Community highlights include "ethical sourcing, production and labelling (e.g. organic farming and Fair Trade), packaging and food waste issues, big environmental and societal challenges (e.g. climate change) as well as all aspects related to sustainable food production and consumption."[16]

Likewise, in the words of Consum's president, "In addition to concern for people, we are concerned about our environment, which is why we focus on local trade and proximity." Since the 1980s, Consum has been implementing environmental sustainability in a variety of ways, and it has been voluntarily measuring and reporting about its work in these areas for over a decade, well before the new European Union mandatory legal requirements for nonfinancial reporting were enacted. Consum has an environmental policy that commits it to reduce environmental impact through continuous improvement processes and raise awareness among all stakeholders. It aspires to "leave no trace" from its operations.[17]

Consum has been an industry leader in the area of measuring and reporting environmental sustainability. Consum was the first grocery distribution company in Spain to register its carbon footprint. Businesses can use a carbon footprint tool to estimate the total amount of GHGs they directly and indirectly produce in operating their business. For example, direct emissions are those controlled by the company (such as refrigerant systems), and indirect emissions are those the company can influence but does not control (such as purchased electricity to power operations). Consum was one of the first supermarkets to remove products that used CFC (chlorofluorocarbon) propulsion. Its store at Calle Historiador Diago in València was the first supermarket in Spain to obtain the ISO 14001 environmental certification. That is an international certification that specifies effective environmental management systems.[18]

Consum's chosen format for sustainability reporting is the Global Reporting Initiative. Through this publicly available report, one can review Consum's progress since 2006. In the reports, Consum clearly calibrates its actions against internationally established Sustainable Development Goals; uses measurable metrics beyond aspirations; and communicates with accessible infographics.

Consum is working to reduce energy demands across all its operations (transporting and distributing products, business trips, and so forth), and it measures and reports on progress. Most of the energy it uses to power its stores is renewable. With a goal of 100 percent, renewables provided 75 percent of the cooperative's electricity in 2017 and 97.7 percent in 2019.[19]

Consum reduced its carbon footprint by 24.3 percent in 2017 compared to the previous year. It primarily accomplished this by recognizing that one of its biggest problem areas was refrigerant gas leaks, which accounted for 45 percent of its carbon footprint. From this awareness, Consum moved to strategically create and implement a correction plan; it has been actively preventing and controlling these leaks and substituting other gases with less harmful global-warming potential. In 2017, it invested €3.5 million to reduce fugitive refrigeration gases at 120 supermarkets. This appears to be a common issue for grocery stores. In the United States, National Co-op Grocers similarly found that when its grocery store members engaged in tracking their carbon footprints, fugitive refrigeration gases were often the biggest culprit. Fortunately, this is an area of operations that is relatively easy to correct through regularly scheduled maintenance and capital investments; however, for small, undercapitalized organizations this can be challenging.[20]

Part of achieving environmental sustainability of the food system is based on reducing the miles between where food is produced and consumed. Consum reports that 99.4 percent of its purchases are from 1,163 national suppliers. Its "local" suppliers are those who are in Spain, with 66 percent of the suppliers from the autonomous communities in which they have stores. Its focus on better transport logistics and more efficient distribution vehicles resulted in driving 4.7 million fewer kilometers in 2017 than the previous year.[21]

Consum's press releases show it promoting social and environmental activities. In 2018, for instance, it announced the opening of its ninth "eco-efficient" supermarket that year—stores that use 40 percent less energy than a conventional supermarket. It advertises that 91 percent of its stores are eco-efficient. The stores have automatic equipment shutdown, LED lighting, and optimized refrigeration, among other features. Even in the midst of the COVID-19 pandemic, Consum announced on the fifth anniversary of the Sustainable Development Goals how its actions advance the goals. Further, as consumers were confined to their homes to reduce the spread of the virus, Consum opened more online stores to serve them safely.[22]

In 2015, the European Commission adopted a Circular Economy Action Plan. The legislation on waste went into force in July 2018, setting clear waste-reduction targets to recycle 70 percent of all packaging waste by 2030. In March 2019 the European Parliament approved a ban on single-use plastic cutlery, cotton buds, straws, and stirrers, to come into force by 2021. Consum is already implementing practices ahead of the European waste regulations and single-use plastics ban. Consum reports that its waste-management model is consistent with the "circular economy"; it uses 100 percent recycled paper and cardboard for packaging Consum brands. In 2017, it started a Zero Discharge project to recover waste and prevent it from going to the landfill. It has involved its customers in environmental campaigns, such as collecting batteries, lightbulbs, and single-use plastic bags. Plastic bags are not part of the regulations, but in 2018 Consum stopped using plastic bags in its online store. However, it continues to use plastic bags and packaging in the stores I visited, so it has more work to do to eliminate single-use plastics.[23]

Consum carries organic products, but they are not a substantial part of its product mix. In 2017, Consum sold thirty-five eco products under the Consum Eco brand, and expanded to fifty-six products in 2018, with plans to continue this trend. Unlike bioTrèmol (see case study below), where organic food is a defining feature of the cooperative, for Consum it is a newer area of growth. Compared to another large Spanish supermarket chain, it is not the leader in this area. The investor-owned Carrefour Market had 1,500 "bio" products in 2018.[24]

However, Consum shows a leading commitment to the broader social aspect of sustainability. In 2017, women comprised 72.4 percent of its 14,364 employees (most of whom are owners). Unlike many other workplaces, women also hold a majority of Consum's management positions (60 percent in 2020). In 2018, it increased paid parental leave to seven weeks, two weeks more than required by Spanish law, partnered with the Red Cross to employ people at risk of exclusion from the labor market, and raised wages by 1.5 percent.[25]

Moving out further to the broader society, Consum declares that as a social economy company, it has "an inherent commitment" to "people and to the social challenges" in the communities where it operates. Its 2017 sustainability report shows that between 2015 and 2017 the cooperative increased its investments in social projects to improve society. In 2017 it spent €14.9 million, with a direct impact on 150,000 people, and donated 6,000 tons of food, and more than 1,000 of its workers volunteered in social services. Examples of the type of social projects Consum supports are donating to the Red Cross to rebuild water and sanitation systems in Cuba after Hurricane Irma and donating to Spanish UNICEF to help Mexican children after an earthquake.[26]

Governance and the Cooperative Structure

Picot observes that Consum's governance documents include all the documents the cooperative has approved and implemented, so the company is managed in an ethical, democratic, transparent, and sustainable manner.

These documents include its articles of organization and bylaws, as well as its strategic plan, annual management plans, Code of Good Governance, Code of Ethics and Conduct, and management policies.[27]

Unlike the other cooperatives discussed in this book, Consum's democratic governance is predominantly representative. Each member has a vote, but they are voting to elect their delegates to represent their interests. Consum is governed by a Governing Council (twelve people) and a General Assembly (150 delegates) that is composed of 50 percent worker-representatives and 50 percent consumer-representatives. The General Assembly appoints the members of the Governing Council for four-year terms. There are a social committee, a management control committee, an appeals committee, and an audit control committee.[28]

The Governing Council is responsible for setting the guidelines for corporate social responsibility. The Governing Council appoints a general director who is in charge of managing the business and environmental sustainability of the cooperative. The audit control committee, guided by internal regulations, regularly assesses whether Consum's governance is fulfilling "its mission of promoting social interest, taking into account the legitimate interests of the Cooperative's different stakeholders." The auditors regularly review the ethics committee's reporting on compliance with the internal Code of Ethics and Conduct. Article 11 of the audit regulations establishes the competencies for corporate social responsibility and sustainability. The committee is charged with monitoring and assessing strategy, practices, and compliance with corporate social responsibility. Separate from this, it is to monitor environmental sustainability policy, strategy, and practices, and assess compliance, with input from the Department of External Relations.[29]

Consum has conducted an annual external financial audit since its founding. It has conducted a nonfinancial audit of its operations and published it in the sustainability report since 2006, well in advance of any legal obligations. In 2018, Spain revised its commercial code with Law 11/2018, which requires more entities (including Consum) to provide nonfinancial reporting on environmental, board diversity, and other matters. Picot asserts that its 2018 sustainability report is already in compliance with the new requirements.[30]

Further, Consum's policy on corporate social responsibility establishes internal procedures for good governance, promoting the ethical behavior of its employees, and monitoring compliance with internal and external regulations. The Code of Good Corporate Governance, approved by the Governing Council on February 23, 2012, and the Code of Ethics and Conduct, approved by the Governing Board on January 25, 2018, are policies aimed to prevent criminal and unethical behavior. The 2018 policy includes an avenue to protect whistleblowers who reveal the possible commission of criminal acts, or the breach of any legal rule or internal policy.[31]

Picot emphasizes that the prestige and reputation of the cooperative are a basic element for its success and sustainability as a business. It wants all stakeholders to be aware of these policies and avenues for reporting, so all of this information is publicly available on its website.[32]

Summary

Created at the dawn of building a new democracy and labor movement in Spain in 1975, Consum has grown to serve three million consumer-members and 14,000 worker-members of its grocery stores throughout the Spanish Mediterranean. Its cooperative values have set it on a path to greater sustainability than required by environmental laws. Consum has consistently promoted environmental sustainability and transparency prior to legal obligations. It has been measuring and reporting on these nonfinancial matters using the Global Reporting Initiative format since 2006 (legally required in 2018). It is the first supermarket chain in Spain to use the carbon footprint and has been using it to strategically target and reduce its impact (entirely voluntary). It is running 97.7 percent of its operations on renewable electricity, with a goal of 100 percent (entirely voluntary). It has adopted a zero-waste policy to increase recycling and avoid landfilling waste (ahead of European Union regulations to recycle 70 percent packaging waste by 2030). While Consum appears to provide stores and prices similar to investor-owned supermarket chains, its cooperative legal structure establishes a sharing of financial surplus with its worker- and consumer-members. Its triple-bottom-line sustainability demonstrates that a cooperative difference is not at odds with being able to compete with large, chain, investor-owned grocery retailers.

BioTrèmol (Grocery Store Chain, Consumer, Spanish Provincial)

I was talking with a new organic berry farmer in the United States about how their sales to grocery stores were going. He said, "Not so great. We're at an odd size where we have too many berries to just rely on direct sales to customers who come to the farm, and too few for the grocery stores to commit to purchasing from us." This seems to encapsulate the tension new organic farmers face when trying to rebuild a local food economy; they find the retail grocery stores are built for larger, industrial-scale operations. Cooperative grocery stores can play an important role in rebuilding local agriculture by providing a market for smaller producers.

The Consum Cooperativa case study introduced a large-scale grocery store chain with more employee-owners in Spain than exist in the United States across every industry combined. Consum shows that cooperative grocery stores can compete with investor-owned grocery chains and still have a strong commitment to environmental sustainability that sets them apart. Despite the many impressive features of Consum, its scale of operation has limitations: it is not primarily selling organic food, it is not connecting consumers to farmers, and it is not able to purchase from small-scale, emerging farmers. For these objectives, a smaller and more local operation is better equipped.

Overview of the Cooperative: History, Type, and Economic Sector

In 2013, 150 founding members launched bioTrèmol, a consumer cooperative grocery store, to provide an alternative to industrial agriculture and chain grocery stores. The start-up years have been a time of growth, and it opened four stores in as many years. By 2019, it had scaled back to three stores in the province of Alacant, in the Valencian Community, had 450 members, and annual revenues of over €1 million. It operates as a nonprofit. The criteria it applies to the food it sells is that it must be:

- organic,
- local, and
- sourced from a small producer or cooperative.

The cooperative views itself as part of the social and solidarity economy and as working for greater *food sovereignty*, which it describes as facilitating local economies and eliminating or shortening supply chains.[33]

To learn more about bioTrèmol's motivations and practices, I interviewed Carmen Llinares (vice president) and Jesus Arnaiz (secretary). Both volunteer to manage the cooperative and are serving as elected leaders. Before they opened their first store, some of the founders were part of a smaller association of people who wanted healthier food, so they started to grow their own crops, operating as MercaTrèmol. Their political motivation was to "break from the consolidated corporate-controlled food system that relied on pesticides, shipping over great distances, and less power for farmers." MercaTrèmol was only available to the members who volunteered. Some of its members wanted to have a broader influence, so they formed bioTrèmol and opened a store to reach more people. Arnaiz explained, "Where MercaTrèmol wanted to be the ecologist in the neighborhood, bioTrèmol wanted to transform the entire neighborhood into ecologists." To have this neighborhood impact, it regularly holds talks in its stores and organizes visits to the farms to connect consumers with producers.[34]

While bioTrèmol only has consumer-members, it views its stakeholders as including employees and farmers. It is committed to paying its employees and farmer-suppliers well. Llinares and Arnaiz reported, "People work in good conditions in the store. They are paid more than at other grocery stores and their work hours are flexible, with many opting to work less than 40 hours a week." They also noted the commitment of the employees to their mission, asserting that many employees also volunteer as activists with the cooperative. As for the farmers, Llinares and Arnaiz say the cooperative pays farmers and producers directly, so they receive better prices without a middle company to cut into the money that goes to those who produce food. Last, they say that consumers receive good-quality organic products.[35]

Industry Context

Food cooperatives were leaders of the movement for organic and natural foods in the United States, starting in the 1970s. In Spain, bioTrèmol entered the organic food movement forty years later, rejecting the dominant food system (production, distribution, commercialization) as generating inequality, injustice, and health problems. It does not want to carry brands that come from consolidated megafirms in agriculture, and only sells organic products.[36]

Spain has the most organic agriculture in Europe, with 2.1 million hectares, most of it in the Community of Andalucia, where almost half the agricultural land is dedicated to organics. While Spain's farmers are leading organic producers, Spain's consumer demand for organics has only recently started to grow. Llinares and Arnaiz explained that Germany and other countries created a consumer demand for Spain's organic products before Spaniards did, so their organic farmers and producers are oriented to an export market and operate at a large scale.[37]

Now consumer demand for organic food within Spain is growing much faster than overall retail food sales. In 2017, organic food sales grew 12.52 percent, and that growth trend continued in 2018, while overall Spanish food retail sales were flat. This has encouraged more large grocery chains to carry organic foods. Industry experts expected Spain to sell €2 billion in organic food in 2018, but noted that small organic stores would have a harder time competing because larger chain stores had entered the organic market.[38]

The trend is similar in the United States. In 2017, the U.S. organic food market increased 6.4 percent while overall U.S. food retail sales only saw slight growth of 1.1 percent. That seems like it could have boded well for cooperatives that have been focused on carrying organic products. Yet, starting around 2013, food cooperatives in the United States have also been feeling this competitive pressure from large retail chains that have started carrying organic and local products. With Amazon's purchase of Whole Foods, and giants like Walmart and Costco carrying organic products, some U.S. food cooperatives are closing. Costco alone had $4 billion in U.S. organic sales in 2015, making it the largest U.S. organic food retailer.[39]

Some cooperatives are offering online sales to compete. Others think that in order for food cooperatives to survive they need to emphasize democratic participation of members and build a sense of community through member involvement in governance, volunteer hours, and community events. Others are using signs indicating the names of specific farms and distance from the store to provide a link between consumers and farmers. Some argue that cooperatives are most needed in places that lack grocery store chains—"food deserts"—and could fill unmet needs by providing inexpensive basic foods instead of focusing on more expensive organic foods.[40]

BioTrèmol's Arnaiz does not see the expansion of organic foods in large retail markets as a positive trend for the environment or the farmer. He says, "These products are not from local suppliers, so they have a bigger carbon footprint even though they are labeled organic." There is no place for small-scale farmers in this large retail market because they cannot reliably produce enough to be a supplier. This is where bioTrèmol plays a distinctly different role, with its commitment to local and small-scale cooperatives or producers. If the organic berry farmer I mentioned at the beginning of this case study had a bioTrèmol-type of store nearby, there would be a store suited to their scale of operations.[41]

How is bioTrèmol different from the large supermarket cooperative Consum, which is also featured in this book? At Consum, Llinares and Arnaiz assert, you'll find products that are similar to those at Aldi, El Corte Inglés, or other big supermarkets. In other words, Consum carries a lot of conventional products. Around 2017, recounts Arnaiz, "the big grocery stores in this region started seeing eco products as a good market, so Consum and others started selling eco products." He thinks this has caused small eco stores to close, including one of bioTrèmol's stores, and so sees this trend as harmful to efforts to create a demand for local organic food and build an economy that is more supportive of small farmers.[42]

Consum reports it purchases from local suppliers in the area where it operates its stores, with 99 percent of its suppliers from Spain. Consum has a limited, but growing, supply of eco products. The difference appears to be the definition of "local" and the scale of farming operations supported by each type of store. For Consum, local means produced in Spain. For bioTrèmol, local means produced in the province where they are based, at a very small scale that would not be acceptable for a large grocery store or export market.

BioTrèmol is working to establish small-scale, local organic farmers to meet a new consumer demand in their province. Through bioTrèmol's relationships with its members, it has encouraged the development of new organic farmers to serve this local market. BioTrèmol negotiates to buy a set number of crops or products from these small-scale farmers. Llinares and Arnaiz gave several examples. One of the coop members had unused land and wanted to start growing organic crops. He was able to start this new farm because the coop agreed to buy a set amount. Another member was similarly motivated to start an organic goat farm because she was able to sell products to the coop. Another entrepreneur, Isabel, sells her organic, Fair Trade chocolates in the coop's stores. Due to the relationship with bioTrèmol, Isabel converted an abandoned small chocolate factory into a functioning business and has hired one employee described as being from a "socially excluded" class. In this way, the cooperative is seeding new businesses that can supply local consumers. This not only creates new opportunities for ownership and wealth, but increases the province's food sovereignty in the face of larger market forces.[43]

BioTrèmol's most active members appear to be motivated by values they do not see present in the larger grocery chains. One of bioTrèmol's activist members urged people to see that small daily purchases can empower certain types of producers by supporting their products. As responsible consumers, they demand compliance with minimal criteria of responsibility: respect for nature, proximity, small production, and short marketing channels. In this call to support local economies, one member did so in a way that was globally aware and connected to a "planetary consciousness" that he described as meaning we share a "common future" from which nobody can escape. Holding global and local awareness as complements to one another mirrors the Cooperative Principles that promote ethical behavior toward the community in which the cooperative operates and in its trade and supply chains with those outside the local area.[44]

Environmental Sustainability

BioTrèmol describes itself as "active agents" promoting "food sovereignty" by developing local economies. It highlights several environmental sustainability practices it is implementing in its stores: it sells only organic products, does not sell bottled water, carries seasonal produce grown in close proximity, is aiming for zero waste and adding more bulk products, and runs 100 percent of its stores' electricity on renewable energy.[45]

Its website contains blog posts written by members, which critique the industrial food system. The posts show their passion for using consumer power to change the food system by purchasing local and organic products to support smaller-scale agriculture. The coop's commitment to local products reduces the carbon footprint of the products it sells.[46]

Llinares and Arnaiz brought these practices to life through their stories of sparking and supporting new entrepreneurial food ventures by buying directly from producers and giving them a store outlet to sell their products. Anything that can be purchased locally is. For items like coffee and bananas that are not produced in this area, the coop takes the complicated step to establish direct links with producers and engage in fair trade without a middle company. (For more on trade, see Chapter 12.)[47]

The dilemma with a business like bioTrèmol is that all these eco and fair payments result in what they call "just prices," which are higher prices for consumers. How do low-income people participate in this cooperative given the "just" prices? Llinares and Arnaiz outlined several practices they are designing to make their cooperative more affordable. BioTrèmol offers a staggered membership fee, with an initial payment of €25 to join and the remaining seventy-five to be paid throughout the year. They asserted that with the discount on food that comes with membership, the price paid is very similar to a conventional store. Second, in 2019 it established a "food basket" with basic items like rice, milk, and other staples. While it will still be sourced according to their commitments, it will have the least expensive options for these healthy foods. Last, it regularly donates perishable items to food banks to provide healthy food to those who cannot afford it and to avoid food waste. If through these practices bioTrèmol can deliver lower prices to make organic and local food more affordable, it could provide a model for those who want to use cooperatives to meet needs in food deserts in rural and urban poor areas.[48]

Another case study in this book is about Som Energia, the consumer cooperative that exclusively sells electricity produced by renewable sources. One of the Cooperative Principles is cooperation with other cooperatives. BioTrèmol and Som Energia are carrying this out in two ways. BioTrèmol purchases renewables from Som Energia to cover all of its electricity use. Beyond this, it created an agreement between the two cooperatives that went into effect at the beginning of 2019. Members of bioTrèmol can also become members of Som Energia without paying the additional membership fee of €100 to join the renewable energy coop. Som Energia has its local group's monthly talks in bioTrèmol's store to raise awareness about energy conservation and switching to renewable sources of electricity. It is too soon to tell how many of bioTrèmol's members will opt to buy renewables for their personal electrical needs, but the two cooperatives have strategically reduced barriers to participation.[49]

Governance and the Cooperative Structure

From the beginning, bioTrèmol hard-wired a commitment to the social and solidarity economy into its governance documents. The first article of its bylaws describes its mission:

> BIOTRÈMOL COOP. V. a cooperative society is constituted, endowed with full legal personality, . . . with the firm will to transform the social reality by means of changing consumer habits, responding to . . . large corporations directing food policies, working to improve relations between people, putting collective interests before any idea of private profit, and contributing . . . to the improvement of the social, environmental, economic, and social condition of its members and of citizens in general, in its double condition of consumer organization and social economy entity.[50]

Following this, it enumerates its commitments in its bylaws, which include developing an ecological, local, and small-scale model of agriculture; guaranteeing transparency in its products; promoting responsible consumption; cooperating with other organizations to pursue these common objectives, while remaining independent and autonomous; guaranteeing democratic control of the entity; and contributing to a truly sustainable economy.[51]

Its social objectives, established in its bylaws, include a commitment to sustainability: to distribute organic ("ecological") products. As a reflection of its focus on building a local food economy, its bylaws commit it to working only in the province of Alacant (Figure 2).[52]

BioTrèmol's founders chose a cooperative form of business organization because they were not motivated by profit and thought it allowed for the

Figure 2. BioTrèmol's Principles as Displayed in Its San
Vicente del Raspeig Store (Credit: Melissa K. Scanlan)

democratic participation they needed for a broader transformation of society. BioTrèmol uses a system of direct and representative democracy in its governance. There is an annual General Assembly of all the members (one member/one vote) and a Governing Council elected by the members. In addition, it has specialized committees for resources and management.

BioTrèmol says its horizontal management structure promotes participation in decision making.[53]

Llinares and Arnaiz explained the different membership categories the coop has established, some required by law and others of its own creation. All the members receive discounts on products, but vary in terms of money and time commitments. The legally required category of "consumer-members" is the most basic membership. The coop charges a one-time fee of €100 to become a consumer-member and then charges €6 per month. "Activist members" pay a one-time fee of €1,000 to join, and volunteer four hours per week. Most of the store's employees have opted to become "activist members," so in addition to their paid work, they volunteer. For those altruistic people who support the mission with donations and not time, and do not need discounted products, the coop has "associate members." Last, inspired by the Park Slope Food Co-op in Brooklyn, New York, it created a new membership category for those who volunteer four hours per month and do not pay a monthly fee. It hopes that this new category of membership will make its healthy food available to more low-income people.[54]

Summary

BioTrèmol consumer grocery store cooperative designed the nonprofit business around a set of values and principles, which it wrote directly into its governance documents and carries out in its daily practices. It is actively working to support the emergence of new small-scale organic farmers in the province of Alacant to supply local consumer demand for organic food. The coop's commitment to small-scale and cooperative farmers means it is providing a market that supports dispersed farm ownership and wealth rather than consolidated agribusiness operations owned by nonfarmer investors.

La Bodega de Pinoso (Organic Wine, Farmer, Spanish National and Export)

Overview of the Cooperative: History, Type, and Economic Sector

Adjacent to the Mediterranean Sea, Alacant boasts a climate that is excellent for wine production. While there are numerous wineries and vineyards, and Spain produces the most organic wine in the world, organic wine is still a small part of the overall Spanish wine market. Talking with local wine merchants in the Central Market revealed there were only a few local organic wines available, and this is where I found wines from the cooperative La Bodega de Pinoso (Bodegas Pinoso).

This cooperative brings together grape farmers to produce high-quality wines, rooted in local agricultural knowledge of how to cultivate grapes with a small ecological footprint. While viticulture and enology are complementary, they are two separate disciplines. In the late nineteenth and early twentieth centuries in the town of Pinoso, in the province of Alacant, there were about 400 farm families who cultivated grapes and had their wine cellars at home. The small farmers in Pinoso were very dispersed, with each producing their own wine and having limited storage capacity. Each farmer had to individually bring their wine to market in the port of Alacant to ship beyond the immediate area, which created imbalances between supply and demand that made prices very unstable.[55]

"LA UNIÓN HACE LA FUERZA" (UNION MAKES STRENGTH)

Then at the turn of the twentieth century, essentially all of the town's grape farmers decided to cooperate in response to a crisis in wine production that affected the main wine-producing countries and threatened their livelihoods. The 400 farmers formed a cooperative that has been producing wines together since 1932.[56]

This is mainly an agrarian cooperative in accordance with Article 2 of its bylaws and with Article 4.2 of the Valencian Agrarian Cooperatives Law. By producing the wine together in a cooperative winery, the small farmers could reduce production costs, reduce costs of intermediaries, increase wine quality, increase the wine storage area, and obtain greater benefits for its member farmers, all while maintaining their independence as owners. This allowed the farmers to gain competitive advantages and be resilient in the face of a structural overproduction of wine they faced at the beginning of the twentieth century. The founding farmers used the popular expression "union makes strength" ("la unión hace la fuerza").[57]

The strength of their union has withstood many decades of changes in culture, law, market demands, and now climate. Their union has allowed

them to innovate and adapt. As more producers wanted to join the cooperative, it expanded its membership in 1969, 1974, and the 1980s. By that point,
it was the largest winery cooperative in Alacant, with 650 grape-growing
members. In 2019, they had 280 farm families as members, and seventeen
employees who were not farmer-owners but who ran the wine production,
bottling, marketing, and financial aspects of the cooperative. The smaller
number of farmers today than historically reflects a problem in many parts
of Europe and the United States that fewer people are becoming farmers.[58]

These farmers were the early adopters of organic grape cultivation and
wine production, having obtained organic certification in 1997. Today the
cooperative is a powerhouse in organic wines, as the largest organic wine
producer in the Valencian Community and also the largest in Spain. Its
organic wines are sold under the labels Vermador, Vergel, and Pinoso. Its
wines that are not certified organic, Cristatus and Diapiro, are still produced, bottled, and distributed using the additional sustainable practices
described below (Figure 3).[59]

Figure 3. Bodegas Pinoso (Credit: Melissa K. Scanlan)

Industry Context

To put Bodegas Pinoso in the context of global and Spanish wine production, caring for vineyards and creating wine are a traditional part of the Spanish culture and economy that has grown in economic importance. Cultivating grapes in this region of Europe dates back 3,000 years. Currently, there are 4,000 wineries in Spain, many of them small- to medium-sized businesses and farmer cooperatives. Of these, in 2016, about 900 Spanish wineries had some organic wine production.[60]

With 975 million hectares planted in grapes in 2016 (97.4 percent for winemaking, 2 percent for table grapes, 0.3 percent for raisin production, and the remaining 0.3 percent for nurseries), Spain has the largest area of land in vineyards in the world. Spain also has the most land in the world dedicated to organic vineyards, with 106,509 hectares in 2016, up from an estimated 84,000 hectares in 2013. Even before the jump in production in 2016, Spain dominated organic wine production, with 27 percent of the world total in 2013.[61]

Most of the organic hectares are dispersed among small Spanish farmers. For instance, Bodegas Pinoso's farmers boast the most organic hectares of grapes in Spain: in 2004 they were collectively cultivating 484 hectares of ecological grapes, and by 2018 they had grown to 600 organic hectares. From this land, Bodegas Pinoso produced approximately 2.75 million liters of organic wine in 2018.[62]

Environmental Sustainability
"SOMOS LO QUE HACEMOS DÍA A DÍA"
(WE ARE WHAT WE DO DAY BY DAY)

Sustainable wineries should be engaged in practices to reduce their environmental footprint in cultivating, winemaking, bottling, and distributing. According to Armand Gilinsky and his colleagues, after the 2007–2008 global economic crisis more wine producers proclaimed themselves to be sustainable as a strategy to gain a competitive edge by differentiating their product or reducing costs. Others are motivated to adopt sustainable practices in order to avoid increasing regulation of agriculture practices such as pesticide spraying and groundwater contamination. Sustainability is a malleable concept, but some certification programs, such as the California Sustainable Winegrowing Alliance, have established benchmarks.[63]

An approach when analyzing the environmental sustainability of wine related to GHG emissions is to use a life-cycle assessment. For wine, the life cycle consists of:

- cultivation—vineyard,
- vinification—winery,
- bottling—packaging, and
- distribution.[64]

Areas where wineries produce the most GHGs are the production of glass bottles for packaging, emissions in the field from tractors and nitrogen fertilizer (conventional wine), and transportation to distribute the wine. The use of new glass bottles is the biggest source of GHGs. In this case study, I use the life-cycle structure but look beyond GHGs to incorporate aspects of the performance areas identified by the California Sustainable Winegrowing Alliance.[65]

Bodegas Pinoso was a certified organic winery a decade before the winery sustainability movement took off in the United States. My interview revealed that the coop's sustainability practices go well beyond those contained in its promotional materials. It lacks an office dedicated to sustainability, a corporate social responsibility officer, and a system of transparently measuring and reporting on sustainability metrics, such as a GHG audit or use of a carbon footprint tool. A perusal of its website will not yield a sustainability report. Yet, environmental sustainability is infused throughout all its operations.[66]

CULTIVATION: ORGANIC VINEYARDS AND WINE

In the 1990s, the cooperative shifted its focus to improve its sustainability and wine quality and moved to get its first fields certified in 1997 as *"agricultura ecológica,"* or ecological agriculture, which is a European Union certification and a Valencian Community certification (ES-ECO-020-CV). It was one of the first Spanish wineries to become certified. To be a certified organic wine in the European Union and Spain, the grapes must be grown without the use of synthetic chemicals or pesticides, and the wines made without additional chemicals or GMOs. Bodegas Pinoso explains the certification means it is improving soil fertility, increasing biodiversity, not using pesticides, and allowing the grape farmers and wine producers to work in a healthier way.[67]

Conventional wine growing, with its use of fungicides and chemical/synthetic fertilizers and pesticides, exposes the farmers, local community (through pesticide drift), local waters, and consumers to chemicals in potentially harmful amounts. When organic wines first entered the market, as "healthier" wines, they were seen as inferior in quality; but by 2019, after improvements in winemaking, many were advocating for the superior quality of organic wines.[68]

By 2010, half of Bodegas Pinoso's vineyards were certified organic, and it is now the largest ecological wine producer in Spain. On its website, the coop declares its commitment to quality, sustainability, and innovation in the description of its roots, history, and development over more than eighty years. The website has a separate page to highlight the coop's environmental commitment, which it describes as its *alma ecológica,*" or "ecological soul."[69]

WATER CONSERVATION

Farming in the dry Mediterranean can require a lot of water, which is in limited supply in this region. Bodegas Pinoso farmers conserve water through their related commitment to preserving their farming traditions using special grape varieties that thrive in the Mediterranean climate. Monastrell grapes, for instance, make up the vast majority of the grapes they grow. Adapted to this climate, these grapes have very deep roots and do not need to be irrigated. Milagros Perez, of Bodegas Pinoso, described these deep-reaching roots as providing part of the terroir of the wine, so you can "taste the earth of the Mediterranean when you drink a Monastrell wine."[70]

VINIFICATION: WINERY FACILITY, WINEMAKING, BARRELS

In practice at Bodegas Pinoso, nothing is wasted. Instead of seeing something as "waste," they see it as a "product" and reuse it in another process, as will be explained. This concept is practiced in their production facility, vineyards, and winemaking.[71]

Bodegas Pinoso has carefully maintained its 1932 production facility, with an original wooden roof. As a cooperative committed to longevity and stewardship of its financial resources, while it has modernized and expanded, there is nothing excessive or flashy about its facilities. Although it does not yet use renewable energy, it conserves energy by using lights on sensors as well as a lot of natural light throughout its production facilities.[72]

The first product that is used in the vineyards is the branches the farmers prune from the vines. The traditional use was for firewood, especially for cooking, because of the particular aroma the woody vines contribute to meat or a regional specialty, paella.[73]

Once the grapes have matured, the farmers harvest them and bring them to the cooperative for destemming and crushing the grapes to start pressing them into wine. The residues or by-products of the vinification (such as skins) could be considered waste, but at Bodegas Pinoso this is another point in the production to return the resources to another use. It delivers these by-products to an authorized distiller who uses them to produce rum, whiskey, spirits, and vinegars; and what remains after that process can be used for fertilizer.[74]

Another element in its winemaking is the wooden barrels used to age some of the wines. When no longer useful to the cooperative, some of the barrels are repurposed as a product for aging whiskey or brandy, and the others are used to make furniture. Thus, from harvest to winemaking, the cooperative is rethinking "waste" and repurposing resources for multiple uses.[75]

BOTTLING: PACKAGING AND DISTRIBUTION

Bodegas Pinoso's website cites additional sustainable practices beyond organic certification: using recycled cardboard to box its wine, using lighter bottles, and encouraging boxes of wine. As noted above, the biggest source of GHGs for wineries is the production of new bottles. Not only does the winery sell some of its wines in boxes (made of recycled material), but all its bottles are recycled so it is not using new glass, significantly shrinking its associated GHG emissions.[76]

Perez has deep roots in this community. Her great-grandfather grew grapes for wine and was one of the early members of the cooperative. She emphasized the farmers' roles caring for the earth that in turn provides for them, describing it as a symbiotic relationship. She clearly loves Bodegas Pinoso and the land and adjacent mountains. She pointed out some of the labels on the wine that feature different animals, which she said were "designed to show a reverence the farmer-members have for protecting those animals."[77]

SUPPORT FOR LOCAL COMMUNITY

The cooperative provides a retail store adjacent to the winery that supports the local community. It sells farm implements and its wine, but also features locally produced bakery items, cheeses, and sausages from area farmers and businesses who are not members of the cooperative. The store is open to the public, but it primarily serves the farmer-members, who get a discount.[78]

All these sustainability practices reflect Bodegas Pinoso's definition of environmental sustainability, which is focused on balance and longevity. For Bodegas Pinoso, according to Perez,

> environmental sustainability is the creation of a harmonious balance between nature (vineyards, fields, nature) and society. They achieve results for the winery, but without threatening natural resources. They will give to future generations a better environment, so they can maintain the wine and agricultural tradition in the area.[79]

FINANCIAL HEALTH AND FUTURE FARMERS

Will there be future generations interested in cultivating organic grapes? The average age of the farmer-members of Bodegas Pinoso is between sixty-five and seventy. This is not an isolated problem. Although rewarding, working in agriculture is not easy. According to data from the European Union's Statistical Office, Eurostat, only 5.6 percent of farmers in Europe are under thirty-five, while about 56 percent are over fifty-five. The Spanish agricultural population is even older than the European average. Only 3.7 percent are under thirty-five, while 63 percent are between thirty-five and sixty-four, and 33.3 percent are over sixty-four.[80]

While Perez thinks that "young farmers are the ecologists of the future," she knows they face high initial costs, even if they inherit the land, because to remain competitive they need to use new and expensive technologies.

In addition, agriculture sometimes has low profitability, as it depends on climatic and environmental factors. Being a farmer is a job that in some sectors is stigmatized as "too hard," since the farmer needs to be dedicated to the work 365 days a year, and many young people opt for better-paid jobs in urban areas that offer greater stability and comfort. Meanwhile, Spain's youth unemployment rate is incredibly high. Between 1986 and 2018, the youth (ages fifteen to twenty-four) unemployment rate in Spain averaged 34.7 percent. Bridging the gap between aging farmers and unemployed youth will be critical to carry on the valuable farming traditions of this region.[81]

Governance and the Cooperative Structure

Bodegas Pinoso has written into its governance documents its mission and the Cooperative Values and Principles, including its commitment to protect the environment. It shares these documents with every member. According to its bylaws, participating farmers will do a variety of things, including:

> promote cultivation practices and techniques for the production and management of waste that respects the environment, especially to protect the quality of water, soil and landscape, and to preserve and/or enhance biodiversity.[82]

This is a larger cooperative, with almost 300 farmer-members and almost 100 years of traditions. The democratic governance it uses is representative rather than direct democracy for most matters. The farmer-members participate through the General Assembly, by voting for the Governing Board and the president who will represent them. Annually, the cooperative holds two General Assemblies in which the members hear about and discuss the strategy, measures, norms, calendars, campaign rules, environmental sustainability, finances, investments, and results. The members then vote to ratify or oppose the management decisions. For day-to-day operations, the Governing Board and president are responsible.[83]

In terms of day-to-day environmental decisions, there is a technical team that collaborates with the Governing Board to make decisions about environmental sustainability goals. The Governing Board decides how to

allocate funds for environmental sustainability. As is done with other matters, the farmer-members learn about and discuss environmental sustainability at the General Assemblies and vote to ratify or oppose.[84]

Financial stability is drafted into the governance documents. The cooperative has a mandatory reserve account that is funded by, among several
sources, taking 20 percent of the net surplus available for each fiscal year.
It also has a mandatory fund to support cooperative formation and promotion. It uses this to support training its members and employees in Cooperative Principles and techniques, cooperation with other cooperatives, and
cultural and social programs in the local community.[85]

Summary

Bodegas Pinoso has brought together farmers to produce high-quality
wines since 1932. It first became certified organic in 1997, and now half
of its fields (600 hectares) are certified, making it the largest organic winery in Spain. Its sustainability practices go beyond organic certification,
and these practices apply to all of its wines (including conventional). Some
significant measures are: growing Monastrell grapes that are accustomed
to the Mediterranean and do not require irrigation; using recycled bottles
and recycled cardboard boxes in the packaging of its wines; sending the
by-products of wine production to another company to produce additional
beverages; and conserving electricity with natural lighting and light sensors. Its commitment to protecting the environment is drafted into its governance documents, along with its overall commitment to the international
Cooperative Principles, which provides a legal durability to the cooperative's
purpose, goals, and day-to-day operations. The farmers' strength in working together has allowed them to innovate and adapt to changing circum-
stances for almost a century.

Cerveses Lluna (Organic Beer, Worker, Spanish National)

Overview of the Cooperative: History, Type, and Economic Sector

Cerveses Lluna, formally Articultura de la Terra Coop.V, is an organic
craft beer company that is owned by its workers. Maria Vicente and David

Seguí cofounded the cooperative in 2008 and moved their craft brewery to Alcoi, Spain, around 2016. For this case study, I was able to meet the co-founders: I interviewed Vicente while Seguí brewed beer. The cooperative currently produces six types of beer annually and adds seasonal varieties, all of which are "artisan, cooperative and ecological." Its annual production is 40,000 liters of organic beer, with the capacity to produce twice this amount.[86]

The name of the brewery, which is in the regional language of Valenciano (a dialect of Catalan), reflects the cooperative's roots in its rural mountain community. Alcoi is a city of 55,000 people, about an hour's drive from an urban hub in Alacant and the Mediterranean Sea. From Alacant, the route up the mountains in February brings you past terraces of flowering cherry and almond trees, interspersed with olive trees. Flanked by two nature pre-serves, the small city of Alcoi is surrounded by green as far as the eye can see, in contrast to the rocky, dry mountains that are lower and closer to the sea. Alcoi has a university, but the town used to host a thriving industry of textile mills. These have mainly closed, leaving behind their empty build-ings. The brewery is in an old textile mill building the cooperative revital-ized to give it another life. Located adjacent to the river and an exercise trail with park benches, the brewery is open to the public for tours and tastings.

The cooperative is on the very small end of the spectrum, and yet it is the largest producer of organic beer in all of Spain. Over its first decade of existence it has experienced growth and decline, and is positioning itself for another period of growth in its second decade. In 2017, it had twenty members who had invested money (and are paid annual dividends in beer), and four worker-owners (members who gave money and time as workers). The cooperative experienced a lot of growth when it had one customer that bought specialty beer for a niche market (beer flavored with aloe vera). This arrangement allowed the cooperative to purchase new equipment for the brewery and expand its brewing capacity. However, when this key customer went out of business, it caused financial problems for the cooperative in 2018. It did not have enough income to pay the monthly salaries of all the worker-owners, so two found new work and left the cooperative.

After the restructuring, the cofounders are the two remaining worker-owners. Still, Vicente is optimistic and as committed as ever. She says,

"They chose the cooperative form to have better conditions for the workers and protect the environment, and they want to grow again to add more worker-owners. They are resilient." While she is working to move the company forward to expand capacity, income, and employees, she reflects that "there is no price for loving your job."[87]

While clearly financially challenged, as a cooperative Cerveses Lluna is independent, and no parent company or corporate headquarters will be able to determine its fate. This ties into why the founders chose to create a cooperative instead of another business form. Vicente had previously worked for a traditional corporation, but the company discharged polluted water in the river in order to save money, and this weighed on her. After management ignored her concerns, she felt the need to do something meaningful by starting a business that wasn't about what she called "brutal capitalism." The cofounders of Cerveses Lluna believe it is important to do something to protect the earth, so everything at this cooperative is oriented toward that goal, and has been since day one.[88]

Vicente says that what motivated them was philosophical and value-driven. Their purpose was to create a democratic workplace where the workers were owners and participated in decisions about strategy. They wanted to create jobs that were improving the world and protecting the environment by design, so they focused on producing craft organic beer.[89]

For a decade, they have been making beer as a cooperative and applying "values of self-management and assembly." They designed the cooperative to "contribute our grain of sand to transform the world a bit, with agro-ecological principles, cooperative work, democratic practices, and the social economy as the cornerstones."[90]

Industry Context

Craft beer? Organic beer? Worker-owned cooperative brewery? You will be hard pressed to find an entity that is all three, but Cerveses Lluna is: craft, organic, and cooperative. When Cerveses Lluna started in 2008, people in Spain were not familiar with craft beer, so the cooperative has been a pioneer for this industry.[91]

At the same time in the United States, the craft beer market was in an early phase, but was gaining momentum and scale far faster than the Spanish market. Between 2005 and 2015, the number of U.S. craft breweries expanded from approximately 1,300 to 4,500, growth of over 200 percent in a decade. By 2017, the U.S. craft beer market was a $26 billion industry with 6,266 craft breweries. The craft beer expansion paralleled a renewed interest in local food and beverages of all varieties. Studies suggest that craft beer consumers reject mass-produced beer because of its homogeneity and low quality.[92]

However, the emphasis on organic foods has not caught on as much for beer. There is a very small segment of craft brewers making beer from organic ingredients. In 2003, the inaugural North American Organic Brewers Festival featured twenty-five breweries, and by 2015 this had grown to thirty-six. During a similar time period, 2003–14, U.S. organic beer sales increased more than tenfold, from $9 million to $92 million. Even with the sales growth, organic beer is not commonly found.[93]

Organic beer in the United Kingdom is also somewhat obscure. However, when you do find a certified organic beer in the United Kingdom, you know that "all the organic raw materials, including malt, hops, and yeast, must be non-GM and grown without reliance on pesticides, herbicides or insecticides," according to an organic certification expert.[94]

In the United States, Wolaver's Fine Organic Ales was the first producer of beer certified as organic by the U.S. Department of Agriculture. Yet, it bottled its last organic beer in 2015. Wolaver's story provides a cautionary tale for those choosing a corporate form. "In 2002, Wolaver's Organic Ales bought the Otter Creek brewery and its brands, relocated to Vermont and took over production of both portfolios. After years of modest growth, Otter Creek, in a move to make the business more financially stable, sold to Long Trail/Fullham & Co. in 2010." Within five years, the parent company shut down Wolaver's and stopped producing organic beer, instead prioritizing expanding conventional craft beer production by Otter Creek. The parent company noted the high cost and limited availability of organic ingredients, as well as the difficulty of separating organic beer production from the rest of its production, which was much larger. In a cooperative like Cerveses Lluna, that scenario is very unlikely because of the democratic control of the members (one member, one vote) and the legal formation documents, which commit it to using organic ingredients.[95]

Cooperative breweries are uncommon. In Spain, there are three cooperative breweries that create organic craft beer. In Vermont, home to a vibrant
and growing craft brewing industry, there are no worker-owned cooperative
breweries. However, Switchback Brewery in Vermont is organized as an
ESOP. Full Barrel Cooperative is in the early formation stages, with a vision
to become a consumer cooperative created to serve beer-drinking members. There are several consumer coops for beer lovers scattered around
the United States. Black Star Co-op and 4th Tap Co-op, both in Texas, are
worker-owned brewery cooperatives, while Black Star also has thousands of
consumer-members. In sum, Cerveses Lluna provides a rare example of a
craft brewery organized as a worker cooperative that is entirely organic and
powered by 100 percent renewables.[96]

Environmental Sustainability

U.S. BREWERS ASSOCIATION SUSTAINABILITY BENCHMARKING

The Brewers Association, which serves small and independent brewers
in the United States, publishes sustainability manuals and provides a tool
for brewers to measure and compare their sustainability, through benchmarking. The annual benchmarking analysis for the industry is relatively
new, started in 2014. They even have a "benchmarking mentor" who can
assist members in their pursuit of sustainability. Of the 4,400 craft breweries, in 2017 only a very small fraction (280) were participating in the Brewers Association's sustainability benchmark. But this is a recent effort that is
gaining traction. Further, as the Cerveses Lluna case study shows, lack of
reporting or benchmarking does not mean lack of sustainability.[97]

The Brewers Association defines a sustainable brewer as one who:

- produces the highest quality beer while minimizing impacts to the
  environment;
- balances profitability with the needs of the planet, our workforce,
  and communities;
- protects the environment for brewing ingredients and future
  generations;
- sources, builds, and operates responsibly, without compromise; and

- uses natural resources in an efficient manner and strives to eliminate waste.[98]

Brewing consumes a lot of water and energy, and the benchmarking reports focus on sustainability in these two areas. Strikingly, however, there
is nothing in the benchmarking about ingredient sourcing and prioritizing
organic ingredients.[99]

To shrink the GHGs associated with creating beer, brewers can use renewable energy systems onsite or offset by purchasing renewable energy
credits. Of the Brewers Association members who are participating in
benchmarking, only sixteen are using renewable energy.[100]

In the third year of benchmarking, the 2017 report highlighted trends
from its case studies of leading breweries. This emphasized the importance of creating an employee culture with a core focus on environmental
sustainability. The Cerveses Lluna case study reinforces this finding, but
goes further back into the DNA of the business structure. Embedded in the
design of Cerveses Lluna, as a worker-owned cooperative, the employees
carry out a vision of using the business they own to protect the environment
and practice democratic participation. The terms Cerveses Lluna uses to
describe its approach are "ecological" and "bio." In the United States this
is understood as "organic" as it relates to USDA certifications. Cerveses
Lluna's website explains the core of its business approach, saying nothing
would be "understandable and coherent if we did not try to minimize the
environmental impact of our work." All the raw materials that it uses come
from organic crops, and it reduces the consumption of energy as much as
possible.

While the coop does not formally measure sustainability metrics and report on this, the absence of such a reporting system does not indicate a
lack of commitment to protecting the environment. It is too small to have
a head sustainability officer, but because protecting the environment was
designed into the fundamental structure of the business, the employees are
continually incorporating this approach. For the employee-owners, this is
their primary motivation, their passion, and not something added onto the
business later to improve marketing.[101]

Cerveses Lluna's business model is based on the circular economy. Sustainability starts with its location, and its focus on reducing waste and repurposing it to become a resource instead of a disposal problem. The building it uses is a 1930 restored textile mill that it has fashioned into a craft brewery. It utilizes natural light to reduce electricity demands, and it built light fixtures from used plastic kegs and beer coasters. It is moving from plastic labels to recycled paper labels. While the law prevents it from using recycled bottles for beer, it uses glass bottles because it can buy them from a local business and customers can recycle them. Further, it can bottle onsite with a small-scale bottling machine. Beer production results in a lot of used grains and hops (picture piles of mushy cereal). Cerveses Lluna does not view this as waste to dispose of, but instead gives it another life by providing it for free to local organic farmers. There it is used to feed chickens and produce compost for olive trees.[102]

Beer is primarily water; the rest is grains, hops, and yeast. The brewery conserves water and energy. It minimizes water use and energy by reusing clean hot water for multiple purposes, such as cleaning. All of Cerveses Lluna's ingredients are certified organic, and the water comes from a source in the mountain nature preserve adjacent to the city of Alcoi. Ideally, it would source all its ingredients locally. However, currently there are no organic farmers in the immediate area who produce the amount of ingredients it needs. Its main suppliers are in Germany and England, and, recently, a Canadian cooperative that grows hops.[103]

With its commitment to local purchasing, Cerveses Lluna is helping develop a local farm economy of suppliers of organic ingredients. Spain produces hops in Galicia and León, which are rainy and cool areas of Spain that some people describe as Celtic in feel. Some organic hops are growing in the immediate area around Alcoi, and some in nearby Catalunya, but these are not in the amounts the coop needs for commercial production. The Alcoi hops farmer produces enough for one batch of beer a year, and Cerveses Lluna hopes the Catalunya farmer will have commercial quantities in a few years.[104]

Cerveses Lluna stands apart from other craft brewers in its focus on climate change. It has a statement about its support for the Paris Agreement and signing onto the Climate Change Solutions campaign on its website.

Beyond talking the talk, it is among the few craft brewers that use electricity produced from 100 percent renewable power. As a member of Seneo cooperative, it purchases renewable electricity from Enercoop (featured in Chapter 8), a renewable energy cooperative in Crevillent, about an hour from its brewery.[105]

Governance and the Cooperative Structure

Cerveses Lluna grafted the Cooperative Values and Principles into its legal formation and governance documents. Vicente pointed out the provisions—written in Valenciano—that explicitly committed the company to produce organic products and follow the Values and Principles.[106]

It holds an annual assembly of all the members (workers and investors), but because of the small size of the cooperative, with only two worker-members this year, collective decision making does not require the formality found at larger entities.[107]

One of the Cooperative Principles is to cooperate with other cooperatives. This is very evident in the web of relationships this brewery has woven. While not exclusively so, many of its business relationships are with other cooperatives. It purchases renewable energy from a cooperative. It makes a special seasonal beer with local organic cherries grown by a neighboring cooperative. After the interview when we tasted the beers, Vicente also served organic olives grown and prepared by a small local company.[108]

This principle also infuses the coop's distribution networks. It focuses on selling beer locally, so it cuts down on environmental impacts from transportation. Margins on beer sales are small so that it prefers to sell directly to consumers from the brewery. It works with distributors to sell beyond the brewery—primarily in Madrid, Barcelona, and València—with a focus on selling through consumer cooperatives. For instance, Eroski, the big grocery store chain that is part of the Mondragón cooperative, carries Cerveses Lluna beer, and happened to show up at the brewery to pick up beer during our interview.[109]

Summary

Cerveses Lluna has been exclusively producing organic craft beer in an employee-owned cooperative since 2008. It has been organic from the beginning and is the largest producer of organic beer in Spain. Additionally, some significant sustainability measures it employs are: purchasing local supplies wherever possible; conserving water and energy; powering itself on 100 percent renewable electricity; rehabilitating a textile mill for its brewery space; and repurposing all of its used grains and hops as feed and compost for local organic farmers. Its commitment to protecting the environment is drafted into its governance documents, along with its overall commitment to the Cooperative Values and Principles, which provides a legal durability to its purpose, goals, and day-to-day operations.

CROPP Cooperative/Organic Valley (Organic Food, Farmer, U.S. National)

Overview of the Cooperative: History, Type, and Economic Sector

Founded in 1988, the CROPP Cooperative has been consistently committed to organic farming, marketing its products under a variety of brand names, including the well-known Organic Valley. I interviewed a founding farmer and former CEO (1993–March 2019), George Siemon, for this research. Since Siemon was no longer employed by the CROPP Cooperative/ Organic Valley (CROPP) at that time, his views are his own and provide a rich reflection on building a mission-driven farmer cooperative from a fledgling group in isolated western Wisconsin into the first organic food company to earn $1 billion in annual revenues.[110]

Siemon was a Wisconsin dairy farmer experiencing a terrible farm crisis in the 1980s and struggling not to fall into bankruptcy. In the midst of this crisis, an idealistic group of fifty produce and seven dairy farmers in the hilly Kickapoo Valley of Wisconsin gathered around an audacious dream: they wanted to form an organic, value-added cooperative to save the family farm. Part of the catalyst for this came from an activist group, the Wisconsin Farm Unity Alliance.

"The original farmer-activists were going to solve problems themselves because the government wasn't helping," remembers Siemon. At the time, the concept to create an organic, value-added farmer cooperative was an advanced idea in the United States. The Kickapoo Valley region of Wisconsin had a lot of tobacco growers, and the founders' original idea to grow organic produce faced an uphill battle trying to replace tobacco. Siemon then took the lead on the dairy side of the cooperative, and dairy is what took off and grew them into a national cooperative with significant brand recognition.[111]

Although the farmer-founders came from a place of disappointment in cooperatives because the existing agricultural cooperatives hadn't helped them during the 1980s farm crisis, they tried to figure out what they needed to do better. They did not reject cooperatives, but wanted a different model for cooperation: one that would save the family farm. As with many of the case studies in this book, the farmers drew inspiration and ideas from the Spanish Mondragón cooperative. Siemon reflects, "We wore out the Mondragón video" while launching this venture. The Spanish cooperative was their "biggest influence." He describes CROPP's founders as "idealistic and community-oriented."[112]

For these farmers, forming as a cooperative was simply a familiar way to organize. "Minnesota and Wisconsin are co-op country, so we had a lot of support right away for this form of business that allows cooperation among the farmer-members to market their products," observes Siemon. He sees their focus on saving family farms as essential to gaining support. He notes, "Everyone in Wisconsin loves family farms, so even people who thought organic farming was nuts supported us because it was a strategy that was working for small farmers." One lesson from CROPP is that they approached organic farming in a way that had a broader appeal, enabling them to overcome those skeptical about the value of organic: they made it a pocketbook issue about better compensation for the farmers who are working the land.[113]

Siemon reflects that this was always a team effort between the activists, the farmers, and the employees of the coop. They steadily grew and in 2019, when Siemon resigned as CEO, CROPP had more than 2,000 organic farmer-members, or 14.4 percent of U.S. organic farms working across thirty-one states.[114]

Industry Context

Siemon believes a "key element" to their success was their decision to be entirely organic. He insists, "The organic market enabled them to grow and succeed." Reflecting on the late 1980s, Siemon explains, "There weren't established prices for organic dairy yet, so they could do it their own way and they could set up a different pricing model that worked for the farmers. The conventional farmer cooperatives had only one marketplace, the one for conventional milk, and that was determined by market forces that were driving farmers out of business."[115]

The U.S. Department of Agriculture uses a complex formula to set minimum prices for milk based on four categories of how it will be used (for a beverage, yogurt, hard cheese, or powdered). But the prices fluctuate and many farmers do not know what they will be paid until a month after they ship their milk to a processor. Imagine a teacher doing the same amount of work each month but never knowing what compensation will be paid until well after the services have been rendered.

Siemon recounts, "We had a collective bargaining attitude: Organic must represent economic sustainability for the farmer *and* environmental sustainability for the earth." In order to work for the farmer, CROPP picked a price for milk that stopped driving farmers into bankruptcy. In 1988, the price the cooperative charged its customers was $17.50 per 100 pounds of milk, when the conventional price was $12.50. He says, "We held to this target price for organic milk. If we had too much organic milk for the market, instead of undercutting our price, we sold it as conventional." In an industry where family farmers are at the mercy of pressures largely beyond their control, this was transformational. Dairy farmers, in particular, are constantly pushed to get big or get out of the way. They have to deal with complex pricing systems and trade wars that can disrupt markets without much warning. In this context, creating a cooperative that could deliver better prices for the farmer and better environmental protections was an incredible feat. This formula worked to launch the cooperative and grow it into a major player in organic foods for three decades. CROPP was the first organic foods company with revenues over $1 billion. In 2017, it was paying its farmers nearly double the price conventional farmers received for their milk.[116]

However, U.S. dairy farmers continue to face tremendous financial strains, and at the start of the 2020s even organic dairy farmers and artisanal cheese makers are facing very difficult market forces and government policies, such as those that encourage farmers to expand milk production, which results in excess supply. Continued excess supplies of organic and conventional milk, along with depressed and volatile milk prices, are the current status quo, leading to more farmers going out of business.

In 2018, the U.S. Department of Agriculture reported that 3,000 dairy farms went out of business. Today the larger conventional factory-style farms, which contain their cows in a barn and have mechanized milking operations, are better able to weather the volatility by negotiating better prices for feed or taking advantage of government subsidies and insurance. Even organic dairy farmers are not immune when milk prices fall unexpectedly and they find themselves facing a glut of milk labeled organic that has come from massive factory-styled farms. Groupe Danone's U.S. organic brand, Horizon, is the largest organic dairy brand in the United States. The independent Cornucopia Institute asserts that Danone has been flooding the market and depressing prices, which is like a "death warrant" for small organic farmers who allow cows to graze on pastures.[117]

After twenty years of operating in the black and grossing over $1.1 billion in sales for two years in a row, Organic Valley posted a loss of $10 million after taxes in 2017. In April 2019, Organic Valley posted a $12 million loss in net profit before tax, even though sales had increased 1 percent.[118]

The cooperative now appears to be in a fragile transition period. Three months after its CEO resigned in March 2019, the cooperative laid off about 5 percent of its employees (thirty-nine of 950), and by fall that number had grown to fifty-nine employees. Either through layoffs or resignations, in that period the cooperative lost its general counsel and chief mission officer, who had sixteen years of experience; its mission executive, who had eight years of experience; and its senior manager for sustainability, who had eight years of experience. Such a broad loss in institutional memory for key leadership positions may be mitigated by well-drafted governance documents that have locked in the mission and purpose of the cooperative. It is too soon to know how the cooperative will navigate through this transition.[119]

Environmental Sustainability

This chapter focuses on how the CROPP cooperative forged a strong sustainability leadership position as an industry pathbreaker before this transition period of 2019. To understand that better, I interviewed Jonathan Reinbold, who had been Organic Valley's senior manager for sustainability until the summer of 2019. At the time of the interview he was no longer an employee of the cooperative, so he did not speak for the business.[120]

CROPP leads the way on a wide variety of environmental sustainability practices that are not required by law. These practices are carried out by the farmer-members on their individual farms, internally in the buildings and factories CROPP controls, and with external partners with whom CROPP contracts. For the latter two areas of focus, the employees are the driving force. Siemon observes, "The CROPP employees are much more idealistic and green than the farmer-members, who are more concerned about the marketplace."[121]

Reinbold led a team of six full-time employees dedicated to achieving a sustainability he describes as making social, environmental, and financial goals "mutually reinforcing." Like Siemon, Reinbold is a proponent of ensuring that environmental sustainability goes hand in hand with solid financial performance. For instance, when working to shift the cooperative off fossil fuels, Reinbold wanted it to use its position as an industrial-level consumer of renewables to get the best possible financial returns, thus pushing the renewable energy industry to be more affordable for future smaller customers too.[122]

The employee management and farmer-member board of directors make decisions about environmental sustainability. The general membership of farmers has not been driving the coop's sustainability focus, in Siemon's experience. He thinks the farmers are proud to endorse sustainability, but a small group of farmers have been complaining that the cooperative should not be spending money this way. With that qualifier, CROPP's farmers, as compared to other U.S. farmers, stand out as sustainability pathbreakers. While cooperatives can function in ways that make them hard to distinguish from investor-owned businesses, cooperatives have greater flexibility to be sustainability leaders because they can have multiple goals beyond profit. Siemon thinks that "cooperatives are not as worried about quarterly returns and are not part of the stock market, so they have more flexibility and freedom to consider multiple stakeholders."[123]

FARMER SUSTAINABILITY

CROPP farmers' sustainability starts with organic and leads to perennial pastures, energy conservation, renewable energy on the farms, and growing seeds to power tractors and trucks with biodiesel. CROPP's farmers have been organic from origins in 1988. The Organic Foods Production Act of 1990 directed the U.S. Department of Agriculture to promulgate regulations for organic products. For dairy, "organic" means the farmers "only feed their cows organically grown crops and do not use antibiotics," according to Siemon. He said, "Organic farming has many benefits for soil health and animal health, so we know it must be healthier for humans; and recently there have been a lot of studies showing the harm pesticides cause to humans."[124]

Many of the farmer-members had been raising cows on organic perennial grass pastures since the beginning of the cooperative. Around 1994 the cooperative made grazing on pasture a requirement for its members. While these practices are "very positive" for animal welfare, improving water quality, reducing energy demands, and sequestering carbon, the cooperative was also responding to the market. Siemon explains simply, "Consumers think of cows on grass, so requiring our farmers to all use pasture worked to differentiate them from conventional dairy farmers." Now, Organic Valley has two pools: farms whose cows graze to a certain level and 100 percent pasture-fed.[125]

As a measure to combat climate change, people have become more aware of the importance of perennial grass pastures to capture carbon. Siemon asserts, "The more roots you have in the soil and organic matter, the less runoff you have and the more carbon you sequester. Depositing manure one patty at a time and not in a slurry of liquid manure is much better for the earth."[126]

The U.S. Environmental Protection Agency provides data that tends to support Siemon's points. The Environmental Protection Agency identifies manure management as releasing large amounts of GHGs (nitrous oxide and methane), which accounts for 16 percent of total agriculture-related emissions. Notably, intensive livestock operations, where cows are concentrated factory style and the manure collected in lagoons, produce the most emissions. By contrast, manure deposited on perennial pastures decomposes aerobically and produces very little methane. Similarly, the agency calculates emissions from agricultural soil management, showing that 74 percent of nitrous oxide emissions come from cropland and only 26 percent come from grasslands for grazing. Further, Ranjith P. Udawatta and Shibu Jose's 2012 study identified perennial agriculture, such as grasslands that do not need to be tilled and planted annually, as a method to lock carbon into plant roots and shoots, creating a carbon sink. Tiziano Gomiero and colleagues' 2011 study comparing conventional and organic practices indicates organic farming increases carbon sequestration in soils and reduces fossil fuel energy requirements. Thus, although dairy production can produce a significant amount of GHGs, the farming methods adopted by Organic Valley's farmers are associated with much lower GHGs than conventional, factory-style, concentrated dairy operations.[127]

The cooperative is also helping its farmer-members reduce their energy demands and power their remaining farm needs with renewables. In 2019, it had 225 farmers with solar, which was about 11 percent of its farmers. This is many times the use of solar among the population at large, according to Reinbold. The cooperative was a catalyst by providing direct services to farmers. The cooperative paid for energy audits and renewable-energy site assessments for farmer-members. Then it paid for grant writers to find money to reduce the costs farmers had to pay for the installations. Some farms have installed solar that pays for itself within two to three years due to a combination of grant money, tax credits, and the price of power and net metering in their areas. By contrast, Reinbold was certain their largest organic dairy competitor, Groupe Danone (owner of Horizon, not a coop), is not giving this type of service to their farmers.[128]

The cooperative pioneered research into the use of vegetable oil for tractors. Its concept was to "grow your own fuel," says Siemon. Organic Valley started this program as an experiment to see if it could grow seed crops, press the seeds into oil, and power its fleet vehicles with this biodiesel. It built a mobile biodiesel plant that could be taken to farms to teach its members to grow oil seeds (sunflowers and canola). Ultimately, the cooperative planted up to 250 acres of farmland in these seed oil crops. It pressed the seeds and used the vegetable oil directly in the vehicles or converted it to biodiesel to be used with any diesel equipment. The cooperative created a group buying program for conversion kits for tractors and trucks, and purchased oil presses to pass on savings to members who wanted to be independent of fossil fuels. However, due to the currently low cost of fossil fuel diesel, farmers have not adopted these practices at a high rate. The cooperative, however, scaled up and built two retail fueling stations to dispense biodiesel to its fleets, and this has been a much more cost-effective approach. When management approved these investments, the expected return on investment was seven years, but the actual payback turned out to be much quicker, as the coop saved money compared to conventional petroleum diesel and produced fewer GHGs, recalls Reinbold.[129]

SUSTAINABILITY OF INTERNAL OPERATIONS

When CROPP built its first new building in 2004, it followed LEED standards and put up solar and wind generation, reflects Siemon. In 2017, the cooperative committed to be 100 percent solar-powered, making it the largest food company in the world to lead on the renewable energy transition. This goal relates to powering its buildings and factories, not its supply chain or the operations of its farmer-members; still, it is a pathbreaking goal.[130]

In August 2019 the solar project went live with 32 MW of new generation. This provides 100 percent of the cooperative's electricity needs and generates renewable power for a wide variety of municipal and business partners. While it reached its goal quickly, the project faced several speed bumps. The coop had been trying to figure out how to reach this goal for years. According to Reinbold, the team looked at an internal carbon tax, efficiency investments, and a $2 million annual investment, but none of these would have led to quick results that were solid financially. The investment tax credit for solar is only available to coops in the year of the investment, and CROPP's business did not have that kind of tax appetite. Ultimately, the cooperative decided to partner with third parties to invest in the 13 MW they projected it would need by 2025. Then in 2018, President Trump put tariffs on solar panel imports, and this made the project more expensive. Still, the cooperative was committed to finding a solution that worked financially and environmentally. It had to scale up the project to 32 MW and find additional partners to buy the extra power and renewable energy credits (RECs).

Dr. Bronners, Native Energy (selling RECs to Clif Bar), and the city of Madison, Wisconsin, among others, signed twenty-five-year REC commitments, and the Upper Midwest Municipal Energy Group, a utility, agreed to buy the power directly. Through a dizzying array of agreements and financial arrangements, CROPP was able to make this large-scale renewable power project a reality, with installations in three states.[131]

SUSTAINABILITY OF EXTERNAL RELATIONS/SUPPLY CHAIN

Siemon says despite all its sustainability progress, the cooperative has more to do. It still needs to address the climate impacts of transporting its products and all the external factories who do business for it. Sustainability is an iterative process that at its best sets up a system of ongoing improvements. Time will tell if CROPP will be able to continue its sustainability momentum to lead in these additional areas where it could have a positive impact across its supply chain. To Siemon, "sustainability is not a luxury, but good business; it leads to cutting costs or increasing sales through brand building."[132]

Governance and the Cooperative Structure

CROPP drafted into its mission a commitment to organic products and improved farmer income, which Siemon explained were the two keys of its founding formula for success. CROPP's mission is "to create and operate a marketing cooperative that promotes regional farm diversity and economic stability by means of organic agricultural methods and the sale of certified organic products." Siemon identified the cooperative's governance documents as:

- mission and goals, written in 1988 and modified slightly later;
- articles of incorporation;
- bylaws;
- board policies; and
- farmer pool policies (such as the policies for the dairy pool or the egg pool).[133]

When Siemon was CEO, CROPP wrote foundational principles and lessons learned, and shared these informal as well as formal governance documents with the employees in their training packets. Yet, he "can't say the farmer-members ever embraced the informal governance documents as guiding tools." In reflecting on this now, Siemon realizes that the informal foundational principles and lessons learned "should have been adopted by the board." He says, "I took it for granted that these were shared principles, and I shouldn't have. This is a lesson for other cooperatives to formalize their principles."[134]

With 2,000 farmer members working across the country, CROPP uses a tiered structure for governance that allows for both direct and representative democracy. There is a Cooperative Board, with an Executive Committee of the Board that connects to various regional membership bases. They hire staff to manage the business and sell their products, and although this staff is quite large (close to 1,000 employees in 2019), they are not members of the cooperative or represented on the board. Farmer-members can participate in the cooperative's governance through a wide variety of committees. About 128 members formally participate in governance through the board and committees, the structure of which is shown in Figure 4.[135]

The cooperative has a variety of ways to involve farmer-members, including the annual three-day general assembly; two regional meetings every year;

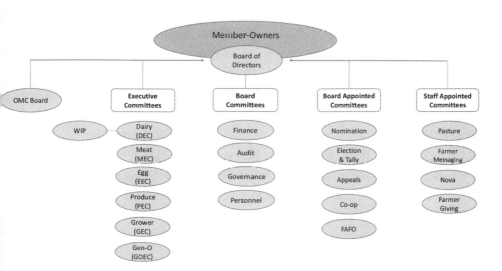

Figure 4. CROPP's Governance Structure (Credit: Courtesy CROPP Cooperative © 2020, all rights reserved. For more information, visit organicvalley.coop or farmers.coop/farmer-owned.)

working in farmer "pools" based on the commodity produced; and serving on committees, as shown in Figure 4. But the primary decision making happens at the seven-member board level, according to Siemon. The general members vote for representatives on the board, and the entire board is composed of farmer-members. Each fall there is a "leadership meeting" of leading members and employees. Siemon observes, "The farmers have a lot of pride in their cooperative, but the big decisions are made by the board."[136]

Siemon noted several tensions in a successful cooperative that is a sustainability leader. First, the board composition of all farmer-members is a weakness because businesses today are now using a multistakeholder approach. He believes the board would be stronger if there was greater diversity. That means, among other things, employee representatives and non-member external professionals who bring a broader stakeholder perspective. Next, he observes, "There is nothing inherent to cooperatives that stops people from being self-centered and want to simply focus on maximizing value for present-day members." This is especially problematic for cooperatives that become financially successful and have not taken steps to protect future members against current members' self-interest. A regular pressure is that farmer-members are worried about the current price they are paid for their products. Valuing current pay can lead to failing to invest in the future, such as in marketing, buildings, management, or renewable energy. Another extreme possibility is that a farmer-member could pressure the board to sell the cooperative to maximize the economic benefits to the current members. Such an argument would be that the board has a fiduciary duty to liquidate the business because the member would make more from the sale of the business than from dairying for the rest of his or her life. Cooperatives need legal structures (either legislation or their governance documents) that prevent this constant pressure to pull value out of the company in a way that privileges current over future members and the cooperative's multiple other stakeholders. The Real Pickles cooperative's governance documents contain an example of how to draft these protections, as described above in Chapter 7.[137]

Summary

An unlikely group of fifty-seven farmers in an isolated area of western Wisconsin founded CROPP cooperative in 1988. Before organic had captured the minds and dollars of U.S. consumers, it pioneered an entirely organic, value-added farmer cooperative to market its products. It got out in front of the curve and was able to establish dairy prices that worked to keep farmers in business rather than waging a losing battle in the push toward conventional mass production. The farmers grew their idea into success-ful national brands such as Organic Valley and became the first entirely organic food company to earn $1 billion in revenues. Organic Valley is one of three national organic fluid milk brands, and the only one the farmers own cooperatively. They committed to a variety of practices both on their farms and at their headquarters and factories that established CROPP as a sustainability pathbreaker. In addition to organic, the cooperative has been committed to raising cows on perennial grasses for twenty-five years, long before it became known as a practice to promote carbon sequestration. It has been a catalyst for its now more than 2,000 farmer-members to adopt sustainable practices, providing them with energy audits, renewable-power site assessments, grant writing for renewables, biodiesel technical support to grow their own fuel, and organic management input. Its farmers have installed solar power at a rate five times that of the American public at large, and its headquarters and factories were entirely powered on renewables as of 2019. Its efforts to adopt a 100 percent renewable electricity goal was the catalyst for a much larger 32 MW solar project that provides renewables well beyond the cooperative to other businesses, cities, and utilities.

The CROPP story provides three important lessons. First, the cooper-ative was successful in attracting early and sustained support because it made people see that organic food was a way to save the family farm. This appealed more broadly to people who were unfamiliar with and skeptical about the health and environmental benefits of organic. Second, when it pursued further sustainability initiatives, such as moving off fossil fuels, it structured the projects to ensure they made sense financially as well as socially and environmentally. These factors needed to be in balance. Third, the cooperative may falter now because it has failed to structure the coopera-tive in ways that protect against members pursuing self-interest over stake-holder interest. Although it has a sizable work force of almost 1,000 people, the workers are not members of the cooperative and lack a governance voice on the board. The cooperative has not included a provision in its governance documents that discourages liquidating the cooperative, so there may be increasing pressure from members to cash out now that the business has become so successful. This would disadvantage future organic farmers who would no longer have the option to participate in the cooperative. Fourth, CROPP demonstrates the value in the private sector leading when faced with government inaction. The founders of CROPP did not see solutions to the 1980s farm crisis coming from the federal government, so they joined with other farmers to find a farm-appropriate solution. This recipe for success worked to provide stable incomes for organic farmers for thirty years while they set the pace for strong sustainability for their industry.

10

Water

THE SPECTER OF CAPE TOWN, a city of four million people in South Africa, running out of water in 2017 and 2018 underscores the fragile situation people around the world will increasingly face in a climate-disrupted world. As they approached day zero, residents' water was restricted to thirteen gallons (49.21 liters) per day, and many had to stand in line for hours to fill up water jugs. In wealthier parts of the city, people dug private wells and filled swimming pools, giving another example of the disparate impact of adapting to climate change. While extreme, Cape Town is not an isolated case. Water-stressed megacities are dispersed across the continents, and include Los Angeles, Mexico City, Rio de Janeiro, London, Beijing, Delhi, Karachi, and Tokyo.

Water is essential for life and is recognized as a human right. But the largely invisible system of buried pipes and infrastructure that delivers it is forgotten when it is easy, clean, and inexpensive to turn on the tap. Even for those with generally reliable water services, however, climate change threatens water security. By increasing flooding on one extreme and drought on the other, systems for supplying water and cleaning sewage before returning it to rivers and lakes are more challenged.

Moreover, water and energy are tied together. The energy footprint of delivering clean water and then treating it as municipal sewage is part of the water–energy nexus. Another part is the amount of water necessary to generate electricity. For the latter, the switch to solar and wind will significantly change the relationship between water and energy demands. Renewables will free up vital water, especially critical in water-stressed cities. In the United States, for instance, the production of electricity from fossil fuels and nuclear energy requires 190 billion gallons of water per day, accounting for 39 percent of all U.S. freshwater withdrawals. This chapter sketches the pathways to sustainable development and deep decarbonization of water services. Then it provides a case study of cooperative exemplars to demonstrate how different aspects of the transition are already under way. The case study includes an overview of the cooperative's history, type, and economic sector; context based on its industry and country; best practices in environmental sustainability; and its governance and cooperative legal structure.[1]

Pathways to Sustainable Development and Deep Decarbonization

Most people would not readily think of water as a priority for deep decarbonization. Yet, pumping, cleaning, and distributing water, and then later treating the wastewater, can be quite energy-intensive. For many water utilities, if they are not powering themselves on renewables produced at their facility, their energy bill is one of their biggest expenses and carries a large carbon footprint. As I have previously calculated, a municipal wastewater treatment facility for a small city of around 70,000 people, with an average flow of ten million gallons per day, uses as much energy as almost 3,000 average U.S. homes. Deeply decarbonizing water services involves a combination of ongoing commitments to energy efficiency improvements, water conservation—including maintenance and upgrades to reduce leaky pipes, and renewable energy to power the operations.[2]

The availability of clean, reliable, and affordable water and sanitation is a Sustainable Development Goal. Water is also fundamental to meeting many of the other Sustainable Development Goals, such as ending poverty and hunger, maintaining good health and well-being, finding decent work, and encouraging economic growth. Further, in recognition that water is essential to the realization of other human rights, in 2010 the United Nations General Assembly articulated a human right to water and sanitation. This followed the United Nations Economic and Social Council's General

Comment No. 15 (2002) stressing that because water is a limited natural resource and a public good fundamental to life and health, the "human right
to water is indispensable for leading a life in human dignity."[3]

For many around the world, the human right to water and sanitation is
an aspiration not experienced in daily life. As I discussed in an earlier article, pressure on the world's water resources is increasing due to a growing
global population, pollution making water unusable, and less reliable supplies due to climate change and its related droughts and floods. The international clean water goal is tied to an indicator of "safely managed drinking
water services," which, according to the World Health Organization, means
drinking water from an improved water source that is "located on premises,
available when needed, and free from fecal and priority chemical contamination." As of 2017, 2.2 billion people lacked such water services, exposing
them to preventable diseases, suffering, and death. An estimated 829,000
people die each year from diarrhea related to unsafe drinking water, sanitation, and hand hygiene.[4]

Even in the United States and European countries where safely managed drinking water service has been widely available for generations, there
are hidden issues. Water infrastructure is literally buried out of sight in
a vast network of pipes, which in the United States often date back to the
nineteenth century and are near or past the end of their useful lives. The
lead contamination crisis in Flint, Michigan, shined a spotlight on how this
aging infrastructure adversely impacts public health. The issues go well
beyond the community in Flint. A 2012 American Water Works Association
report says the U.S. water infrastructure will cost at least $1 trillion over the
next twenty-five years to maintain and update.[5]

A controversial global response to inadequate water infrastructure has
been to privatize water services. When Flint's water supply was found to be
contaminated by lead, people similarly advocated privatization of the water
service. The internationally orchestrated push at the end of the twentieth
century for governments to privatize water services and allow bulk water
delivery (via bottled water, water tankers, and water bags floated across
the ocean) created a highly consolidated industry. After legal reforms in
the 1980s and 1990s, a handful of multinational investor-owned corporations grew increasingly wealthy by privatizing water through a number of

avenues, including the management and control of municipal water distribution systems and taking water out of the public commons—at no cost—and bottling it for sale around the world. By the early 2000s, multinational companies providing water services were a more than trillion dollar per year industry, not including the more than $19 billion-per-year U.S. bottled water industry.[6]

Also, by the early 2000s, a small group of investor-owned corporations controlled much of the international water market. Two French-based corporations, Suez and Vivendi, owned or had controlling interests in water companies in over 130 countries serving more than 100 million people. That hegemony has continued, and currently the top two corporations, based on revenues, remain Suez and Veolia (changed name from Vivendi in 2014). As of 2014, private water corporations were serving more than 280 million people.[7]

In an earlier article, I wrote about the water commons and pushes to privatize water services. A misperception about the commons is that they must be privatized in order to be protected (the so-called tragedy of the commons). The government could manage the commons, but when the government fails to build and maintain a clean, reliable, and affordable water infrastructure, privatization may look like the best option. Yet, privatization of the water commons carries with it the risk of unjustly enriching the few, while depriving the many; it allows the commons to be taken out of the public domain and given to the few who have the ability to enclose and profit off them.

An alternative model for government management of the water commons that has deep legal roots in ancient Roman, English, and U.S. law is management based on principles of the public trust doctrine, whereby the government holds the common water resource and, by extension, water service infrastructure in trust for the public, and regulates in the public interest. For the trust relationship to function well, there must be transparency of the trustees' actions; adequate regulations governing shared access and use of the trust property; and public participation by the trust beneficiaries (the public), including the ability of the civil society to enforce and to call for routine accounting of the trust to ensure the government is managing it in the public interest. Coupling these trust principles with a cooperative enterprise structure, democratically governed by water consumer-members, presents a middle ground to the private-versus-public water service debate that has been waged over the past thirty years. It could mitigate the worst excesses of the purely public and the purely private systems while addressing the climate crisis. Cooperative water services can provide reliable and affordable water services, while deeply decarbonizing this sector. The sustainability pathbreakers featured in this chapter show how they put this theory into practice.[8]

Water Cooperative Pathbreakers

This chapter features case studies of two water cooperatives: Olesana Cooperative, a historic cooperative that has been providing the water supply for a municipality in Spain for generations, and Aigua.Coop, a modern water-supply consulting cooperative providing services around the world.

Olesana Cooperative (Water Supply, Consumer, Spanish Local)

Aigua.Coop (Water-Supply Consulting, Cooperative, International)

Overview of the Cooperative: History, Type, and Economic Sector

This chapter weaves two related water cooperatives into one narrative. One of the cooperatives has a 150-year history and the other is newly formed: the Olesana Cooperative and Aigua.Coop, respectively. I interviewed Joan Arévalo i Vilà, who is the president of both social enterprises.[9]

This story starts in the middle 1800s, when a group of 114 neighbors decided to form a common enterprise to provide clean water to their community in Vila d'Olesa de Montserrat in Catalunya, Spain. In 1868 they founded the Mining Community Olesana to manage the integrated water supply in the municipality, and although its legal status has changed, it has continued to provide uninterrupted water service as the community population has expanded.[10]

The original water-service entity was legally organized as a group of co-owners who were providing water services to themselves. After Spain

changed its laws in 1985 to allow investor-owned corporations to manage
water supplies, the leaders of this historic water enterprise worried their
legal structure was not strong enough to guarantee they could continue to
offer community management of their water. They thought a multinational
would try to come in and obtain the contract to manage the water with
the municipality because they were seeing this happen all around them.
So in 1992, the Mining Community SCCL Olesana (Olesana Cooperative)
changed its legal, corporate, and administrative structure to convert into a
cooperative society of consumers and users. It currently has eleven employees and supplies water to 24,000 Olesans. It promises to offer high-quality
water while practicing values of the "social economy." It converted into a
consumer cooperative to achieve a variety of objectives, including:

- produce high-quality water;
- provide the same water service to people regardless of their socio-
  economic level;
- use a management model in which all the members have the same
  power using the democratic principle; and
- reaffirm that water is a communal and democratic asset.[11]

Since municipal water service is necessarily a monopoly service (in the
sense of a natural monopoly) and cooperatives are voluntary associations of
members, it is curious the two can coexist. Arévalo i Vilà explained that anyone who lives within the municipal water service boundaries is eligible to
join the cooperative, but joining is voluntary, and the cooperative provides
water to nonmember water users as well. While there are two classes of
users (members and nonmembers), 99 percent are cooperative members.
Arévalo i Vilà thinks they have such a high rate of membership because
people in the town understand the importance of water in their daily lives.
Nonmembers are typically short-term water users. The nonmembers pay a
higher price for water and do not participate in the governance of the cooperative. The members pay a one-time fee of €400 to join the cooperative.
This money goes into the "social capital" fund, where it grows over time. If
the member decides to leave the cooperative, they are able to get back their
membership fee and any interest on the money.[12]

According to Arévalo i Vilà, the cooperative model of water service provides better quality and prices to members than a corporate private service. He explained that the Olesana Cooperative's surplus revenues are not
"profits" earmarked for shareholders and directors, but are invested in the
community for better and more efficient service. The cooperative embraces
transparency and gives an overview of its accounts on its website, including
accounting for its social actions. It reports that it has saved its members
more than €1 million compared to water costs in neighboring towns. The
cooperative provides water to its members that is 40 percent less expensive
than in neighboring municipalities managed by investor-owned corporations, and 50 percent less expensive than the metro area of Barcelona that is
managed by a corporation. Barcelona's water service contract with a subsidiary of the French-based Suez has been the subject of heated debate about
whether to return to a publicly managed system.[13]

In addition to managing water, the cooperative is active in supporting
the surrounding community and building a cooperative movement. It uses
social funds to help those struggling with social exclusion and energy poverty, to promote cooperativism through Aigua.Coop, and to participate in
cooperative federations.[14]

The Olesana Cooperative sees water as a vital element of a community's
identity, which must be managed as a common good for the public interest,
democratically controlled by the water users. The cooperative is motivated
by the recognition that water is a vital resource. The Olesana Cooperative
is unique in Spain, as the only water cooperative providing the public municipal water service. Based on its long-standing experience with community water management and its positive benefits, the Olesana Cooperative
wanted to create a new cooperative platform to share its experience and
highlight the substantial differences between the management models to
show their various social and economic implications.

In 2016 it launched Aigua.Coop, a second-degree cooperative (that is, a
cooperative composed of at least two legal entities, of which one is a cooperative), to spark development of new water cooperatives. Aigua.Coop is
designed to serve social economy entrepreneurs and groups that want a
cooperative model for municipal water service. Aigua.Coop is a consulting
entity that advises about how to establish a cooperative to manage municipal

water supplies; and about environmental sustainability, administrative, legal, and technical issues. The European Social Fund, managed by CEPES, provides 50 percent of the support for Aigua.Coop.[15]

Industry Context

Similar to other parts of the European Union and the United States, in 1985, Spain opened up its municipal water systems to private competition. Spanish Law 7/1985 of April 2, Regulating the Basis of the Local Regime, allowed municipalities to choose between public management or private management for their water service. Today more than 80 percent of the Catalan population receives its water supply through a mixed public/private arrangement or a totally private one, offered by investor-owned corporations.[16]

Advocates for private water corporations argue that government water utilities are nonresponsive to customers, undercharge ratepayers for political reasons, and therefore underfund infrastructure. Challengers to private systems contend that investor-owned utilities gain the benefits of massive infrastructure investments previously made by the public. Further, they assert that private corporations charge consumers more for water and deliver lower quality because the corporations need to borrow money at a higher cost and pay investors a profit. For instance, water infrastructure is cheaper when a government uses municipal bonds with an average of 4 percent interest. In the United States, one investigation showed private water corporations charged consumers 59 percent more for water service than local government utilities. While proponents of privatization highlight the positives of competition leading to more efficiencies, detractors counter that the water services industry is highly concentrated, so there are few firms competing for contracts. Thus, the only point of competition occurs at the time of issuing the contract.[17]

Aigua.Coop emphasizes another part of this debate. It is concerned with democratic governance and asserts that neither municipal nor investor-owned management guarantees that the majority of water users own and participate in managing the water service. It further contends that municipal employees are not necessarily motivated to provide efficient service and that there is a conflict of interest when the municipality evaluates service it is providing. On the private end of the spectrum, it alleges that investor-owned management prioritizes service and benefits, but creates a structure that is often disproportionate to justify costs that are not proper to the service or its quality.[18]

Utilizing water cooperatives for municipal water services avoids the worst of government-only and investor-owned corporate water service. Aigua.Coop argues the way to guarantee responsive public management of a service owned by the users is to use joint management, where the city council is the collaborating partner overseeing the cooperative, which manages integrated municipal water service, without seeking a profit. The main features of the cooperative management model are:

- Maintain the municipality as the trustee responsible for guaranteeing a public service as important as water.
- Municipality partners with consumer cooperative and retains public oversight of water management.
- Consumer cooperative manages water as a common good, not a commodity.
- Cooperative is operated as 100 percent social-commitment, non-commercial venture, under the principles and values of cooperativism and the social economy.
- Water management is sustainable and protective of the environment.
- Cooperative plans activities to engage the public around the importance of water, nature, and culture.[19]

A critical distinction with the cooperative model is its support of democratic participation of its members—the water users. Investor-owned entities are not designed to include the participation of water-consumer stakeholders in any meaningful way. The investors are most likely not using the service in which they are investing. This distinction is crucial, as the United Nations stresses that water service should follow good practices to implement the human right to water. Accordingly, human rights standards insist on "active, free and meaningful" participation by civil society in decisions about water and sanitation. Related, the United Nations urges that society should be provided with information so people can monitor contracts, budgets, and other water-service compliance issues. This democratic participation value is directly aligned with the core values of cooperatives and is present in the water cooperative case studies here.[20]

Yet, even international water experts fail to distinguish cooperative water enterprises from investor-owned corporations when discussing the private sector. The United Nations' special rapporteur on the human right to safe drinking water and sanitation produced an in-depth report, *On the Right Track: Good Practices in Realizing the Rights to Water and Sanitation,* in which water service providers were simply referred to as "public or private." This level of generalization misses an important element about the private sector: a social-enterprise water cooperative can be more easily designed to fit the goals and principles inherent in the human right to water than an investor-owned corporation that has a primary obligation to produce a profit for shareholders and lacks a mechanism to democratically engage water users.[21]

After thirty years of experiencing privatization, Aigua.Coop sees an emerging opportunity coming soon because many municipalities that have privatized water service will have to open new tenders or reverse the service to the municipalities to recover the direct management of their water supplies. There is a broader movement across Europe advocating for a return to public ownership and management of water systems. This stems partly from an ideological position that water is so vital it should be controlled by the government, and partly from empirical studies that show privatization has resulted in higher prices and poorer quality.[22]

Aigua.Coop works to raise awareness about the water cooperative model. This is necessary because when a public utility is under pressure to privatize, the cooperative form is often missing from the debate. The absence is not due to a lack of water cooperative examples. Like other public utilities, such as electricity, water cooperatives have a long history in many countries of the world. For instance, Argentina has hundreds of water service cooperatives, many of which have been operating since the 1960s and 1970s: most towns with fewer than 50,000 people rely on cooperatives for their water, while private corporations dominate urban areas. Because many Argentinian cooperative members participated in the establishment and con-

struction of their water systems, they have resisted efforts by private corporations to gain management contracts in these smaller towns. Around fifty
water cooperatives joined together to form the Buenos Aires Water Supply and Sanitation Cooperatives Federation (Spanish acronym FEDECAP).
Shortly after FEDECAP's formation in 2000, the province of Buenos Aires
regained public control of its water supply.[23]

Finland and Denmark have long relied on water cooperatives, and these
are in fact more common than public water services. For hundreds of years,
cooperatives have provided water supply services in small towns throughout Finland. In 2009, there were 902 consumer cooperatives and only
400 municipal utilities providing water supply services in Finland. Similarly, there were 2,575 consumer cooperatives and only 165 municipal utilities providing water supply services in 2001 in Denmark.[24]

Consumer cooperatives also provide water utility services in the United
States, where there are almost 3,300 water cooperatives. Most often they
are found in suburban and rural communities with smaller systems. In
certain parts of the western United States, "water mutuals" are cooperatives
that have dominated the provision of irrigation water for generations. Today
water mutuals serve 85 percent of all farmers and 20 percent of irrigated
acres in the western United States. Further, in 1990, EJ Water Cooperative
launched in the state of Illinois to provide water in an area that saw its economic development limited by lack of a reliable water supply. It has grown
over time, added a wastewater treatment subsidiary cooperative, and currently serves 10,000 members. So, while the Olesana Cooperative may be
one of a kind in Spain, countries ranging from the United States, Finland,
and Denmark to Argentina have used water cooperatives to deliver water
supply services to thousands of communities.[25]

The rise of investor-owned water corporations followed 1980s and later
changes in international and national laws and policies that promoted water
privatization. Similarly, international and national laws and policies could
be amended to promote water cooperatives by encouraging social enterprises to compete for water service contracts and deliver water in a way that
is consistent with deep decarbonization and the Sustainable Development
Goals. For instance, a municipal request for proposal (RFP) could require
applicants to be social enterprises that are not organized for profit, are

member-owned, and are governed according to the democratic principle of one member, one vote. Such an RFP could set clear requirements for environmental sustainability and deep decarbonization of water services. The Olesana Cooperative shows how a cooperative has led in these areas to meet aggressive sustainability goals.

Environmental Sustainability

Arévalo i Vilà makes the case that cooperatives are focused on environmental sustainability. He is concerned about climate change in this century and is leading the Olesana Cooperative to create a water plan to account for climate disruption. He says its plan contains infrastructure projects to diversify its water sources and build more resilience into its water system.[26]

The carbon footprint of water services can be reduced if the water is not wasted before it reaches its ultimate end users. The cooperative's efforts to promote water conservation and reduce demands are multifaceted. One of the big areas where water services waste a lot of water is by using a system of old, leaking pipes that lose water before it is ever delivered to the user. Because the Olesana Cooperative does not need to pay shareholders, it is always reinvesting in its water infrastructure and replacing pipes, asserts Arévalo i Vilà. Public health crises, like that encountered in Flint, Michigan, originate from lead pipes leaching into the water and harming people's health. The Olesana Cooperative has replaced all its lead pipes. It minimizes chlorine use in the water and filters with active carbon.[27]

Another area of environmental progress for the Olesana Cooperative is in convincing members to reduce their water use and be conscious consumers. Twenty years ago, each person used an average of over 200 liters per day (52.8 gallons); by 2019 they almost cut this in half, to 106 liters (28 gallons) per day. This is less than the average per person usage in Spain, which in 2014 was 132 liters per day. As a point of contrast, the average per person water usage in the U.S. in 2019 was estimated to be 303–79 liters (80–100 gallons) per day. The Olesana Cooperative accomplished this reduced usage through a combination of efforts. It is using smart meters so people can see and adjust their water use. It has focused on creating awareness about water conservation, and this became especially important during and after a drought in 2008. It uses conservation water pricing, as does the entirety of Catalunya, where water use is charged in rising price blocks. For instance, after one uses twenty-seven cubic meters of water during a three-month period, they move into a more expensive price block.[28]

The Olesana Cooperative has also been working to reduce its electrical demands associated with delivering clean water. In the past twenty years, it has reduced its electricity use by 37 percent, primarily by making capital investments in more efficient water pumps. Further, it strategically times its electricity usage to obtain the lowest price. There are periods of time in the day with different prices, and the cooperative focuses all of its activities in the lowest price block in order to reduce the cost of water service to its members.[29]

The remaining energy required to run the water system is powered by 100 percent renewable energy, which it purchases through Factor Energia. Thus, it has taken significant steps to reduce the carbon footprint of its water services through conservation and using renewable energy. Last, it has an agreement with its green electricity supplier that allows the cooperative members to also have contracts with this company for personal use. Coop members receive a discount of 13–14 percent on their personal electric bills for using renewable energy.[30]

The Olesana Cooperative is sharing its environmental sustainability knowledge by offering consulting services through Aigua.Coop to other cooperatives. Aigua.Coop provides environmental consulting for:

- water engineering projects related to the capture (surface or ground), treatment, distribution, and sanitation of water;
- decontamination of aquifers;
- sustainable energy/water projects, such as using smart meters and optimizing networks to offer real-time water quality information;
- implementation of environmental impact studies for water infrastructure projects;
- analysis of the life cycle of products and services based on the norms UNE EN ISO 14040 and 14044; and
- calculation of the carbon footprint according to the carbon footprint standardization UNE EN ISO 14064.[31]

Aigua.Coop has received a lot of interest in its consulting. Arévalo i Vilà noted that Cuzco, Peru, may establish water cooperatives soon. Aigua.

Coop collaborates with Engineers Without Borders on projects in Peru and other countries in the Global South. It currently provides technical support to Peruvian indigenous communities, where it is also raising awareness about water as a human right and commons that should not be commercialized.[32]

Arévalo i Vilà says there is a rising interest in these consulting services that is primarily coming from small- to medium-sized communities with populations between 30,000 to 70,000. In 2019, it was working with an even smaller town of 3,000 people that may convert to a water cooperative. He explained that Sant Hilari de Sacalm, in Spain, had a private water contract that just ended. The town's experience with the private water service included a situation where the corporation was selling bottled water from the municipal supplies, while forcing the people who lived there to go without running water for dry periods in the summer. This situation led the community in search of a better water management model with a cooperative water service.[33]

Governance and the Cooperative Structure

COOPERATIVE VALUES AND PRINCIPLES

The Olesana Cooperative's bylaws declare the community's commitment to "harmony" with water. It further states that the cooperative respects environmental sustainability, while advancing the social economy in economic development. The coop pursues this commitment by providing water services and access to better prices for renewable energy. Further, its objective to educate the public manifests in promoting and publishing books and images and supporting scientific research related to the natural environment, water, and the socio-cultural environment of the town in which it operates. The bylaws require that "at least 10%" of the net surplus each year go into the Cooperative Education and Promotion Fund to train members and workers on the Cooperative Principles, entrepreneurship, and economics. The fund also supports relationships between cooperatives; fighting social exclusion; payments to join coop federations; promoting cultural activities; corporate social responsibility; and launching new and growing existing cooperatives by donating money.[34]

The cooperative built in a disincentive to members to try to extract wealth from the cooperative. Even in the event of liquidating the enterprise, nei-

ther this Education Fund nor any surpluses or other obligatory funds can
be seized and distributed among the members. Instead, if they cease to exist, the Education and other Funds will go to a cooperative federation in the
region. Like the model provided by Real Pickles, this structure protects the
rights of future generations of Olesana Cooperative members to the con-
tinuance of the cooperative even when current generations may see a more
immediate financial benefit to them to cash out and sell the enterprise.[35]

Aigua.Coop's bylaws directly reference the Cooperative Values and Principles to make this part of their legal design. Article 2 of the bylaws states
that in accordance with the principle of cooperation, Aigua.Coop is focused
on promoting, training, and disseminating cooperativism as a business
model, as a relationship between people, and as a commitment to the environment. Like the Olesana Cooperative, it actualizes this commitment by
setting up a fund for cooperative education and requiring that at least 10
percent of net surplus annually go into this fund. The money is used to pay
for training members and workers on the Cooperative Principles, entrepreneurship, and economics. The fund further supports building relationships
between cooperatives. Last, it is dedicated to social impact and the "struggle
against social exclusion."[36]

DEMOCRATIC PARTICIPATION

In its bylaws, the Olesana Cooperative declares the community must promote and protect the use of democracy in its operations. The Olesana Cooperative uses a mix of direct and representative democracy in its governance.
The members can participate in the annual General Assembly, where they
review and approve the budget, discuss the quality of water, their services,
the audit of the financial accounts, and the management report. The General Assembly must also approve any modification of the bylaws or internal
regulations, as well as any substantial modification of the economic, social,
and organizational structure of the enterprise. Since there are 11,000 members, they cannot all come together to speak at one assembly. The members
elect representatives to the Governing Council, as is typical of cooperative
governance for this size of entity, for day-to-day management oversight.
The Governing Council is composed of up to eleven members, serving
five-year terms, and any member of the cooperative can be a candidate for
this office.[37]

Members have many other avenues to participate in the Olesana Co-operative, as explained by Arévalo i Vilà. The cooperative holds events for members and communicates with them through WhatsApp groups and the general media. Arévalo i Vilà insists that cooperatives have a duty to explain the advantages of the cooperative system, as it motivates and encourages the participation of members, and makes them more responsible for the management and sustainability of the cooperative.[38]

Aigua.Coop, whose members are other cooperatives, also uses an annual General Assembly of all the members and an elected Governing Council. Each member has one vote in the General Assembly. The types of matters that require a discussion and vote in the General Assembly include approval of the annual management plan; selecting and removing members of the Governing Council and the auditors; modifying any bylaws or internal regulations; new financial contributions from members; and any substantial modification of the economic, social, organizational, or functional structure of the cooperative. The Governing Council is a smaller group of three to six elected officials from the general membership. They serve five-year terms and meet monthly. The members are expected to attend the meetings, and the bylaws establish that an unjustified failure to do so is a violation. Violations are punishable by reprimand, sanction, and expulsion from the cooperative.[39]

Summary

Aigua.Coop describes itself as a small republic of water. As water co-operative consultants, it promotes a water service model that is free from multinational investor-owned utilities. Instead, it advocates for water services controlled by local water users, on a democratic basis. It is motivated by the belief that the water commons belongs to the people. So, it links the concept of water as a social good belonging to the people and held in trust by the government to the choice of business form for the water utility. It concludes the most apt business form is a cooperative water service in partnership with municipalities.[40]

The Olesana Cooperative has been deeply decarbonizing and working toward the Sustainable Development Goals in its day-to-day operations. It meets the needs of the municipality for clean, affordable, and reliable water service, consistent with the Sustainable Development Goal of safely managed drinking water service. Its consumer-members pay 40 percent less for water services than residents of neighboring municipalities managed by investor-owned corporations. It has been able to deliver lower costs while implementing strong sustainability best practices, as follows:

- moved away from fossil fuels by powering water service with 100 percent renewable electricity;
- offered group purchase discount of 13–14 percent on individual electric bills of cooperative members who switch to renewables;
- reduced energy consumption by 37 percent over the past twenty years by using more efficient equipment;
- created a climate change plan to build resilience into water supply;
- reduced water usage per person by almost half (usage lower than national Spanish average) over the past twenty years through smart meters, member education, and conservation water pricing;
- replaced all lead pipes and avoided water loss and leaks through ongoing investments in infrastructure improvements; and
- minimized chlorine by using active carbon filters.

Last, both cooperatives spur cooperative education and the launch of new cooperatives, thus enhancing a supportive cooperative ecosystem. Both allocate 10 percent of net surplus to these efforts. The Olesana Cooperative incorporated environmental protection into its bylaws, which has set the purpose for its day-to-day sustainability leadership. It also includes a dissolution clause that protects against current member self-interest.

11
Finance: Capitalizing the Cooperative Movement

AT THE BEGINNING OF THE 2020s, the world faced the triple challenge of recovering from a major economic crisis related to the global coronavirus pandemic, needing to rapidly reduce GHGs over the decade, and adapting to climate disruptions already occurring. Each challenge will take enormous sums of money, and to date financial institutions have not aligned their investments with these realities. In simplest terms, climate finance—investments devoted to mitigation and adaptation—needs to grow, and fossil fuel financing needs to end. Yet, according to the Climate Policy Initiative, in 2018 climate finance fell by 11 percent from the prior year (from $612 billion to $546 billion). Meanwhile, investments in fossil fuel industries effectively canceled out the benefits of these greener investments.

Then in 2020, a subcommittee of the U.S. Commodity Futures Trading Commission issued a report sounding the alarm that "U.S. financial regulators must recognize that climate change poses serious emerging risks to the U.S. financial system" and that "[r]egulators should recognize that the financial system can itself be a catalyst for investments that accelerate economic resilience and the transition to a net-zero emissions economy." The subcommittee also found that some banks, especially those serving low- to moderate-income communities, are particularly at risk of climate-related shocks, which could disrupt financing for small businesses, farmers, and borrowers.[1]

As a major part of the financial system, banks play an important role in the world's deep decarbonization and work toward Sustainable Development Goals. This chapter sketches the pathways to sustainable development and deep decarbonization of finance, with a focus on banks. Then it provides a case study of a cooperative exemplar to demonstrate how different aspects of the transition are already under way. The case study includes an overview of the cooperative's history, type, and economic sector; context based on its industry and country; best practices in environmental sustainability; and its governance and cooperative legal structure.

Pathways to Sustainable Development and Deep Decarbonization

It is essential to understand the relationship between environmental sustainability and banking. Unlike an organic farmer or a renewable energy company that has a direct and clear impact on the environment, banks' roles are more obscured, yet they play a defining part in the global system that has produced climate disruption. They will need to be a clear leader in climate solutions for the world to keep warming within a safer range.

Bank finance determines whether we take the path of business as usual or the path of the new economy and sustainable development. Banks establish access to capital, with significant implications for the growth and development of institutions, communities, and cultures. Banks directly impact environmental sustainability when they provide capital for businesses and homeowners. Even after 197 nations signed the Paris Agreement in 2016, banks actively undermined it by increasing the amount of money available for fossil fuels. In data from 2016–19, the *Banking on Climate Change* report shows thirty-five investor-owned banks funded $2.7 trillion in fossil fuels in the four-year period. The World Resources Institute reports that the U.S. bank with the best ratio of investments in renewables versus fossil fuels is the Bank of America; yet even the best bank has an annualized sustainable finance commitment of $27.27 billion, compared to $36.56 billion in fossil fuel finance. Whether banks invest in fossil fuels or renewables, provide low-interest loans for energy efficiency improvements, or finance green bonds for climate mitigation, banks strongly influence environmental outcomes globally.[2]

Moreover, one of the limits on growing cooperatives is access to sufficient amounts of capital to scale up their business enterprises. Because they do not place investor returns above social mission, they often cannot attract investors. If low-income people are trying to start a new worker cooperative, they may not have enough credit worthiness for a bank loan. Without access to significant funds from investors or commercial banks, values-based stakeholder banks geared toward capitalizing cooperatives are key to growing these enterprises.

Banking Cooperative Pathbreaker

This chapter features a case study of Caixa Popular, a Spanish stakeholder bank owned by its employees and other cooperatives. The bank has a values-based purpose, implements strong environmental sustainability, did not engage in the speculative lending practices that resulted in the global financial crisis, did not foreclose on any homeowners during the crisis, has total transparency in salaries, and has a CEO who is paid only four times more than the lowest-paid employee.

Caixa Popular (Bank, Hybrid Second-Degree Cooperative and Worker, Spanish Regional)

Overview of the Cooperative: History, Type, and Economic Sector

Imagine a stakeholder bank owned by its employees and by other cooperatives; has a values-based purpose; reduces waste, operates on renewable power, and has no loans to fossil fuel companies; did not engage in the speculative lending practices that resulted in the global financial crisis; did not foreclose on any homeowners during the crisis; has total transparency in salaries; and whose CEO is paid only four times more than the lowest-paid employee. Let me introduce you to Caixa Popular.

Founded in 1977, Caixa Popular is a cooperative bank, with worker and other cooperatives as members, headquartered in València, Spain. By 2019 it had grown to have seventy-two offices throughout the Valencian Community and was the largest cooperative bank in the community.[3]

Eight people and four founding cooperatives developed the idea of creating Caixa Popular after a visit to the Basque Community, where they had learned about Mondragón. Inspired, they brought back these concepts to build a cooperative movement in València. In the early 1970s, leaders in the Valencian cooperative movement first created cooperatives like Consum (grocery store) and Florida (agriculture). The capital structure and incentives for investing in cooperatives present more challenges than a traditional business. The early Valencian cooperators realized they needed a bank to capitalize their cooperatives and fuel the growth of new cooperatives for a society-wide movement to be possible.[4]

Similarly, in Mondragón's early years, it established a cooperative bank to serve this purpose. However, the Valencian movement diverged from the Mondragón model. Unlike Mondragón, which is an umbrella cooperative that connects all its related businesses, the Valencian cooperatives work together and with cooperative federations, but maintain their individual autonomy.[5]

One of those early Valencian cooperative pioneers, Josep María Soriano Bessó, the founder and former president of Caixa Popular, connected me to Francesc "Paco Alós" Alabajos, the bank's director of social responsibility, whom I interviewed for this research.

Today, Caixa Popular is owned by its workers and by 150 Valencian cooperatives. In the latter sense, the bank is a second-degree cooperative, formed when cooperatives join together or federate to undertake common functions. In this way, they utilize capital (minimum investment of €3,000) from cooperatives in different economic sectors, such as consumer, furniture, textiles, services, education, housing, construction, metals, and agriculture. In turn, they offer special benefits to cooperative members, such as a high interest rate on deposits; support for cooperative training; an annual external audit; and annual membership fees for the Valencian Federation of Worker Cooperatives (Spanish acronym FEVECTA).[6]

In 2019, Caixa Popular had 350 employees, of whom 326 were worker-owners. Their articles of association establish that employees can become member-owners after a probationary period of nine months to one year, which then entitles them to participate in the "economic and social" activity of the bank, receive part of the distributable annual surplus, vote in the general assemblies, and be elected to social positions.[7]

Caixa Popular is a bank with a strong purpose, which it has imbedded in its legal formation documents and daily practices. It seeks to improve Valencian society through its financial activity and collaborations. It uses its surplus to fund educational, social, and cultural activities, sports, and festivals to enrich Valencian life. Yet, more fundamentally, Alós explained, the bank's mission is to transform Valencian society through cooperative values. It sees supporting employee-owned cooperatives and their lending and donating practices as activities aimed toward that overarching aspiration.[8]

The bank's articles of association establish its "social purpose" as receiving funds from people in the form of deposits and so on, and then using these funds for loans to meet the financial needs of its members. While it may have transactions with nonmembers, as limited by law, it will give preferential attention to the financial needs of its members. It will limit transactions with nonmember third parties to less than 50 percent of its total resources. Thus, it has financed, advised, and helped grow other cooperatives as a primary social purpose. But it also serves individuals, self-employed people, and noncooperative businesses.[9]

Caixa Popular annually allocates 10 percent of its net profits to the Cooperative Training and Promotion Fund, and its member cooperatives determine how to spend this. Further, the bank collaborates with FEVECTA to support the creation of new worker cooperatives and encourage cooperation among cooperatives. Caixa Popular also collaborates with the Confederation of Cooperatives of the Valencian Community, which is the representative of the Valencian cooperative sector more broadly.[10]

The ecosystem of support for worker cooperatives provides fertile conditions for growth. The Valencian Community now has a thriving worker cooperative sector. In 2016, there were 2,037 worker cooperatives with 21,378 worker-owners. To put that in perspective, the entire United States has approximately 350 worker cooperatives with over 4,500 worker-owners.[11]

Industry Context

Caixa Popular recorded profits of €12 million in 2018, growth of 20 percent, and growth of 7 percent in deposits over the prior year. With almost

200,000 customers, it is the largest cooperative bank in the Valencian Community and a medium-sized cooperative bank in all of Spain. Throughout
Europe, cooperative banks are quite common. The European Association of
Cooperative Banks asserts that in 2011 there were 4,000 cooperative banks
serving more than 181 million customers, representing 50 million members,
employing 777,500 people, and with an average market share of 20 percent.[12]

In contrast to commercial banks that are organized to maximize returns
to their shareholders, cooperative banks are stakeholder institutions. They
were originally created to serve people and places that lacked commercially
provided financial services. There are four types of stakeholder financial
institutions: cooperative banks, credit unions, community development
finance institutions, and savings banks. These are also known as values-
based banks. In addition to serving its stakeholder members (workers and
cooperatives), Caixa Popular is conscious of additional stakeholder groups
as it carries out its business operations: clients, public administration, suppliers, professional associations, lobbies, public opinion, the environment,
and Valencian society.[13]

The existence of a cooperative bank with a mission to serve as a force to
capitalize cooperatives is a key element in building the cooperative sector.
According to Melissa Hoover, executive director of the Democracy at Work
Institute, it is difficult for cooperatives to access capital in the United States
because "we don't have the kind of values-aligned capital that understands
how to finance conversions, understands the risks, or understands the up-
sides of cooperatives."[14]

Caixa Popular has changed that dynamic in the Valencian Community,
which is now home to a thriving and growing system of worker-owned and
other cooperatives. The U.S. corollary to Caixa Popular is Shared Capital
Cooperative, which is similarly organized as a second-degree cooperative
made up of 237 cooperative members in thirty-five states. Shared Capital
is a national loan fund and federally certified Community Development
Financial Institution, based in Minnesota, with a mission to build a "just,
equitable and democratic economy" by investing in cooperatives, especially
those in low-income communities. Shared Capital, unlike Caixa Popular, is
not also a worker-owned cooperative.[15]

With a forty-year history, Caixa Popular has survived when other banks
failed during economic crises. After the global economic crisis, Spain

restructured its banking system. Since 2011 it has reduced the number of banks, branches, and employees, and has transformed savings banks (nonprofit organizations) into commercial banks (joint stock firms). Cooperative banks fared better than other banks in Spain during and after the crisis. Cooperative banks usually have a high level of capitalization, stable incomes, and diversified credit portfolios. Their credit ratings are at the high end, between AA and AAA. However, some of Spain's cooperative banks also closed during this time, with seventy-four in 2011 going down to sixty-eight in 2012.[16]

The Co-op Alliance has Guidance for cooperatives on how to implement sustainable development as part of Cooperative Principle 7 (concern for the community). The Guidance points to the global financial crisis of 2007–2008 as proof that cooperatives are resilient and sustain local communities when put to the test. It observes that because cooperatives' economic activities focus on meeting their members' needs, they are less inclined to engage in financial speculation. According to the Co-op Alliance, "member-control and deep local roots" lend themselves to cooperatives avoiding the "excesses that can take place in investor-owned businesses, the systemic nature of which is to seek to obtain the greatest possible profit for investors often at the expense of the community in which the business operates."[17]

That distinction holds true for Caixa Popular. There are several differences between its operations, which are influenced by cooperative values, and those of commercial banks. A remarkable example that Alós was proud to recount is that during the global economic crisis, they did not lay off any employees, and in fact have been adding employees and growing.[18]

One may wonder if an employee-owned cooperative would tilt in favor of employee stakeholders when deciding whether to spend money on environmental sustainability. Discussing this with Alós provided another strong distinction from commercial banks. He explained they do not have conflicts like this because there is total transparency, and levels and criteria for employee salaries are established according to their responsibilities. Caixa Popular has a strict salary policy, based on the criteria of solidarity, responsibility, and work performance. Fundamental to its values, it has written into its bylaws a requirement that remuneration be based on the pay scale. Alós ties this clarity and transparency to removing tension between paying salaries and investing in projects that would protect the environment.[19]

Its salary policies have also resulted in successes in two additional areas that distinguish it from large commercial banks: the gender pay gap and the CEO–worker pay gap. Through its salary policies, Caixa Popular successfully eliminated problems with disparities in pay between men and women in the same level of remuneration. However, Caixa Popular continues to show disparities in women holding management positions. Women make up 60 percent of the workforce, but only 35 percent of the management positions, so it is still working on correcting this imbalance. By comparison, in 2018, major U.S. commercial banks Wells Fargo and Citi self-reported that, on average, they paid women less than men in similar positions. This disparity persists at Wells Fargo and Citi despite laws requiring equal pay.[20]

CEO-to-worker pay ratios in the banking industry can be quite extreme. However, at Caixa Popular the highest-paid employee receives a salary only four times higher than that of the lowest-paid. The result is that the lowest salaries are higher than the market norm, but the highest salaries are lower than the market. By contrast, in the United States' ten largest banks, the average pay gap was 265 to 1 in 2017. Looking at the smaller U.S. banks, their average pay gap was 154 to 1 in 2017. This gap persists in U.S. banks despite pay reforms in the 2010 Dodd-Frank Wall Street Reform and Consumer Protection Act. Legal requirements can be weak medicine to cure structural problems emanating from the core purpose and values of a business.[21]

Some experts tie high levels of bank CEO pay and wage gaps to encouraging risky speculative practices, such as those that led to the global financial crisis and banks that foreclosed on the homes of millions of people. Incredibly, Caixa Popular had no mortgage foreclosures during the global economic crisis. Why? Alós says they "refused to speculate in their home loans." When other banks were doing the opposite, especially with subprime lending, Caixa Popular followed a strict policy to not loan more money than the cost of the house and would not allow a loan repayment amount that was more than 30 percent of a person's monthly income. As a result, it didn't grow as fast as the other banks that were speculating leading up to the housing bubble, but it didn't crash as hard when the bubble burst, and, importantly, their customers did not lose their homes. Imagine how this would have changed the course of history and averted the global financial crisis had more banks been organized and managed like the cooperative Caixa Popular.[22]

Environmental Sustainability

The relationship between environmental sustainability and banking is essential to understand. Of course, banks can pursue internal sustainability measures typical of most businesses, such as powering their offices on renewable energy, recycling, and encouraging less-polluting commutes for their employees. Or they can set aside funds to donate to support environmental causes. However, banks have a much more direct role in environmental sustainability because they provide capital for businesses and homeowners. Bank decisions to fund fossil fuels or renewables, and more broadly to assess the environmental impacts of their financing decisions, have ripple effects across the global economy. Turn off the financing spigot for fossil fuels and turn it on for renewable energy, and corporations will stop developing climate-disrupting fuels and capitalize the renewable transition. For instance, banks could use a carbon footprint of businesses and the impacts of their products or services as a contributing factor in loan approval or savings rate. The Equator Principles is another approach some large banks have endorsed to assess and manage environmental and social risks of projects.[23]

Caixa Popular gets at sustainability directly and ties it back to the Cooperative Values and Principles. The bank explains the Cooperative Principles and describes Principle 7 (concern for the community) as: "The cooperative works for the sustainable development of its community through policies accepted by the people that form it." It further declares it is "committed to the environment" and developing "attitudes and activities" for a sustainable planet. Caixa Popular implements this commitment to the sustainability principle in a variety of ways. Alós describes its internal practices as covering everything from waste to energy. It is reducing and trying to eliminate the use of paper; recycling; conserving energy; and making all the offices more energy-efficient whenever it renovates. It has arrangements with local renewable energy cooperatives and with the national Som Energia (featured in Chapter 8) to purchase renewable electricity to power its operations. With a goal of 100 percent, it is incrementally adding more renewables every year. Currently, 30 percent of its offices are running on renewable energy. As the offices' long-term energy contracts end, it is switching to renewable energy contracts.[24]

Additionally, Alós explained that Caixa Popular will start measuring CO_2 emissions related to all its operations because its stakeholders have asked for this. This underscores the role its stakeholders play in influencing sustainability behavior in the absence of legal requirements. As one example, the UK National Centre for Business and Sustainability for Cooperatives provided a case study of a cooperative bank that measured CO_2 emissions. The bank measured emissions based on energy consumption and then calculated emissions per customer. The first year of measurement set the baseline. It then set a goal of reducing per customer emissions by 70 percent by 2001, which it met. The case study highlighted that cooperative banks can reduce emissions by improving energy efficiency and switching to renewable energy; and can offset emissions through planting trees.[25]

Caixa Popular supports a variety of nonprofit projects, including those that protect natural resources and reduce GHGs. For example, it has partnered with the Sant Jordi Foundation to create the Calderona Environmental Center in Sierra Calderona. This will be self-sufficient, with water from an onsite well and energy from onsite renewables. The center uses environmental education programs to connect children to nature. Another example is a project with Samaruc Digital, which provides photos and descriptions of natural areas in the Valencian Community to promote public understanding and appreciation of these landscapes.[26]

Unlike large commercial banks, Caixa Popular does not work with any fossil fuel companies and is instead loaning money to renewable energy companies. In this way, one could hail them as a climate champion; however, as a regional bank they play a small role so we need to move beyond the case study to better understand the dynamics between banks and climate change.[27]

Globally, banks are providing more financing to fossil fuels than renewable energy, a situation that is taking the world in exactly the wrong direction on GHGs. Thus, a major system influencer in the transition off fossil fuels is obtaining enough financing that renewables can replace fossil fuels. Do cooperative banks have a better foundation from which to advance environmental sustainability, if they choose to do so? Answering this is complex. Let's compare Spain's largest investor-owned bank to the world's largest cooperative bank to assess their roles financing fossil fuels versus renewable energy.[28]

The biggest cooperative bank in the world, France's Crédit Agricole, had net banking income of €3.6 billion in 2017. The biggest commercial bank in Spain, Banco Santander, had net banking income of €8.96 billion in 2017. In 2019, Rainforest Action Network identified both Crédit Agricole and Banco Santander on its list of the top banks financing fossil fuels, giving both a C on their bank policies to stop financing the expansion of fossil fuels and commit to phasing out fossil fuel financing. Crédit Agricole had $32 billion and Santander had almost $15 billion of financing in fossil fuels from 2016 to 2018. Based on that metric, the cooperative bank is not looking very green. A year later, Rainforest Action Network reported that Crédit Agricole has the strongest overall fossil fuel policy of all the banks they analyzed.[29]

Crédit Agricole is France's leading bank financing renewable energy. In 2016, it financed 27 GW of renewable energy (enough power for twelve million French households) and in 2017 committed €100 billion in new green financing by 2020. While it has not divested from fossil fuels, that fossil fuel financing is significantly smaller, measured in euros, than its quickly growing green-economy financing. In June 2019 it announced new policy commitments that earned it the top ranking by Rainforest Action Network in 2020. It will no longer finance companies developing or planning to develop any new coal infrastructure and pledged to phase out coal from its portfolios in the European Union by 2030. Further, it will not finance any companies involved in Arctic drilling.[30]

Turning to the commercial banking example, confusingly, a bank can be considered an environmental sustainability leader as judged by the Dow Jones Sustainability Index and *Newsweek*, and still have invested almost $15 billion in fossil fuel companies from 2016 to 2018. The Dow Jones Sustainability Index gave Banco Santander the best possible score (100) for internal environmental management and *Newsweek* Green Ranking listed it as the best bank based on environmental management. Banco Santander publishes an annual sustainability report. One of the purposes of such reporting is to identify and disclose risks of a nonfinancial nature so investors can make informed decisions. In the report, Santander appears to be a corporate leader in addressing climate change, and according to its self-description, it "supports the recommendations of the *Task Force on Climate-related Financial Disclosure* of the *Financial Stability Board*" and hopes transparency will "raise awareness about both the financial risks, and the opportunities, that climate change brings." Further, Banco Santander reports that 100 percent of the electricity it uses in Spain comes from renewables, but does not disclose the percentage used to power its global operations. It also highlighted financing 3.39 MW of renewable energy in 2017 and being the lead lender for renewables in the United States.[31]

Neither Santander nor Crédit Agricole mentions fossil fuel financing in its sustainability reports. This disconnect shows a significant weakness in sustainability reporting. Crédit Agricole, however, at least provides financing information for renewables, while Santander does not even disclose this amount in its sustainability report, obscuring an assessment of its role in supporting these industries. On these limited measures of transparency, the cooperative bank fares better. Crédit Agricole's financing of 27 GW of renewable energy in France in 2016 is orders of magnitude beyond Santander's 3.39 MW, so on this metric the cooperative bank is doing more to finance the critical transition to renewable energy. This points in a direction favoring cooperative banks, but clearly requires a more rigorous data-intensive study to better understand whether different financial actor types (cooperative versus commercial banks) show different investment patterns in fossil fuels and renewable energy.[32]

Governance and the Cooperative Structure

Caixa Popular explains the Cooperative Principles that mark a difference between it and investor-owned banks. It highlights that as a worker cooperative, it aims to provide its members with jobs through which they provide banking services to others. It asserts that "people who work in a cooperative have an ethical commitment to honesty, transparency, social responsibility and concern for others." Its purpose is infused with a focus on collaborating for a common benefit: this union of wills "is the cornerstone to ensure a society based on the well-being of citizens."[33]

According to Caixa Popular's governance documents, Article 2 mandates that the structure and operation of the bank follow the Co-op Alliance's Principles. Cooperatives cannot be members of Caixa Popular if they operate "contrary to the cooperative principles" or social purpose. Although being profitable is a goal, that is in balance with its social purpose to promote employment, social equity, and equality. It describes the differences in this business model as follows:

- It is a stable employment formula: the partners come together to meet their work needs in the best possible conditions.
- People and the value of work are above the capital contributed. The cooperative is a democratic organization where decisions are made in an equal manner.
- The workers are at the same time owners and managers of the cooperative. That is why there is greater motivation and identification with the company and its future.
- It is an expanding business formula that offers great opportunities for entrepreneurs.[34]

On a global level, Caixa Popular's commitment to triple-bottom-line banking is shared by members of the Global Alliance for Banking on Values. Founded in 2009, its member banks (cooperatives, credit unions, and other community development banks) are motivated to finance new business models that have people, planet, and prosperity in balance. Its research of financial results from 2010 to 2017 compares value-based banks to other large banks throughout the world (such as Barclays). It found that values-based banking consistently delivers better financial returns.[35]

Caixa Popular's governance model is based on democracy and cooperation, and it is governed by its employees and cooperative members. By its governance documents, each employee and cooperative member has one vote. Members have the right to participate in the assemblies and committees and run for elected positions. Alós reports that, informally, the bank has a very fluid and nonhierarchical structure where employees can easily talk with management regularly. The general director holds regular breakfasts with the employees to also facilitate communication.[36]

Summary

The largest cooperative bank in València, Caixa Popular is a worker-owned and cooperative-owned bank that serves its stakeholder interests.

With a mission to infuse Valencian society with cooperative values, it has focused on providing financial services to build worker cooperatives and support families. In addition to offering favorable rates to cooperative members, Caixa Popular contributes to a supportive ecosystem for cooperatives by paying membership dues in the worker-cooperative federation, providing annual audits for its member cooperatives, and setting aside 10 percent of net surplus for cooperative training and education.

A hallmark of this cooperative bank is an employee compensation ratio of the highest-paid employee making only four times the lowest, eliminating a gender gap in salaries for the same positions, and allowing total transparency in salaries. Caixa Popular refuses to engage in speculative lending, and did not need to foreclose on a single home through the housing and global financial crisis. It has drafted the Cooperative Principles into its governance documents and carries out Principle 7 in support of environmental sustainability by reducing the environmental impact of its offices through waste management, energy conservation, and a goal to meet 100 percent of its electrical demands with renewables. It donates to environmental programs. It has no fossil fuel projects in its portfolio of financing, and instead finances renewable energy development. In these ways, it is implementing sustainability broadly defined and capitalizing the cooperative movement in the Valencian Community of Spain.

12

Trade

IN THE POST–WORLD WAR II ERA, international leaders emphasized trade as a way to build prosperity and promote peaceful relations. Since 1947, the GATT (General Agreement on Tariffs and Trade) has established the framework for trading goods. In 1995, the World Trade Organization (WTO) replaced the GATT and expanded trade from goods to services and intellectual property among its 164 members (representing over 98 percent of international trade). Subsequent free-trade agreements reduced barriers; international trading and global supply chains have expanded dramatically and come to dominate the global economy.

International trade brings positives, such as greater variety of markets, lower prices, opportunities for both importing and exporting countries, and global relationships between producers and consumers. However, the international trade agreements that have facilitated this were not primarily concerned with protecting the environment or broadly sharing wealth generated from trade. As trading escalated after the formation of the WTO, so too did unsustainable use of the climate, fresh water, forests, and farmland, as well as the gap between the ultrawealthy and the rest of the world. Further, addressing climate change through new laws and taxes may be limited by existing trading rules. The WTO cautions countries to take into account the trade impact of climate mitigation measures. It is unclear what the WTO means, but it is certainly not a directive to mitigate GHGs consistent with the IPCC recommendations.[1]

Part of the complexity of addressing the benefits and harms of international trade is manifest in the repercussions of U.S. president Donald Trump's four-year tariffs on imported solar panels. Starting in 2018 with a 30 percent tariff, and dropping 5 percent each year, the U.S. Solar Industries Association asserted the restrictions on trade will result in 62,000 fewer jobs and 10.5 GW of solar energy that will not be installed. The case study above about CROPP Cooperative/Organic Valley explains how this tariff impacted their solar installation. While it was initially a major setback, the cooperative creatively redesigned the project with more partners and ultimately installed greater renewable generation capacity. That story should not be misunderstood as favoring the solar tariff, but as showing the resilience of sustainability pathbreakers and the strength of the business case for solar, even in the face of policy obstacles.[2]

Recognizing the complexity of the topic, this chapter sketches the pathways to sustainable development and deep decarbonization of trade. Then it provides a case study of two interrelated cooperative exemplars to demonstrate how different aspects of the transition are already under way. The case study includes an overview of the cooperatives' history, type, and economic sector; context based on its industry and country; best practices in environmental sustainability; and its governance and cooperative legal structure.

Pathways to Sustainable Development and Deep Decarbonization

Trade is included in the sectors addressed in this book due to a combination of the climate impact from transportation-related emissions and the relationship between trade and the Sustainable Development Goals. Achieving the broader Sustainable Development Goals of alleviating poverty and inequality and cultivating decent work and economic growth, among others, requires a trading policy geared toward sustainable economic development. SDG 13, to take urgent action to combat climate change, means United Nations member countries need to address the links between trade and climate disruption. The UN Conference on Trade and Development wants trade to be part of the solution to deep decarbonization and urges countries to see the enormous cooperation potential to share benefits among trading partners.[3]

To date, however, climate action and trade appear not to have followed a coordinated policy approach. To take one example, trade relies on the shipping of goods between suppliers and customers. Under the Paris Agreement, which followed the Kyoto Protocol, global GHG emissions accounting does not attribute any emissions from the international shipment of goods to any country. In essence, these international transportation-related emissions are phantom emissions for which no state or party is accountable. If shipping were a country, in 2015 it would have been the sixth-largest emitter of GHGs in the world, between Japan and Germany.[4]

Further, international transportation-related GHG emissions grew by 90 percent (international aviation bunkers) and 64 percent (international marine bunkers) between 1990 and 2013. Experts project world trade will continue to increase, and the International Maritime Organization expects emissions from international shipping to increase 50–250 percent by 2050. Thus, these trajectories are headed in exactly the wrong direction to meet climate goals to deeply reduce GHGs.[5]

Deep decarbonization and meeting the Sustainable Development Goals require a rethinking of local production and the necessity of trade. International shipping will need new, yet-to-be-commercialized technology to switch from using heavy fuel oil to something that is carbon-neutral. The world's largest shipping company, the Danish conglomerate Maersk, is not waiting for government action; it has set a goal to be carbon-neutral by 2050. Hydrogen fuel cells, if made with water, hold promise, but it remains unclear how the company will reach the goal. In a deeply decarbonized world, while there will still be international trade using carbon-neutral shipping, trade will likely be at a lower level and countries will emphasize local production. If more places are effective at building resilient and diverse local food systems that have a much smaller carbon footprint, for instance, there will be less need for the volume of food trade that currently exists.

The trading that occurs should be aligned with meeting climate goals. There are a wide variety of policy ideas to affiliate trade with climate objectives, an in-depth discussion of which is beyond the scope of this book. However, at the core of the issue is that the international approach to trade agreements has been out of sync with the climate imperative. Current trade agreements are seen as potential obstacles to achieving deep decarbonization. New agreements will, at a minimum, need to be in harmony with meeting climate goals and allow governments to quickly move off fossil fuels without the threat of being stymied in a trade tribunal. New agreements will also need to address, reduce, and mitigate GHG emissions related to shipping. Finally, new trade agreements should recognize whether and how they advance the Sustainable Development Goals so trade will yield multiple co-benefits.[6]

The European Union has staked out a leadership position on these issues. In December 2019, in advance of the climate negotiations in Madrid, the European Parliament approved a resolution declaring a climate and environmental emergency, and the European Commission released a European Green Deal. The parliament's resolution stated that all legislation is to align with the climate imperative of limiting warming to 1.5°C. The members of the European Parliament urge all countries to include emissions from international shipping and aviation in their national contribution plans under the Paris Agreement. The European Green Deal, among other things, calls for more effective carbon pricing paired with a carbon border tax adjustment to make sure the price of imports reflects their carbon content. The Green Deal also notes that the European Union's most recent trade agreements include binding commitments to ratify and implement the Paris Agreement; that all chemicals, materials, food and other products imported into Europe "must fully comply with relevant EU regulations and standards"; and that this facilitates enhancing environmental protection and climate mitigation in the exporting countries. As one of the largest markets in the world, the European Union's climate leadership could influence many private and public entities across the world.[7]

Private Governance of Fair Trade

Trade should be conducted in ways that support countries meeting the Sustainable Development Goals. Not willing to wait for reform of the existing trade and investment regimes, about a decade after the formation of the WTO, advocates developed a system of private governance around concepts of fair trade. In 2009, the World Fair Trade Organization adopted a Charter of Ten Principles of Fair Trade, which is now seen as an international codification of the fair trade movement's principles. Cooperative Coffees, one of the exemplars introduced below, says it adheres closely to these principles, which are as follows:

Principle 1: *Creating Opportunities for Economically Disadvantaged Producers.* Poverty reduction by making producers economically independent.

Principle 2: *Transparency and Accountability.* Involving producers in important decision making.

Principle 3: *Fair Trading Practices.* Trading fairly with concern for the social, economic and environmental well-being of producers.

Principle 4: *Payment of a Fair Price.* Paying producers a fixed price by mutual agreement, ensuring socially acceptable wages depending on the location.

Principle 5: *Ensuring No Child Labour and Forced Labour.* Adhering to the United Nations (UN) Convention on children's rights.

Principle 6: *Commitment to Non-Discrimination, Gender Equity and Women's Economic Empowerment and Freedom of Association.* Respecting the trade union rights and rejecting discrimination based on gender, religion or ethnicity.

Principle 7: *Ensuring Good Working Conditions.* Providing a safe and healthy working environment for producers and workers in line with the International Labour Organization (ILO) conventions.

Principle 8: *Providing Capacity Building.* Seeking to develop the skills of producers and workers so they can continue to grow and prosper.

Principle 9: *Promoting Fair Trade.* Raising awareness for the need of greater justice in world trade by trading fairly with poor communities.

Principle 10: *Respect for the Environment.* Caring for the environment by maximising use of sustainable energy and raw materials while minimising waste and pollution.[8]

The World Fair Trade Organization certifies enterprises based on the ten principles. It says fair trade is a different business model that challenges the conventional way of putting shareholders ahead of all others. It emphasizes the importance of the structure of business: ownership, profit strategy, management, relationship with stakeholders, and core purpose. And it asserts that the "new economy" is defined by certified Fair Trade enterprises, circularity, upcycling and recycling, and organic farming.[9]

The purpose of fair trade is to address the imbalance of power in trading relationships where farmer/producers are not paid well for their work and face unstable markets and poverty. In the late 1990s, national fair trade organizations banded together to create the Fair Trade Labeling Organizations International, which provided certification in compliance with international standards. However, individual countries may have specific fair trade labels for products sold within their borders, which has made navigating the labels and certifications more confusing for consumers. For instance, U.S. consumers see four competing fair trade labels, each certified by different third parties based on divergent standards: Fair Trade USA, Fairtrade America, Fair for Life, and the Small Producer Symbol. Two of these certifications (Fair Trade USA and Fairtrade America) require no minimum level of fair trade purchases in order to use their seal, and the other two require only 10 percent and 5 percent, respectively, of total volume. The push to make fair trade mainstream resulted in conventional brands such as Dole, Nestlé, Starbucks, and McDonald's obtaining certification. While this has expanded the scale of Fair Trade certifications, some argue that it has undermined the brand as a useful distinguisher for consumers.[10]

In the cooperative exemplar story that follows, fair trade as a concept (lower-case) and as a certification (capitalized) will be discussed. The story also provides a perspective from which to think about how cooperatives can advance and build strong trading relationships that are ethical, transparent, and environmentally sustainable.

The Co-op Alliance's Principle 7 understands the concept of community in interconnected tiers. Cooperators' commitment to local development is at the center of the concept, but it is not the entirety of the meaning. Local commitment is not at the expense of and isolation from global development. According to the Co-op Alliance's Guidance for implementing Principle 7, cooperatives should:

- use "ethical" supply chain contracts,
- engage in Fair Trade,
- promptly pay suppliers,
- trade with and support other cooperatives,

- use green consumerism, organic agriculture, and renewable energy, and
- advocate for governments to implement environmental policies and initiatives.[11]

This melding of intense commitment to the local and a concern with connecting and trading globally in a fair and ethical manner is the balance that is missing from some current internationally important debates. In a climate-stable world, trade should occur when necessary, in accordance with the World Fair Trade Organization's Charter of Ten Principles of Fair Trade, and taking into account the carbon footprint of the transaction.

Trade Cooperative Pathbreakers

This chapter features a case study of two cooperatives involved in the coffee trade, Just Coffee, a U.S. worker-owned coffee roaster and distributor, and Cooperative Coffees, a North American green coffee (meaning unroasted) importer owned by coffee roaster members. I chose coffee for the trade case studies because it is ubiquitous throughout the world as a drink many feel they cannot live without, yet it only grows in certain regions in the Global South, so the Global North imports it in abundance. Coffee is one of the world's largest commodities: in the next twenty-four hours, people around the world will consume around 2.5 billion cups. How coffee is grown and the price farmers receive for their work make a big difference to the coffee-growing environment and ability to have a reliable farm income.

Just Coffee (Coffee Roaster and Wholesale Distributor, Worker, U.S. and International)

Cooperative Coffees (Green Coffee Importer, Coffee Roaster, International)

Overview of the Cooperative: History, Type, and Economic Sector

Matthew Earley and Mike Moon formed Just Coffee in 2001 in Madison, Wisconsin. The founders created an entity to purchase coffee from farmer cooperatives in the Global South, roast it in Madison, and distribute it to customers in the United States. They offer wholesale coffee that is organic and fair trade to grocery stores, coops, restaurants/cafes, and offices, and directly to customers online. In creating this cooperative, they were attempting to directly link small organic farmers with their U.S. customers.

Just Coffee began by working with one group of Mayan Zapatista organic coffee farmers in Chiapas, Mexico. It had to incorporate as a limited liability corporation (LLC) until it had five worker-owners and could then legally reorganize in 2006 as a cooperative under Wisconsin law. By 2019, Just Coffee had expanded to six worker-owners and over thirty employees, who partner with twenty-five farmer cooperatives growing coffee in fourteen countries in Latin America, Asia, and Africa.

I interviewed Matthew Earley for this research because Moon was no longer working with the coop. At the time of the interview, Earley was president of the coop's board of directors, and general maven of farmer and community relations and sustainability. Reluctant entrepreneurs, the cofounders of Just Coffee were trying to assist a cooperative of Zapatista farmers in Mexico gain access to the U.S. market for their coffee so the farmers could obtain better prices. None of the existing roasters and importers in the Midwest were interested in working with this farmer cooperative because, according to Earley, the roasters didn't understand the value added by purchasing directly from an organic farmer cooperative in Mexico and did not want to deal with the difficulties of doing business with a group of farmers who had no formal organic certifications and had never exported coffee.[12]

At the beginning of the new millennium, coffee commodity prices were very low, so the farmers were interested in finding a way to obtain better prices for their products. When Earley returned to Mexico with the bad news that none of the midwestern coffee roasters would buy the farmers' coffee, they were undaunted and simply said, "No problem. We'd like you to do it." And it was with that hope and strong nudge that Earley returned north to set up a new coffee roaster to provide a market for the Zapatista farmer cooperative.[13]

Just Coffee started out distributing coffee on a very small scale: the employees biked to farmers markets in Madison to sell coffee and raise awareness about fair trade. Twenty years later, they've exchanged the hyperlocal delivery and direct sales for a national distribution network. One can now find the original delivery bike as a conversation piece in the coop's roasting factory café in Madison. While growth comes with downsides, such as a larger carbon footprint, it has allowed the coop to build relationships with more farmer cooperatives in the Global South and provide markets in the United States for their coffee, at better prices than the farmers would otherwise find.[14]

Just Coffee participates in a supply chain of cooperatives. It obtains green coffee through Cooperative Coffees, the importing cooperative of which it is a member. Cooperative Coffees imports green coffee for its roaster-members and is committed to supporting and partnering with small-scale organic coffee farmers and their exporting cooperatives. Cooperative Coffees says its goal is to make coffee growing sustainable and beneficial for farmer families and their communities. At the time of the interview, Earley, of Just Coffee, was on the board of Cooperative Coffees, serving as the vice chair. Without the importing cooperative, Just Coffee would have been buying coffee off a menu from an importer. Using a traditional importing business makes it really hard to form a relationship with your trading partner, observed Earley. Cooperative Coffees had been around for about three years when Just Coffee was getting started, and Cooperative Coffees was key to importing the original Zapatista coffee. Figure 5 shows the supply chain for the coffee cooperatives.[15]

Cooperative Coffees only works with coop farmers who are producing organic or organic-transition coffees, and it is focused on building relationships with the farmer/producers to create a fairer, more transparent, and environmentally sustainable trading system. The importing cooperative then provides this coffee to members, the majority of which are small busi-

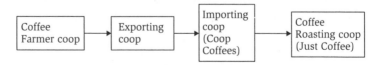

Figure 5. Supply-Chain Relationships Between Coffee
Cooperatives (Credit: Melissa K. Scanlan)

nesses that roast and sell coffee in North America. Just Coffee is the only U.S. roaster in Cooperative Coffees' membership that is a worker-owned cooperative. There is only one other North American worker-owned roaster, and it is in Quebec, Canada.[16]

Cooperative Coffees' volume of imports has grown a lot over its twenty-five years. It imports approximately half as much coffee as the Equal Exchange cooperative, estimates Earley. In 2016, Cooperative Coffees imported almost 4.5 million pounds of organic and Fair Trade specialty-grade green coffee to the United States.[17]

In 2017, Just Coffee became a certified B Corp, so now it has a third-party certifier to assess and rate its practices. B Corp certification does not look at the transactions with the coop's trading partners (the farmers), but reviews employment practices, community impact, environmental practices, and governance structure. Earley reported it was a rigorous process to obtain certification from B Lab, and the coop will need to repeat the audit every three years. B Corp certification adds a level of mission durability on top of its cooperative structure. Earley sees it as "locking them in differently to these commitments and requiring regular quantification of whether they're meeting their commitments." He had wondered if the coop had slipped away from its mission as it scaled up to national distribution and selling to large chain stores, and was happy to confirm through the certification that it is still on target.[18]

Industry Context

Fair Trade, as a system of private governance, has its flaws. The Fair Trade price for coffee was established in 1988 and was not indexed to inflation or rising producer costs, so it quickly lost meaning. The price was $1.26 per pound of green coffee, which Jaffee and Howard calculated as only 96 cents per pound in real value by 2014. By contrast, in 2017, Cooperative Coffees paid a minimum base price of $2.20 per pound on all organic, green coffee contracts.[19]

In 2004, Just Coffee stopped focusing on Fair Trade certification because it felt the certifications were too corporate-friendly and not protective of the coffee farmers. Earley cited the low price per pound of coffee for the farmers as a specific reason he thought the certification had limited value, as it set the bar too low. That said, many of the farmer cooperatives that supply Just Coffee and Cooperative Coffees are Fair Trade–certified.[20]

But Just Coffee is not content with simply relying on the certification, and goes further in its practices. Committed to trading fairly, Just Coffee says its business values include transparency, social activism, environmental sustainability, human dignity, quality of coffee, and quality of life. Instead of relying on what it sees as a weak third-party Fair Trade certification, Just Coffee uses transparency and B Corp certification as strategies. It puts all its coffee contracts online to show people how it does business and provides evidence of the fairness of its trade. It offers a coffee tracker on its website that allows one to enter a code from each bag of coffee to see its source, its organic certification, and the contract terms with the farmer.[21]

"The traditional business model for tropical commodities trade, such as coffee, is one based on exploiting the farmer," says Earley. The cofounders of Just Coffee wanted to create a business that challenged assumptions about the traditional trade model. They wanted to develop and participate in a supply chain that directly links farmers to consumers and is rich on developing relationships of mutuality rather than exploitation.[22]

Both Just Coffee and Cooperative Coffees are focused on developing long-term relationships with small-scale coffee farmers and their cooperatives. Often Just Coffee and Cooperative Coffees visit the farmers together. They also support community development projects, and through these efforts Earley sees the quality of the coffee and their relationships with farmers improving. Just Coffee asserts, "Every day when we drink our coffee we are supporting the lives of the people who planted, tended, harvested, processed, and shipped it to us."[23]

Although Just Coffee has expanded over the past twenty years from a single relationship with a Mexican cooperative to trading with cooperatives in fourteen countries in Central and South America, Africa, and Asia, it has maintained this relationship-building as a core part of its business model. In addition to providing information about the farmer cooperatives with which it is trading, it has organized a trip each year to take its customers—usually those who are purchasing for grocery stores or restaurants—to meet with the farmers who are producing the coffee.[24]

Environmental Sustainability

Just Coffee's fundamental commitments include environmental sustainability, which Earley defines as "not harming the environment, supporting regenerative activities, and thinking about all operations and aspects of the supply chain in terms of carbon, with the goal of a net negative impact." This definition is not simply about refraining from harm, but about being an active force for positive improvements. It sets the bar much higher than that required by law or private governance of Fair Trade certifications. This sustainability purpose infuses the coop's day-to-day decisions and efforts at ongoing improvements, reports Earley.[25]

Just Coffee "supports agroecological and organic practices, carbon neutrality, worker health, and low-impact business practices." This is not a small portion of their business, but the entirety of it: all the coffee it roasts and distributes is organic and sourced from farmers who are working in producer cooperatives. Similarly, Cooperative Coffees is exclusively importing organic and small-farmer coop Fair Trade coffee to distribute to its roasting business members.[26]

CARBON FOOTPRINT

Since Just Coffee is a member of the importing cooperative, it has access to a carbon-footprint tracking tool. Cooperative Coffees has offered this tool to its members since 2017 (using 2016 emissions) as part of its initiative to track and fund carbon reductions within its supply chain. With this tool, members track emissions associated with employees' daily commutes, employee travel, importing green coffee from origin to roaster, energy consumption at the roasting factory, and shipping the roasted finished product to distribute it. The accounting tool does not capture all the supply chain emissions, however. For instance, among other things, it does not measure emissions associated with coffee cultivation and processing, nor with packaging. Nonetheless, the fact that an importer is tracking emissions related to the supply chain of all of the companies that obtain the product from it is a model for other importers and can have a sustainability ripple effect on the companies it supplies.[27]

Carbon-tracking tools can help focus a company's attention on the carbon footprint of its activities and supply chain and provide ideas for reductions.

Just Coffee's 2016 data showed that each pound of coffee is associated with 1.25 pounds (0.57 kg) of CO_2; this is lower than the industry average, but it still has opportunities to make greater reductions. The vast majority of these emissions are from shipping (importing the coffee to a warehouse, shipping it to the roaster, shipping it to consumers), which made up 73 percent of its total CO_2 emissions. Recall that international shipping is a major contributor to global emissions that is not associated with the emissions accounting for any country, so I called them "phantom" emissions at the outset of this chapter. There are some adjustments that can be made to reduce phantom emissions by purchasing from countries that are closer to the importing country and from producers that are located closer to an exporting point, but the overall strategy is to impose a carbon tax to fund offsets, as explained below.[28]

A smaller part of Just Coffee's carbon footprint is due to energy consumed in its roasting factory and offices. It has been working to reduce energy demands through production efficiencies and reducing its energy-related CO_2 emissions by switching to renewables. All the electricity Just Coffee uses to power its roasting factory and offices in Madison is renewable because it purchases renewable energy credits from the local utility. However, the utility does not extend the renewable credits to the coop's natural gas usage. The coop identified that it used the most natural gas running the coffee roasting machines. When the opportunity to replace came up, it switched to a Loring high-efficiency roaster in 2017 and reduced its annual gas usage by about 70 percent.[29]

CARBON "TAX"

Each roaster-member of the importing cooperative takes this carbon-tracking initiative one step further and pays three cents per pound of green coffee into a fund to offset its carbon footprint. This "climate resiliency" charge is akin to a voluntary internal carbon tax on its members. Cooperative Coffees then uses the fund to pay its coffee producer partners to engage in practices that sequester carbon, such as forest restoration and good organic farming and composting practices. Cooperative Coffees seeks to avoid top-down climate mitigation approaches and wants to collaborate with its farmer partners to develop "regenerative solutions in farmer communities."[30]

As implemented by Just Coffees in 2017, the "climate resiliency" payment was $15,166, or a bit over $64 per metric ton of CO_2. Cooperative Coffees insists the fund is intended not to buy off environmental debts, but to soften carbon footprints, to acknowledge the environmental services that producer/farmers provide, and to encourage more results in carbon sequestration.[31]

One such climate resiliency project funded by the "tax" is a pilot to measure the carbon footprint of coffee farmer cooperatives. The coop is using the Cool Farm tool to measure this on Peruvian farms. Further, Cooperative Coffees works with a farmer coop in Nicaragua that is ahead of the curve on agro-ecological practices, and the fund pays for farmers to attend workshops and share their knowledge. Cooperative Coffees has hosted farmer-to-farmer learning exchanges for Latin American farmer cooperatives representing more than 12,500 coffee farmer families.[32]

To put this in the context of the larger literature on carbon taxes, putting a price on carbon is a standard economic answer to the problem of climate change. The cooperatives who are voluntarily doing this are out in front of what needs to happen across the economy as a matter of public law. JPMorgan Chase is the largest financier of fossil fuel developments, and even its investment services arm, JPMorgan, describes climate change as "a global market failure in the sense that producers and consumers of CO_2 emissions do not pay for the climate damage that results." The JPMorgan *Risky Business* report goes on to explain that a carbon tax is a standard solution and it is "best if the CO_2 tax is global to deal with the free rider problem." It repeats the International Monetary Fund's suggestion that a global carbon tax should be introduced immediately and rise to $75 per ton of CO_2 by 2030. Along these lines, a subcommittee to the U.S. Commodity Futures Trading Commission issued a report in 2020 in which it found the financial system faces significant risk from climate change and "financial markets will only be able to channel resources efficiently" to reduce GHGs "if an economy-wide price on carbon is in place at a level that reflects the true social cost of those emissions."[33]

In the absence of a global carbon tax, critics of carbon pricing worry that it puts domestic industries at a competitive disadvantage. One way to mitigate that disadvantage is to use a border tax adjustment on importers of products from countries that do not use a carbon tax. For coffee imported into North America, as is the business of Cooperative Coffees, there is no significant domestic industry disadvantage concern (with the exception of Hawaiian-produced coffee), so there would unlikely ever be a push for such a border tax adjustment on coffee even if the United States initiated a national carbon tax. Clearly, Cooperative Coffees is not using this form of private governance as a pragmatic way to get out ahead of regulation. Instead, it appears motivated to go above and beyond the law by the core purpose of its business to be a responsible and ethical trading entity that wants to reduce and offset any negative environmental impacts of its business.[34]

There are two additional interesting aspects of Cooperative Coffees' climate resiliency initiative: the level of taxation and how the funds are spent. In 2017, fifteen countries used a carbon tax as part of their climate policy. Sweden's is the highest: it taxes electricity consumption at about $120 per ton of CO_2. In general, however, other governments' taxes range from $2 to $30 per ton of CO_2. The U.S. Energy Information Administration, in 2014, used a model tax of $25 per metric ton, increasing 5 percent per year above inflation. In a separate effort, the Council of Economic Advisors and the Office of Management and Budget estimated the "social cost of carbon" to be $42 per ton. They calculated the social cost to inform federal agencies to conduct cost-benefit analysis. Compared to these measures, Just Coffees' 2017 payment of more than $64 per metric ton of CO_2 is well above the level commonly discussed in policy debates. Earley expressed no hesitation about paying such a high carbon "tax" on the impacts of the coop's operations and supply chain.[35]

Further, there is the issue of how to spend the revenues generated by a carbon tax. Policy debates about carbon-tax revenues revolve around three overarching categories. Revenues could be used to reduce individual or corporate income taxes (with lots of variations having differing regressive or progressive impacts), be put into a general fund for government expenses, or be targeted to fund actions that accelerate deep decarbonization. Cooperative Coffees has chosen the last approach: to use the revenues to fund farmer projects that could offset carbon impacts by sequestering carbon. While there are many uncertainties about how much carbon such projects actually sequester, Cooperative Coffees has chosen the most targeted of the revenue-spending approaches to multiply its impact on deep decarbonization.[36]

Additionally, Just Coffee partners with the nonprofit On the Ground to raise awareness and funds supporting the work of small-scale coffee-farming communities. Just Coffee sets aside $10,000 per year to contribute to farmer cooperative projects. For instance, it is funding work with a women's coop in Nicaragua that is diversifying with honey, wine, and tea. Driven by long-term drought and climate change, the move to diversify is a push for climate resiliency and recognition that coffee may no longer be grown in Nicaragua in about a decade. This example shows the depth of the relationship building between Just Coffee and its trading partners—a relationship built on coffee that is strong enough to move beyond coffee.[37]

Given all of the measuring and reporting Just Coffee and Cooperative Coffees do to inform their operational practices and ensure they are making a positive impact on the environment and climate resilience, one might expect to see all this documented in a glossy report for their customers and stakeholders. However, Just Coffee simply produces a one-page annual report to the public that is not really a sustainability report. There is nothing on Cooperative Coffees' website about its climate resiliency initiative to track carbon with all its members, tax it on each pound of coffee, and use the fund for carbon sequestration by its farmer partners. So, if one used the existence of a sustainability report to indicate a sustainability leader, one would completely miss these cooperative pathbreakers.[38]

Governance and the Cooperative Structure

Earley reflects that in the beginning of Just Coffee, the Cooperative Values and Principles were part of the founders' instincts. In Earley's words, "The DNA of the business is to be a force for good in the world." He thought they formalized this purpose in their governance documents when they reincorporated as a cooperative in 2006. Their lawyer was a key factor in explaining the Cooperative Values and Principles and thinking through how to reflect them in their legal documents, recalls Earley.[39]

Just Coffee further revised and updated the governance documents in 2016. Upon reviewing Just Coffee's revised articles of incorporation and bylaws, however, it was apparent that they did not include a specific purpose reflecting its mission or the Cooperative Values and Principles in its governance documents. Within two months of my initial interview with Earley, he left the cooperative. Just Coffee will continue to sell organic Fair Trade coffee without its cofounders, but because of the lack of specificity in its governance documents, it could be more susceptible to mission creep.

Its B Corp certification, with audits every three years, may mitigate this danger. The absence of care in drafting the cofounders' purpose into the governance documents makes the third-party certification more valuable for mission stability. This will be something to revisit in future years to see how the cooperative evolves in the absence of its cofounders.

The six worker-members of Just Coffee make up the board of directors, each with one vote. They meet monthly to discuss financial analysis, bank loans, and projects. Thus, unlike many of the other cooperative exemplars in this book, Just Coffee is small enough to function entirely with direct instead of representative democracy for the worker-members.[40]

The cooperative has three classes of stock: one for members (the employees), one for capital stock investors, and one for community stock investors. The latter two are considered nonmembership stock. These investors typically earn a 3–4 percent dividend per year on their investment, as determined by the board of directors. Investors don't have a representative on the board, but have the right to vote on any potential decisions to sell or dissolve the cooperative. In the event of Just Coffee's liquidation, the worker-members are the last in line to be paid and the capital stock investors are first. However, the worker-members are allowed to also be capital stock investors, so this line of payment does not necessarily serve as a disincentive to "cash out" of the cooperative by voting to dissolve it. Further, it has not included any restrictions in its governance documents that would prevent the cooperative from selling to another entity, such as a larger cooperative, or even Starbucks. The determination would be one the members make, in accordance with the bylaws.[41]

Just Coffee's worker-owner structure has changed as the business has grown. Originally, all the worker-owners were doing production work roasting coffee, but between 2009 and 2012 the coop engaged in a deep transformation. After a lot of investigation and work with outside consultants, it determined that it needed to scale up the cooperative and specialize the employees' roles. When it expanded, all of the original worker-owners transitioned into management positions. Alongside this stable membership, it has expanded the number of employees. In 2019 it had more than thirty employees who were not members. These employees are primarily production workers in their twenties who stay a few years and move on to other jobs. This is a tension because the short-term employees don't participate in governance. Meanwhile, the worker-owners in management are aging, and it is somewhat difficult to attract new worker-owners. In 2019, one employee became an owner. The barriers to becoming an owner are high for these production jobs because they are repetitive, physically demanding, and there's a limit to remuneration at about $15 per hour. Of the production crew, only the current production manager is a worker-owner. So, although Just Coffee has expanded the number of jobs it provides, this has not kept pace with expanding the number of owners fully participating in cooperative governance.[42]

Summary

Cooperative Coffees imports organic and Fair Trade green coffee grown by farmer cooperatives. Its members are coffee-roasting businesses in the United States and Canada, of which Just Coffee is the only one that is also organized as a U.S. worker cooperative. These cooperatives work exclusively with farmer cooperatives in the Global South that are certified organic or in transition to organic. They build their trade practices around relationship building and transparency in contract terms, while providing the farmers with access to the North American market of coffee drinkers.

Cooperative Coffees' climate resiliency initiative encourages its member roasters to track their carbon footprints in order to assess where to make changes to reduce the carbon footprint of coffee. It additionally imposes a carbon tax of 3 cents per pound of green coffee imported, and uses the funds for carbon sequestration by its cooperative farmer partners. This provides a model for other importers, showing how they can track and tax emissions related to the supply chain of all of the companies that obtain products from the importer to have a sustainability ripple effect on the companies it supplies.

Just Coffee used the carbon tracker to identify areas for reductions, and then invested in a more efficient roaster that reduced its natural gas use by 70 percent. The electricity Just Coffee uses for its roasting factory and offices is 100 percent renewable. As a member of Cooperative Coffees, in 2017 its carbon tax was a bit over $64 per metric ton of CO_2, well above the level commonly discussed in policy debates about the level at which to set a carbon tax or the social cost of carbon. These cooperative pathbreakers are going above and beyond the legal requirements of international trade in many ways. They use a business model based on fair prices to farmers and transparency in contracts, only organic products, renewable electricity, carbon tracking, and an internal carbon tax that is used to fund deep decarbonization projects by their trading partners.

Findings and Lessons for a Livable Planet

WE LIVE AT A TIME THAT IS both overwhelming and exciting. The choices we make today about how to develop our political economies will have lasting and defining consequences for those we will never meet. If we can collaborate across nations to transform our global economy and the institutions that are the primary economic actors to deeply decarbonize and reach the Sustainable Development Goals, we have the potential to create a much more vibrant, equal, and livable planet. We can mitigate climate disruption, reduce suffering, and prosper if we create conditions that allow social enterprises to flourish (Figure 6).

The 2018 study by Guillermo Montt and others, using the now-outdated +2°C global-warming goal, calculated that the United States could create one million net jobs by 2030 if it rejected business as usual and, among other things, adopted measures to move energy off fossil fuels. We are on the verge of a very rapid movement of capital into more sustainable businesses, as heralded by BlackRock's 2020 announcement of its change in investment screening and priorities. The businesses that are poised to participate in this shift will be the largest players in this new political economy. If we want to create a more democratic and participatory politics and economy, which promotes widespread and dispersed ownership and greater equity in incomes, business institutions need to be structured to reflect that commitment.[1]

Transitions of this magnitude, speed, and scope demand that governments at all levels create public laws that set clear boundaries of protection

Figure 6. *The Embrace* Statue in Madrid, Based on *El abrazo* (1976)
by Valencian artist Juan Genovés (Credit: Melissa K. Scanlan)

for ecosystems and human health, and reform the investor-owned corporation and its blinding norm of shareholder primacy. However, we have no
time to wait for legal changes. The case studies of cooperative sustainability
pathbreakers in this book exemplify the power of the private sector to effect positive changes in advance of legal requirements. Their stories hold
lessons for others who seek to replicate and scale their successes and adapt
what is compelling to other places. They teach about the importance of care-
ful drafting of governance documents to include clear purposes and direc-

tion to advance multiple sustainability goals, as well as how to advance the Sustainable Development Goals and deeply decarbonize energy, food and agriculture, water, finance, and trade.

Drafting Governance Documents

Sustainability pathbreakers tend to use governance documents to articulate their purpose and set a trajectory for their day-to-day operations. The strongest articles of organization/incorporation and bylaws include provisions that address the following:

- weave Cooperative Values and Principles into the core purpose;
- elaborate on the meaning of Cooperative Principle 7 as it relates to the particular cooperative, such as a clear commitment to environmental sustainability, circular economy, biodiversity, organic and regenerative agriculture, renewable energy, decarbonization, and the like;
- include a dissolution clause that directs financial resources to similar cooperatives, cooperative federations, and likeminded nonprofits in the social economy;
- specify democratic governance of one member, one vote, with possibility of diverse board membership to include stakeholders who are not members;
- allow hybrid memberships of various classes (for example, workers and consumers);
- if including an investor class, offer limited or no voting authority in order to continue to separate capital from control;
- establish a fund for education, technical, and promotional support for cooperatives to build the cooperative ecosystem (some set aside 10 percent of net surplus);
- establish a fund for external charities (some set aside 10–40 percent); and
- articulate a salary policy that may specify caps between highest and lowest paid, transparency, and a commitment to gender equality for equal work.

Renewable Energy

The Apollo Project or New Deal of our time is to transform our electric system and economic activity. We need to create electricity with renewable energy and shift heating, transportation, and industry to renewables in this decade, and completely decarbonize economic activity by 2050. Addressing the climate emergency requires a disruption of the business model for energy, not only in who has access to energy and how, but in who owns and controls it. Bringing greater equity in wealth between the Global North and South means opening up platforms of participation for an additional 1.1 billion people to access clean, efficient, and affordable renewable electricity. At the same time, for those who already have access, we need to deploy comprehensive energy efficiency and conservation programs, including enforceable energy performance standards for buildings. Additional parameters of this new reality involve the need to:

- provide for increased electricity demands due to electrification of transportation (mass transit and private) and heating;
- incentivize use of electric vehicles, emphasizing fleets and sharing rather than individual ownership;
- retire all fossil-fuel electricity generation plants that do not capture and store the GHGs they emit;
- develop renewable energy generating capacity in the least environmentally destructive ways available (for example, on already built surfaces and brownfields, compatible with multiple land uses);
- adopt widespread energy conservation practices and building standards;
- invest in energy storage to smooth out variability in renewable generation; and
- incorporate a diversity and abundance of customer-sited generation, storage, and demand management.

In 2018, the IPCC warned that the majority of models do not show humanity meeting the climate challenge without addressing inequality. The instability of unequal societies has intensified during the coronavirus pandemic, with greater numbers of people experiencing insecurity and loss of income. If we are to take this to heart, then we need to promote and incentivize the business structures that are most apt to deliver shared prosperity. Cooperatives, with their emphasis on democratic ownership and participation, are poised to reduce energy poverty, ensure a just transition, and equitably share the benefits of the renewables revolution.

Unprecedented in scale and time frame, the shift to renewables offers a chance to build a new energy model that is not only environmentally cleaner, but more resilient technically, and with greater equity and ownership among a broader group of stakeholders and investors. The modular nature of some renewables, such as rooftop and community solar, makes this new form of energy generation more readily available to shared ownership than the monopoly model of large fossil fuel generating stations.

Why have renewable electricity cooperatives grown so rapidly in Spain relative to the United States? Spain set a national renewables goal in law that by 2010, at least 29 percent of total generation of electricity would be from renewables. Spain coupled this goal with government subsidies to increase the affordability of renewables. Spain passed the Electricity Sector Act of 1997, amended in 2007, that opened electricity generation, transmission, distribution, and marketing to any entity and guaranteed access to transportation and distribution networks. Spain's national law:

- facilitated a wholesale market pool to fix the price of electricity across the country;
- made the grid operator (previously government) a publicly traded company;
- established a national system of certifying "guarantees of origin" of renewable electricity, prohibiting double use and fraud, that allowed a national market for renewables; and
- created a regulatory body to supervise the entire system.

These legal reforms reduced the barriers for average consumers to switch to renewables and created an opening for new renewable electricity businesses of all types, including cooperatives. Two of the exemplars featured in this book were founded in the wake of these reforms, in 2010.

The cooperative pathbreakers show how they can make these shifts to a deeply decarbonized and sustainable energy model, with significant roles in production, distribution, and commercialization. In less than a decade, Som Energia grew from its founding 150 members to 64,000 members supporting the production of 17 GWh per year of renewable electricity. Som Energia shows that a powerful source of capital for renewable energy projects is from thousands of small consumer-members of the cooperative. Som Energia's members contribute €100 to join the cooperative and have the option to invest more. Som Energia uses this capital to fund new renewable energy generation. In 2017, during a seven-day period it raised €5 million from its members to build three solar projects. With these small amounts of capital, the cooperative opens up opportunities for more people to invest in renewables.

A national cooperative that appears to effectively engage members in democratic governance, Som Energia also provides an inspired example of participation that pairs in-person contacts in local groups with utilization of technology to communicate across larger geographies. The local groups meet regularly to discuss and debate the issues and share best practices in energy conservation and generating renewable electricity in homes and businesses. In a website dedicated to member participation, it articulates a multimonth sequence of events, from surveying members to making proposals, to discussing and debating, to finally defining the issues before a vote of the annual meeting of the General Assembly. It uses this local-national process to elect committee members, develop budgets, make strategic plans, and create internal regulations. It even provided a video tutorial on how to participate in the 2019 process. On its member participation website, it transparently includes all the proposals that have been discussed, the number of votes, and the status of the proposals in its defined process of democratic decision making.

The U.S. electricity distribution cooperative Cobb EMC demonstrates that even without the favorable national laws that exist in Spain, and in fact in the absence of any state laws encouraging renewables, consumer cooperatives can add renewable generation and maintain low electricity rates. The business case for solar is strong even without subsidies. By 2019, Cobb EMC had invested in the most solar generation in terms of megawatts of any cooperative in the United States. It is delivering the benefits of renewable electricity to its 180,000 members in rural and suburban Georgia. Yet, part of the lesson of Cobb EMC is that this rate of transition is too slow to meet the climate emergency. Overall, the almost 900 U.S. electric cooperatives are heavily reliant on the most polluting of fossil fuels, coal. Many are using facilities that are near the end of their useful lives. Federal policy is needed to strongly incentivize cooperatives to bring the co-benefits of renewable electricity and democratic ownership throughout rural and suburban America. The original New Deal brought electricity to rural America through electric cooperatives. We now need an updated version of this to transition those cooperatives off fossil fuels. Further, sixteen investor-owned power producers in the United States are responsible for 50 percent of the nation's GHGs from electricity. The most direct way to reduce GHGs and bring the benefits of clean renewable energy to America is to simultaneously set enforceable reductions for those sixteen companies to retire their fossil fuel generating facilities within the next decade.

Food and Agriculture

Agriculture for a deeply decarbonized planet uses practices that sequester carbon in soils and plant roots, minimize chemical and fossil fuel–powered mechanical inputs, and support local and diverse food systems. Sustainable agriculture saves the family farm by providing fair wages for farmers and reduces hunger by keeping prices reasonable for consumers.

CROPP Cooperative/Organic Valley pioneered organic farming in the United States and was able to successfully market its farmer-members' products, delivering payments to farmer-members that were well above, in some years double, that of conventional farmers. It gained widespread support for its new idea by drawing on traditional values, such as saving the family farm. Its long-serving CEO insisted that sustainability is simply good business sense: it reduces costs and/or gains customers. He also was uncompromising about requiring that organic food production had to be both financially sustainable for the farmer and environmentally sustainable for the Earth and the wider community.

CROPP demonstrates the value of the private sector leading when faced with government inaction. The founders of CROPP were not seeing solutions to the 1980s farm crisis coming from the federal government, so they joined with other farmers to find a farm-appropriate solution. This set the pace for them to consistently get out ahead of the law in adopting leading sustainability practices. Organic Valley is one of three national organic fluid milk brands, and the only one farmers own cooperatively. It committed to a variety of practices, both on its farms and at its headquarters and factories, that established it as a sustainability pathbreaker. In addition to organic, it has been committed to raising cows on perennial grasses for twenty-five years, a practice now being recognized to sequester carbon and reduce water pollution. The cooperative has been a catalyst for energy efficiency and renewables. The cooperative has provided its more than 2,000 farmer-members with energy audits, renewable-power site assessments, grant writing for renewables, biodiesel technical support to grow their own fuel, and organic management input. Its farmers have installed solar power at a rate five times that of average Americans, and its headquarters and factories were entirely powered on renewables as of 2019. Its efforts to adopt a 100 percent renewable electricity goal was the catalyst for a much larger 32 MW solar project that provides renewables to other businesses, cities, and utilities.

There are some features of the CROPP cooperative that make it more vulnerable than other cooperatives and provide a lesson for them to avoid similar potential pitfalls. Its membership is entirely composed of farmers, despite its large workforce of almost 1,000 employees. Without a membership class or board of director seats for workers, that group of stakeholders does not have a voice in governance. Further, CROPP has a board that is entirely composed of farmer-members, rather than diversifying its representation to include broader stakeholders and professionals who could strengthen the enterprise. This, combined with not including a provision in its governance documents that requires the cooperative's assets to be distributed to similar cooperatives or nonprofits upon dissolution, may not insulate the cooperative against members' self-interest. At this point, these vulnerabilities are purely theoretical, and hopefully they will not manifest in actually undermining the success of this organic foods pathbreaker.

A much older farmer cooperative, one started in 1932 in Spain, shows that the cooperative form has allowed its members to adapt to changing market conditions and modify their farming practices and governance documents to emerge as a leader in organic wines. Bodegas Pinoso first began producing wines from certified organic grapes in 1997. By 2019, half of its fields (600 hectares) were certified, making it the largest organic winery in Spain. As for the farmers in the CROPP Cooperative, the appeal of organic agriculture is grounded in traditional values of preserving small family farms and the knowledge that has been passed from generations to reduce their environmental footprint. For instance, most of its grapes are the Monastrell variety, which have deep roots that do not require irrigation, an important feature in the dry Mediterranean. Its wine production's sustainability practices follow a circular business model. They include using recycled bottles and recycled cardboard boxes in the packaging of the cooperative's wines, sending the by-products of wine production to another company to produce additional beverages, and conserving electricity with natural lighting and light sensors. It has updated its governance documents to explicitly commit its farmer-members to follow environmentally protective farming practices and to endorse the international Cooperative Values and Principles.

A worker-cooperative that, like the farmers of Bodegas Pinoso, places its commitment to environmental protection and to the Cooperative Principles in its governance documents, can maintain and continually enhance its sustainability, despite market difficulties. Cerveses Lluna has been producing 100 percent organic craft beer in an employee-owned cooperative since 2008. It has expanded and contracted in response to the market, but through these financial difficulties it has maintained its commitments to organic beverage production, working with other cooperatives, and reducing its GHGs. The largest producer of organic beer in Spain, it is actively trying to spark organic hops farming in Spain in order to create a local supply of organic ingredients for its products. It meets all its electricity needs with renewable electricity, provided by an electricity cooperative. Because its commitment to protecting the environment is drafted into its governance documents along with its overall commitment to the Cooperative Values and Principles, its founders have provided a legal durability to its purpose, goals, and day-to-day operations.

Unless one buys directly from farmers, grocery stores are the primary intermediary between agricultural producers and consumers. Food cooperatives can organize at different scales, with hybrid memberships, and with different lessons for sustainability. On one end of the spectrum, bioTrèmol is a nonprofit grocery store cooperative of consumer-members designed around a set of values and principles, which it wrote directly into its governance documents and carries out in its daily practices. It is actively working to support the emergence of small-scale organic farmers in a defined local province. It is nimble enough that it can contract with new organic farmers to provide a secure market for their products. Such farmers would not have access to larger grocery chains because their supplies are too limited. This model creates a platform for interested consumers to support local, diversified, organic agriculture produced by small farmers.

On the other end of the spectrum, Consum Cooperativa shows a proof of concept for a large chain grocery store that serves worker- and consumer-members and leads in sustainability. Created in 1975, Consum in 2019 had three million consumer-members and 14,000 worker-members at over 700 grocery stores throughout the Spanish Mediterranean. Consum has provided ownership opportunities to more workers than exist in all sectors of all worker cooperatives in the United States. It shows that a cooperative can compete with large investor-owned grocery chains.

Yet, Consum is not just another chain store. As a cooperative, it has embedded values and principles into its business design that have led it to pursue greater sustainability than required by law. Consum has been measuring and reporting on nonfinancial matters, using the Global Reporting Initiative format, since 2006. As the first supermarket chain in Spain to measure its carbon footprint, it has been able to strategically reduce its environmental impact and save money. It set a 2019 goal of meeting all its electricity needs with renewables. It has adopted a zero-waste policy to increase recycling and avoid landfilling waste. It is able to advance these environmental protections in a financially viable way. When Consum does well financially, it shares the "surplus" with its members and broader stakeholders rather than send profits to investors. For instance, in 2017, it donated €14.9 million, directly impacting 150,000 people; it shared €31.4 million with its worker-members and €35.6 million with its consumer-members.

Consum's triple-bottom-line sustainability demonstrates that a cooperative difference can exist even when an enterprise is operating at the scale of large chain grocery stores.

Water

After legal changes in the United States and European Union in the 1980s and policy changes at the World Bank that promoted water service privatization, investor-owned water corporations saw a strong period of growth. Now, with increased recognition of the need to design business to rapidly decarbonize the economy and meet the Sustainable Development Goal and human right of providing clean and affordable water, international and national laws and policies could be amended again—this time to promote water cooperatives by requiring that social enterprises compete for water service contracts and deliver water in a way that is consistent with deep decarbonization and the Sustainable Development Goals. For instance, a municipal request for proposal could require applicants to be social enterprises that are not organized for profit and are member-governed on the democratic principle of one member, one vote. The main features of the cooperative management model to be included in legal reforms are:

- maintain the municipality as the trustee responsible for guaranteeing a public service as important as water;
- build partnership between the municipality and the consumer cooperative, and retain public oversight of water management;
- clarify that the consumer cooperative manages water as a common good, not a commodity;
- ensure that the cooperative is operated as a noncommercial venture, and follows the Cooperative Values and Principles;
- draft a municipal contract with the cooperative that requires water management to be sustainable and protective of the environment through ongoing infrastructure improvements, energy efficiency, running operations on 100 percent renewable energy, and reducing water usage;

- require the cooperative to engage the public around the importance of water; and
- set goals and strategies for water supply resilience in the face of increasing climate disruptions.

The economic repercussions of the coronavirus pandemic are just starting to be felt. However, it is clear that municipal water utilities will face an increased financial strain. This situation greatly increases the risk that some municipalities will seek to sell their water systems in pursuit of short-term fiscal relief. In light of this pressure, it is even more important to understand the co-benefits that cooperative water utilities can bring for greater water security aligned with deep decarbonization and sustainable development.

Finance

Cooperative and stakeholder, values-based banks are essential to capitalizing cooperatives and other social enterprises. To deeply decarbonize the economy, financial institutions need to turn off the money spigot for fossil fuels and heavily invest in renewable energy, energy efficiency, and climate adaptation. The largest cooperative bank in València, Spain, Caixa Popular, is a worker- and cooperative-owned bank that serves its stakeholders' interests. It has no money invested in fossil fuels, and instead has focused on renewable energy.

The larger lesson it offers is its commitment to building a strong ecosystem for the emergence and endurance of worker cooperatives. In addition to favorable rates for cooperative members, it pays membership dues in the worker-cooperative federation, provides annual audits, and sets aside 10 percent of net surplus for cooperative training, education, and promotion. The autonomous community in which it operates has grown from little cooperative presence in the 1970s to a hub of cooperative enterprises today. The Valencian Community is home to more worker-owned cooperatives than the entire United States, and this cooperative bank has played a significant role in this success.

Another important aspect of this cooperative bank is that it has a hybrid membership that also creates a class of worker-owners in addition to the cooperative-owners. It has drafted into its governance documents a commitment to equity in its salary policies. It carries out this strong value in its policies, and the results set it apart from other banks. Employees work within a compensation ratio of four to one. It has eliminated a gender gap in salaries for the same positions. Unlike many private-sector companies, it offers total transparency in salaries, which in turn is a ready compliance mechanism for its salary policies.

Trade

In a deeply decarbonized world, there will be stronger, more resilient and diversified local economies, and less trade in things that can be produced with a smaller carbon footprint closer to home. International trade agreements will need to be aligned with the Paris Agreement and the Sustainable Development Goals to provide frameworks for trade that produces co-benefits of peace and broadly shared prosperity for trading partners. In the meantime, the coffee cooperatives discussed in this book provide examples of how to deeply decarbonize trade while promoting the Sustainable Development Goals through private-sector commitments to fair trade. The featured coffee importing and roasting cooperatives in North America work exclusively with farmer cooperatives in the Global South. They are building trading relationships based on ethics of fairness and transparency, with a commitment to supporting organic agriculture, agro-forestry, and reforestation to sequester carbon.

By disseminating a carbon-tracking tool and utilizing an internal carbon tax with its coffee-roaster members, the importing cooperative is able to influence the amount of GHGs associated with coffee. The roasting cooperative used the carbon footprint tool to identify the largest sources of GHGs at its facility and then made strategic investments in more efficient equipment to reduce its use of energy. Also its internal carbon tax program creates a fund the importing cooperative uses to pay its cooperative farmer-members to engage in reforestation, agroforestry, and other practices to sequester carbon. Combining reductions in GHGs at the roasters and investing in carbon sequestration with the growers models a best practice for deep decarbonization of trade in one of the most common global commodities. This should set a benchmark for others involved in the coffee trade and provides a model for other trading relationships.

Hope

There are inherent difficulties in generalizing from case studies, especially to other political, historical, and legal contexts. Even without the complexities of transboundary application, simply because one identifies a cooperative leading the way on renewable energy in Spain does not mean all electric cooperatives in Spain are renewable energy pathbreakers. Likewise, there is nothing inherent in the cooperative form that mandates environmental sustainability or compliance with any of the other international Cooperative Values and Principles. These are encouraged and part of the cooperative identity and private governance, but unless founders and members press for these priorities, they could certainly be dormant. What we can see from the case studies is that the cooperative form of business provides the optimum flexibility to pursue multiple Sustainable Development Goals and deep decarbonization.

The pathbreakers offer a proof of concept. They show examples of best practices in maximizing the benefits of the cooperative form of business. My research draws out how legal design at the legislative and institutional level is critical for empowering people to articulate and lock in a strong triple-bottom-line purpose. The legal structures that support and encourage sustainability innovation by the pathbreakers can be modified and replicated to build strong purposes into social enterprises. The world needs social entrepreneurs. I hope these stories will spark the creativity of innovators who seek to build resilient businesses for shared prosperity in the next economy. We need you now, more than ever.

There is much greater instability in highly unequal societies. Our current global economic system and the investor-owned corporation that dominates it generate inequality as part of their design, not as a glitch. As we see greater gaps in wealth and security between the Global North and South, and internally within countries based on race/ethnicity and gender, violence and dysfunction will be constantly seething under the surface. In this type of system, which reinforces discrimination and inequality based on race, gender, and class, eruptions of anger and civil unrest come to the surface. Ignited by extreme and chronic police brutality against Black people, riots and protests spread across the United States during the coronavirus pandemic. As this book goes to press, many are pushing for police reforms. Antiracist policing is absolutely necessary. A strong theme in these reforms centers on democratic power: the concept that the most impacted need to have a strong voice in shaping laws, policies, and institutions meant to serve the people. This reclaiming of democratic power, in particular by people of color, is reinforced by building a democratic economy to address the underlying economic system that is perpetuating greater disparities.

In the United States, people of color have developed cooperatives and systems of mutual aid as prosperity strategies. Starting in 1969, Black farmers in Georgia secured their land through New Communities, Inc., which became the first community land trust in the United States to promote access to affordable housing. The Federation of Southern Cooperatives represents many Black farmers who are building local and resilient food systems. In 1997, cooperative-members of the Roanoke Electric Cooperative in North Carolina supported the first Black CEO of a U.S. electric cooperative.

Cooperatives offer a possibility to build businesses for a new system based on values of equality, sustainability, solidarity, and democracy. When business leaders implement the best practices in legal design highlighted in these pages, they are able to harness the power of this business form. Cooperatives readily allow people to advance multiple benefits ahead of reforms to substantive environmental and corporate laws. Not only can cooperatives be successful organizational models for the transition to a fossil-free economy, they provide important complementary benefits. Cooperatives offer the opportunity to reinvest surplus revenue in the community instead of siphoning such revenue to shareholders. They create more owners and enable them to exercise democratic governance of businesses. And ultimately, they can provide greater equality in incomes generated from work. The motivation of the farmers who created Bodegas Pinoso almost a century ago, who knew that their union made them strong, transcends time and borders; it is just as true in the fossil-free global economy we are building in this decade. Choosing to cooperate is our clearest path to prosperity as we work to mitigate climate disruption.

APPENDIX: RESEARCH METHODS

MY RESEARCH METHODS FOR THIS BOOK project were varied, crossed disciplines, and bridged three languages (English, Spanish, and Catalan/Valenciano). Regarding the languages in this book, I have used the official language for every government entity, hence you see Alicant (Valenciano) instead of Alicante (Spanish). As seen in Part I of the book, I used traditional legal analysis of constitutions, statutes and regulations, judicial opinions, and secondary sources written by scholars, journalists, and nongovernmental organizations. Further, I drew on the work of political scientists, economists, climate scientists, and ecologists. These research methods informed the first part of the book, which takes the reader on a journey through environmental and corporate law and efforts to reform it; the history, development, private governance, and laws of cooperatives; and the conceptual framing of sustainability and how it is measured and reported.

I used social science research methods for the case studies in the second part of the book. Through qualitative research interviews with sustainability leaders within these cooperatives, one can discern how they understand sustainability and democracy and how it impacts the purpose and operations of the business. This type of research is used to identify and describe themes in the interviewees' lived worlds. I coupled this with document analysis of what interviewees understand as their businesses' governance documents, usually their bylaws, articles of incorporation, and internal policies. I also conducted content analysis of their publicly available information to assess how they were measuring, describing, and reporting about their sustainability.[1]

I've written the case study narratives in language that is not overly technical in the hope that this will make them available for a broader audience. I attempted to bring to life what distinguishes each cooperative, given its particular industry context, from investor-owned corporations. I described their environmental sustainability practices, their governance, and assessed the role of law and policy in their success. Through this transatlantic research, the overarching question is how we might utilize cooperative ownership more effectively to transition the world to the next system, one that prioritizes environmental sustainability integrated with promoting democratic participation, employment, dispersed ownership, and equitable compensation.

Identification of Study Sites

In the search for cooperative sustainability pathbreakers, there is no directory labeled "exemplars" or even a database where one may find ranked sustainability scores for cooperatives. If searching within investor-owned corporations, a starting place could be Global Reporting Initiative's database of sustainability reports, since reporting is almost universal for the largest corporations. In Esty and Winston's 2006 case studies of corporate sustainability leaders, they too encountered some difficulty identifying environmental leaders and described their methods of using sustainability scorecards, including the Dow Jones sustainability index. Given the dearth of cooperatives using the Global Reporting Initiative reporting system and the limitation of the Dow Jones index to publicly traded companies, this was even more complicated for a cooperative-centered study.[2]

I combined utilizing general cooperative directories and individual company website searches with snowball sampling of experts in each country. I was unable to find a national searchable database that identifies all cooperatives, with contact information, and that combines all types of cooperatives (worker, producer, consumer), in either Spain or the United States. Even if one existed, there would be no easy way to use that data to identify sustainability leaders, given the nonexistence of a sustainability reporting database for cooperatives.

B Lab has a searchable database of all certified B Corps, so that could have been one way to narrow my search. However, it does not provide the ability to search by type of enterprise, so there was no publicly available way to find cooperatives that are B-certified. I obtained lists of B-certified cooperatives in Spain and the United States from helpful B Lab employees in each country. In Spain, as of January 2019, there were only three that fit that category, but none were in economic sectors of interest to this study.[3]

In Spain and the United States, the cooperative federations keep membership lists of their particular type of cooperative (for example, all worker cooperatives in X geographic area). I used these lists to find businesses in the economic sectors that are high-priority for deep decarbonization. I then searched hundreds of websites for these cooperatives and conducted content analysis to determine if they included public communications about their sustainability practices.

Simultaneously, during this phase of the research, I used snowball sampling by talking with experts and leaders in academia, government, cooperative federations, and market vendors in Spain and the United States to identify cooperatives the experts thought were implementing sustainability practices that made them stand apart. If I found these recommendations produced names that I had also identified through my website content analysis research, they went on a short list to request an interview. Many of my interview requests went unanswered. For some, I needed to make several requests, some did not want to participate, and some never responded. I found it very helpful to work with cultural brokers with excellent connections in the cooperative sectors in the locations I was studying. Some of these enthusiastic cooperators made helpful introductions that led to research interviews. This book would have been greatly diminished without their assistance. Having enough time and connections for introductions was critical to gaining access to the interviewees' worlds and nonpublic documents.

Interview Process

My research interview process was substantially similar in the United States and Spain. However, conducting the interviews in Spain, which is a foreign country to me, required a level of language fluency that allowed me to speak and understand, review documents, understand a different legal system, and correspond in writing with the interviewee. I used a variety of supports and processes to verify the accuracy of all the interviews. Before meeting, I provided all interviewees written questions in their native language. This included an advance notice of confidentiality and introductions to me and the overall research purpose. I gathered their written responses to questions and conversational responses, which I recorded only when the interviewee granted permission. For document review, I used translation software and compared my interpretation with the software's translation. For some of the Spanish interviews, I also had a native Spanish-speaking translator so I could check my understanding. This proved especially helpful with interviewees who switched seamlessly between Spanish and regional languages of Catalan and Valenciano.

Member Checking with Research Participants

After reviewing all company-related documents and writing the case studies, I provided each interviewee with a draft of the case study and an opportunity to identify any inaccuracies. After receiving input from the interviewees on the drafts, I finalized the case studies for this book. The end result would not have been possible without the many cooperators who contributed to the research. If any errors remain, they are regrettably mine and entirely unintentional.

ACKNOWLEDGMENTS

I COULDN'T HAVE WRITTEN THIS book without support, encouragement, introductions, and suggestions from many people. I owe debts of gratitude and appreciation to many, and I will not be able to name them all. I especially thank my family: Martin Scanlan for feedback on drafts, navigating hard-to-reach research locations, and cooking countless meals while I wrote; my children (Dominic, Lucas, and Clara) for relocating to Spain during my research and sharing my enthusiasm for cooperatives and Spanish culture.

Muchas gracias y abrazos to my Spanish colleagues who generously offered to assist in so many ways. Prof. Germán Valencia Martin was my host in Spain and ensured I had an institutional home at the University of Alacant's Law School. He made key introductions and accompanied me on trips for research meetings with officials in València and energy cooperatives, offering insights into Spanish environmental and energy law along the way. Special thanks also to Prof. Gemma Fajardo García at the University of València, Prof. Alexandre Peñalver i Cabré at the University of Barcelona, and Prof. Fernando de Rojas at the Miguel Hernández University.

The U.S. Fulbright Commission and the staff at Fulbright España supported my research with a Senior Fulbright Scholar award, which allowed me to live in Spain for the spring 2019 semester. The Hanover Consumer Cooperative Society in New Hampshire honored me with the Gerstenberger Scholarship to pursue this research.

I found the cooperative community full of people who don't hesitate to help. I give special thanks for useful contacts, references, and introductions to Meegan Moriarty at the U.S. Department of Agriculture, Prof. George Cheney, Prof. Joseph Blasi, Esteban Kelly, Mo Manklang, Prof. Tom Webb, David R. Cook, Coro Strandberg, Ricardo Nuñez, Rodrigo Gouveia, Michael Peck, and Prof. Hagen Henrÿ. Of course, all the cooperatives that agreed to have their staff and leadership participate in my research interviews were essential to sharing their lessons with us. All of you are true cooperators!

I wrote much of the book in the ultimate shared commons, libraries. I could often be found in the Biblioteca Pública Sant Joan d'Alacant in Spain and the Sister Bay Public Library in Wisconsin. The Boston College Law School and Vermont Law School librarians were outstanding, and VLS Library director, Jane Woldow, found me excellent student research assistants.

Three student research assistants appeared when I needed them. Marco Sánchez, a Mexican researcher in Spain, got hooked on cooperatives and volunteered as my interpreter and research assistant for interviews with bioTrèmol and Aigua.Coop, gracefully bridging three languages. Vermont Law School's Caitlyn Kelly (VLS '19) helped with early-stage research. Hunter Sutherland (VLS '21) provided critical and tireless fact and citation checking for the entire manuscript, while navigating online learning during the coronavirus pandemic. True stars!

The critical readers and subject-matter experts who reviewed my early stage ideas, draft chapters, and blog posts helped improve this book in so many ways. Special thanks to Prof. Gabriel Pacyniak, Prof. Emeritus Michael Dworkin, Antonio Oposa, Prof. Kevin Jones, Eva Moss, Adam McGovern, Laurie Ristino, Prof. Emeritus Richard Oliver Brooks, Terry Appleby, Andrew Ganserberg, Prof. Carlo Soncini, Prof. David Wirth, Prof. Patrick Parenteau, Prof. Zygmunt Plater, Max Lyons, Colleen Scanlan Lyons, Annabel Stattelman-Scanlan, Brigit Stattelman-Scanlan, and Jim Westrich. My wonderful editors at Yale University Press did a fabulous job fine-tuning my work. For inspiration along the journey and connecting me to so many ideas and people, special appreciation goes to the inimitable Gus Speth.

NOTES

Chapter 1. Our Present Challenge

1. Intergovernmental Panel on Climate Change, *Summary for Policy Makers*; Dernbach, *Introduction*, 2-6 (focusing on United States).
2. Intergovernmental Panel on Climate Change, *Summary for Policy Makers*.
3. Intergovernmental Panel on Climate Change, *Summary for Policy Makers*.
4. Ritchie & Roser, *CO₂ and Greenhouse Gas Emissions*; Mackie & Murray, *Risky Business*, 16 (for data from 1960 to 2016).
5. Heede, *Tracing Anthropogenic Carbon Dioxide*, 234; Van Allen et al., *Benchmarking Air Emissions*.
6. Intergovernmental Panel on Climate Change, *Summary for Policy Makers*.
7. Dernbach, *Introduction*, 8. [All of the ways to construct an economy for deep decarbonization are based on "'three pillars of energy system transformation.' These are (1) energy efficiency and conservation across all sectors of the economy, including power generation, transportation, buildings, industry, and urban design; (2) low-carbon electricity from replacement of fossil fuel based generation with combinations of renewable energy, nuclear energy and the use of carbon capture and storage at fossil fuel based generating facilities; and (3) switching from more carbon intensive fuels to less carbon intensive fuels, and eventually switching from fossil fuel use to decarbonized energy carriers, principally electricity, in all economic sectors."]; Intergovernmental Panel on Climate Change, *Summary for Policy Makers* [According to the IPCC net zero is achieved when "anthropogenic CO₂ emissions are balanced globally by anthropogenic CO₂ removals over a specified period.]
8. Intergovernmental Panel on Climate Change, *Summary for Policy Makers*; International Energy Agency, *Global Energy Review 2020*.

9. Williams, *The Cooperative Movement*, 19; Cartwright, *Feudalism*. [Feudalism in Europe was the dominant political economy from approximately the 9th to the 13th century (England). Historians dispute the time-frames and variation among European countries, but appear to agree that feudalism used a perpetual divide between the few landed aristocracy (monarchs, lords, and some tenants) and the vast majority of the population who worked the land as free or unfree laborers.]

10. Shaik, *Capitalism*; Kubiszewski et al., *Beyond GDP*, 57–68; Wilkinson & Pickett, *The World We Need*, 61–82.

11. Oliver & Shapiro, *Black Wealth/White Wealth*, 5 [income is distinct from wealth; income is money earned through one's labor while wealth is money earned through ownership of assets, such as land and stock]; Cassidy, *Picketty's Inequality Story in Six Charts*.

12. Urban Institute, *Nine Charts About Wealth Inequality in America*; Reeves & Krause, *Raj Chetty in 14 Charts*.

13. Gonick & Kasser, *Hypercapitalism*, 15; Reich, *Higher Wages Can Save America's Economy*.

14. Piketty, *Capital and Ideology*; Reich, *Saving Capitalism*; Gonick & Kasser, *Hypercapitalism*.

15. Global Justice Now, *69 of the Richest 100*.

16. Kelly & Howard, *The Making of a Democratic Economy*, 6.

17. Esty & Winston, *Green to Gold*, 11–21, 30–31 [similarly, over a decade earlier, Unilever, one of the world's largest purchasers of fish, committed to sourcing 100 percent of its fish from sustainable sources by 2005 because of its enlightened self-interest of not destroying fish stocks]; RE100, *Companies*.

18. Williams, *The Cooperative Movement*, 22–23.

19. Kelly & Howard, *The Making of a Democratic Economy*, 80–81 [quoting Millstone (2019)].

20. University of Wisconsin-Madison Center for Cooperatives, *History*; Kelly & Ratner, *Wealth Creation in Rural Communities*.

21. International Co-operative Alliance, *Cooperative Identity*; Fici, *An Introduction to Cooperative Law*, 33.

22. Orr, *Dangerous Years*, 14–21, 76–98; Kelly & Howard, *The Making of a Democratic Economy*, 3–4.

23. García, *Spain*, 702; H.R. 561 (United States).

Chapter 2. Corporate Purpose and Governance for a Livable Planet

1. Richardson & Sjåfjell, *Capitalism, the Sustainability Crisis, and the Limitations of Current Business Governance*, 5; Global Reporting Initiative, *Sustainability and Reporting Trends in 2025*, 5; Scanlan, *Climate Change, System Change, and the Path Forward*, 7.

2. Global Reporting Initiative, *Sustainability and Reporting Trends in 2025*, 5; Dernbach, *Sustaining America*, 32.

3. Light, *The Law of the Corporation as Environmental Law*, 137, 159, 164.
4. Sjåfjell et al., *Shareholder Primacy*, 86–87, 88; Greenfield, *The Failure of Corporate Law*, 75.
5. Sjåfjell et al., *Shareholder Primacy*, 81.
6. Sjåfjell et al., *Shareholder Primacy*, 79; Greenfield, *The Failure of Corporate Law*, 43.
7. Dodge v. Ford Motor Co., 170 N.W. 668 (Mich. 1919); Greenfield, *The Failure of Corporate Law*, 41, 42.
8. Liao, *Limits to Corporate Reform and Alternate Legal Structures*, 277, 278; Sjåfjell et al., *Shareholder Primacy*, 79–80.
9. Liao, *Limits to Corporate Reform and Alternate Legal Structures*, 280; Light, *The Law of the Corporation as Environmental Law*, 181–82 (citing Milton Friedman, A Friedman Doctrine—The Social Responsibility of Business Is to Increase Its Profits, N.Y. Times Mag., Sept. 13, 1970).
10. Liao, *Limits to Corporate Reform and Alternate Legal Structures*, 282.
11. Frank Bold, *Responsible Companies*; Frank Bold, *What Is the Purpose of the Corporation?*; Greenfield, *The Failure of Corporate Law*, 16.
12. Kubiszewski et al., *Beyond GDP*, 57–68.
13. Sjåfjell et al., *Shareholder Primacy*, 85; Greenfield, *The Failure of Corporate Law*, 25; Frank Bold, *The Purpose of the Corporation*, 3, 5.
14. Frank Bold, *The Purpose of the Corporation Project*, 8; Modern Corporation, *Statement on Company Law*.
15. Light, *The Law of the Corporation as Environmental Law*, 181–85.
16. Sjåfjell et al., *Shareholder Primacy*, 125, 145–46.
17. Sjåfjell et al., *Shareholder Primacy*, 137.
18. Research interview with Prof. Tom Webb.
19. Light, *The Law of the Corporation as Environmental Law*, 201, 205.
20. Light, *The Law of the Corporation as Environmental Law*, 207, 210.
21. Sjåfjell et al., *Shareholder Primacy*, 90, 91.
22. Greenfield, *The Failure of Corporate Law*, 125–52.
23. British Academy, *Future of the Corporation*; British Academy, *Principles for Purposeful Business*, 8, 13, 16, 20–21; Lipton, *The Future of the Corporation*.
24. Frank Bold, *Responsible Companies*; Frank Bold, *The Purpose of the Corporation Project*, 9.
25. Frank Bold, *The Purpose of the Corporation Project*, 7.
26. Alliance for Corporate Transparency, *2018 Research Report*.
27. Business Roundtable, *Statement on the Purpose of a Corporation*.
28. Sjåfjell et al., *Shareholder Primacy*, 94, 146–47; Richardson & Sjåfjell, *Capitalism, the Sustainability Crisis, and the Limitations of Current Business Governance*, 31–34.
29. Richardson & Sjåfjell, *Capitalism, the Sustainability Crisis, and the Limitations of Current Business Governance*, 1, 31–34; Greenfield, *A Skeptic's View of Benefit Corporations*, 18–19.
30. Morgan, *Legal Models Beyond the Corporation in Australia*, 183 (citing Galera and Borzaga, 2009, 213, 215); Fici, *A European Statute for Social and Solidarity-Based Enterprise*, 6, 7.

31. Fici, *A European Statute for Social and Solidarity-Based Enterprise*, 9.
32. Fici, *A European Statute for Social and Solidarity-Based Enterprise*, 10.
33. Morgan, *Legal Models Beyond the Corporation in Australia*, 182–83; Fici, *A European Statute for Social and Solidarity-Based Enterprise*, 7.
34. Fici, *A European Statute for Social and Solidarity-Based Enterprise*, 19, 20; Liao, *Limits to Corporate Reform and Alternate Legal Structures*, 293–310; Möslein, *Certifying 'Good' Companies*, 681.
35. Möslein, *Certifying 'Good' Companies*, 681; B Corps, *Home*; Liao, *Limits to Corporate Reform and Alternate Legal Structures*, 303–4, 305. [Carol Liao notes that in 2014 there were 1,045 B Corporations in 34 countries, but overwhelmingly in the U.S. By January 2019, that number had grown to 2,655 companies in 60 countries. While the growth is outstanding, the total is quite small when compared to 27 million businesses in the US alone; citing, US Census from 2009.]

Chapter 3. Social Enterprise Design

1. B Corps, *About B Lab*; Reiser, *Benefit Corporations*, 594; Möslein, *Certifying 'Good' Companies*, 670; Liao, *Limits to Corporate Reform and Alternate Legal Structure*, 303, 304. [Further, in states that lack Benefit Corporation or Constituency Statutes, B Lab allows companies to "build stakeholder interests into a signed term sheet, with an understanding that if the company's resident state eventually creates a benefit corporation, the company will adopt benefit corporation status by the end of their two-year certification term."]
2. Kurland, *Accountability and the Public Benefit Corporation*, 520.
3. Möslein, *Certifying 'Good' Companies*, 677.
4. Liao, *Limits to Corporate Reform and Alternate Legal Structures*, 305, 306 [quoting Reiser, Governing and Financing Blended Enterprise, 642].
5. Kelly & Howard, *The Making of a Democratic Economy*; Kelly & Stranahan, *Employee Ownership and Ecological Sustainability*; Stranahan & Kelly, *Mission-Led Employee-Owned Firms*, 21.
6. Kurland, *Accountability and the Public Benefit Corporation*, 519–20; Benefit Corporation, *State by State Status of Legislation* [this web page also provides links to all of the state statutes related to benefit corporations, their effective dates, bill sponsors, and key supporters]; Reiser, *Benefit Corporations*, 596; Benefit Corporation, *International Legislation*.
7. Clark, *Model Benefit Corporation Legislation*, § 102, § 201.
8. Clark, *Model Benefit Corporation Legislation*, § 301, § 301(a)(3); Greenfield, *A Skeptic's View of Benefit Corporations*, 18–19; Reiser, *Benefit Corporations*, 598–99.
9. Clark, *Model Benefit Corporation Legislation*, § 101.
10. Clark, *Model Benefit Corporation Legislation*, § 102, § 302, § 301(3), § 304, § 401(a)(1)(i).
11. Clark, *Model Benefit Corporation Legislation*, § 401(c).
12. Reiser, *Benefit Corporations*, 600–601; J. W. Hampton, Jr. & Co. v. United States, 276 U.S. 394 (1928).

13. Clark, *Model Benefit Corporation Legislation*, § 102, § 301(d), § 305(c).
14. Kurland, *Accountability and the Public Benefit Corporation*, 520, 521 [citing Reiser's research from 2011; citing Hemphill and Cullari's research from 2014 and Nass's research from 2014].
15. Möslein, *Certifying 'Good' Companies*, 669–70.
16. Möslein, *Certifying 'Good' Companies*, 670, 681.
17. Benefit Corporation, *Delaware*; Kurland, *Accountability and the Public Benefit Corporation*, 520, 521 [citing State of Delaware, 2016 S.362; citing State of Delaware, 2016 S.367]; British Academy, *Principles for Purposeful Business*, 8, 13, 16, 20–21.
18. Kurland, *Accountability and the Public Benefit Corporation*, 522, 522–25, 526–27, 528 [quoting Gilbert & El-Tahch, 2012; Her thinking on this builds on earlier work on "adaptive learning frameworks" by Ebrahim (2010), Ebrahim & Weisband (2007), and Cummings (2012)].
19. U.S. Federation of Worker Cooperatives, *Worker Ownership*.
20. U.S. Federation of Worker Cooperatives, *Worker Ownership*.
21. International Co-operative Alliance, *What Is a Cooperative?*; Henrÿ, *Basics and New Features of Cooperative Law*, 197.
22. H.R. 561 (United States); Henrÿ, *Basics and New Features of Cooperative Law*, 198.
23. Sustainable Economies Law Center, *Cooperatives*.
24. Packel, *The Law of Cooperatives*, 1, 2 [Packel elaborates on what he means by entrepreneur profit. "In a cooperative, all the members assume, in a broad sense, the economic risk, and they contemplate no return for the undertaking of the risk. In cooperatives there may be a return for the use of capital investment and even for the risk of loss, but there is no contemplation of an additional return on capital based upon the potentialities or the actualities of successful operation"]; Liao, *Limits to Corporate Reform and Alternative Legal Structures*, 287.
25. Packel, *The Law of Cooperatives*, 6–7.
26. Packel, *The Law of Cooperatives*, 12–27; University of Wisconsin-Madison Center for Cooperatives, *Types of Co-ops*; Birchall, *A Comparative Analysis of Co-operative Sectors in Scotland, Finland, Sweden, and Switzerland*, 11, Table 1.1; Fici, *A European Statute for Social and Solidarity-Based Enterprise*, 12–13, 18.
27. International Co-operative Alliance, *About Us* [Globally, the Alliance represents 317 cooperative federations and organizations in 110 countries, which totals 1.2 billion members of cooperatives]; International Co-operative Alliance, *Guidance Notes to the Co-operative Principles*, ix, 1, 71 ["The Alliance is now the largest non-governmental organisation in the world in terms of membership and has significant reach, recognition and influence as a formal consultative body with the United Nations (UN), the International Labour Organization (ILO) and the UN Food and Agriculture Organization (FAO)"]; International Co-operative Alliance, *Blueprint for a Co-operative Decade*, 39.
28. U.N.G.A. Res. 64/136, *Cooperatives in Social Development* (proclaiming the year 2012 the International Year of Cooperatives).
29. H.R. 561 (United States), 4; National Cooperative Business Association CLUSA, *Congressional Caucus*.

30. H.R. 561 (United States), 9; National Cooperative Bank, *The 2017 NCB Co-op 100*.
31. H.R. 561 (United States), 4–5.
32. H.R. 561 (United States), 8; U.S. Federation of Worker Cooperatives, *Home*; U.S. Federation of Worker Cooperatives, *Worker Ownership*.
33. Fici, *Italian Co-operative Law Reform and Co-operative Principles*, 24; García, *Spain*, 702; COCETA, *What Is Coceta?*; Generalitat de Catalunya, *Classes of Cooperatives* [The worker cooperative federation in Catalunya provides a list of different types of cooperatives and the laws that correspond to them.]; Generalitat de Catalunya, *Conditions that Cooperatives May Have*; Generalitat de Catalunya, *Cooperative Regulations* [This is another list of cooperative laws in Catalunya.]; Generalitat de Catalunya, *What Is a Cooperative?*
34. COCETA, *What Is Coceta?*; FEVECTA, *Valencian Work Cooperatives Already Exceed Pre-crisis Levels of Business and Employment*.
35. Birchall, *A Comparative Analysis of Co-operative Sectors in Scotland, Finland, Sweden, and Switzerland*, 16.
36. Birchall, *A Comparative Analysis of Co-operative Sectors in Scotland, Finland, Sweden, and Switzerland*, 15.
37. University of Wisconsin-Madison Center for Cooperatives, *History*.
38. Co-operative Heritage Trust, *Who We Are*.
39. International Co-operative Alliance, *Guidance Notes to the Co-operative Principles*, 1, 2 [Alliance members have revised these Principles three times after special commissions and consultation with Alliance members: Paris in 1937, Vienna in 1966, and in Manchester in 1995.].
40. University of Wisconsin-Madison Center for Cooperatives, *History*.
41. García, *Spain*, 702.
42. International Co-operative Alliance, *Guidance Notes to the Co-operative Principles*, 3.
43. International Co-operative Alliance, *Guidance Notes to the Co-operative Principles*, 1.
44. Co-operative Heritage Trust, *Who We Are*; International Co-operative Alliance, *Cooperative Identity, Values & Principles*.
45. International Co-operative Alliance, *Cooperative Identity, Values & Principles*.
46. International Co-operative Alliance, *Guidance Notes to the Co-operative Principles*, 2; Study Group on European Cooperative Law, *Draft Principles of European Cooperative Law*, 12.
47. International Co-operative Alliance, *Guidance Notes to the Co-operative Principles*, ix, 3.
48. International Co-operative Alliance, *Cooperative Identity, Values & Principles*.

Chapter 4. The Cooperative Difference, Private Governance, and the Law

1. García, *Cooperative Finance and Cooperative Identity*, 2.
2. Fici, *An Introduction to Cooperative Law*, 17–18.
3. U.S. Federation of Worker Cooperatives, *Worker Ownership*; Vermont Employee Ownership Center, *Who We Are*.

4. Fox, *A Rose by Any Other Name.*

5. Henrÿ, *Basics and New Features of Cooperative Law*, 213–23; Fici, *An Introduction to Cooperative Law*, 33.

6. Henrÿ, *Basics and New Features of Cooperative Law*, 213–23; Fici, *An Introduction to Cooperative Law*, 26–27.

7. García, *Cooperative Finance and Cooperative Identity*, 2, 4 (referencing the ICA's Statement on Cooperative Identity).

8. Fici, *An Introduction to Cooperative Law*, 32–33, note 108.

9. Fici, *An Introduction to Cooperative Law*, 33–34.

10. Vandenbergh & Gilligan, *Beyond Politics*, 9–10; RE100, *Progress and Insights Annual Report*, annex 1.

11. Light & Vandenbergh, *Private Environmental Governance* 254–55, 261.

12. Light & Vandenbergh, *Private Environmental Governance*, 257–60.

13. Light & Orts, *Parallels in Public and Private Environmental Governance*, 26–27.

14. Fici, *An Introduction to Cooperative Law*, 9–11.

15. Fici, *An Introduction to Cooperative Law*; Henrÿ, *Basics and New Features of Cooperative Law*; García et al., *Principles of European Cooperative Law.*

16. Henrÿ, *Basics and New Features of Cooperative Law*, 199, 200 [noting that the ILO recommendation is the "first and only instrument of universal applicability on cooperative law adopted by an international governmental organization"]; Fici, *An Introduction to Cooperative Law*, 66.

17. Henrÿ, *Basics and New Features of Cooperative Law*, 199; Fici, *An Introduction to Cooperative Law*, 5, 66.

18. Henrÿ, *Basics and New Features of Cooperative Law*, 201, 203 (In this paper, Hagen primary argues that ILO R. 193 constitutes binding international cooperative law); Fici, *An Introduction to Cooperative Law*, 15.

19. Henrÿ, *Basics and New Features of Cooperative Law*, 209–10.

20. Henrÿ, *Basics and New Features of Cooperative Law*, 210 [quoting ILO 193(7)(2)].

21. Henrÿ, *Basics and New Features of Cooperative Law*, 211 [quoting ILO 193(6)(d)].

22. Henrÿ, *Basics and New Features of Cooperative Law*, 212.

23. Henrÿ, *Basics and New Features of Cooperative Law*, 212.

24. U.N. Conference on Environment and Development, *Rio Declaration on Environment and Development.*

25. García et al., *New Study Group on European Cooperative Law*, note 3.

26. Henrÿ, *Basics and New Features of Cooperative Law*, 224; García et al., *New Study Group on European Cooperative Law*, 9; García et al., *Principles of European Cooperative Law.*

27. Study Group on European Cooperative Law, *Draft Principles of European Cooperative Law*, 18.

28. Study Group on European Cooperative Law, *Draft Principles of European Cooperative Law*, 26.

29. Study Group on European Cooperative Law, *Draft Principles of European Cooperative Law*, 28, 37.

30. Study Group on European Cooperative Law, *Draft Principles of European Cooperative Law*, 38, 41.
31. Study Group on European Cooperative Law, *Draft Principles of European Cooperative Law*, 89–91, 94 [The comments explain that some countries, such as Spain, waive the audit report requirement for small cooperatives. The legally required Spanish audit report is annual and includes non-financial indicators such as measures to promote cooperative development].
32. Study Group on European Cooperative Law, *Draft Principles of European Cooperative Law*, 95, 101.
33. Creation of author, with factors adapted from Henrÿ, *Basics and New Features of Cooperative Law*, 209–12; Study Group on European Cooperative Law, *Draft Principles of European Cooperative Law*, 89–91; International Co-operative Alliance, *Cooperative Identity, Values & Principles*.

Chapter 5. Sustainable by Design

1. Richardson & Sjåfjell, *Capitalism, the Sustainability Crisis, and the Limitations of Current Business Governance*, 22–23.
2. Richardson & Sjåfjell, *Capitalism, the Sustainability Crisis, and the Limitations of Current Business Governance*, 1; International Co-operative Alliance, *Guidance Notes to the Co-operative Principles*, 87.
3. International Co-operative Alliance, *Guidance Notes to the Co-operative Principles*, 86.
4. Richardson & Sjåfjell, *Capitalism, the Sustainability Crisis, and the Limitations of Current Business Governance*, 21; International Co-operative Alliance & Sustainability Solutions Group, *Sustainability Reporting for Co-operatives*, 5.
5. Sustainable Development Solutions Network, *Mapping the Renewable Energy Sector to the Sustainable Development Goals*, 6–10, 12.
6. Paris Agreement, Art. 4; Intergovernmental Panel on Climate Change Report, *Special Report on Global Warning of 1.5°C*, 24.
7. Millstone, *Frugal Value*, 2.
8. Richardson & Sjåfjell, *Capitalism, the Sustainability Crisis, and the Limitations of Current Business Governance*, 2; Millstone, *Frugal Value*, 190.
9. Millstone, *Frugal Value*, 185.
10. Millstone, *Frugal Value*, 4–5; Esty & Winston, *Green to Gold*, 11–21; Khan et al., *Corporate Sustainability*, 1703 (citing Friedman, 1970; Aupperle et al., 1985; McWilliams & Siegel, 1997; Jensen, 2002).
11. Eccles et al., *The Impact of Corporate Sustainability on Organizational Processes and Performance*, 2849; Khan et al., *Corporate Sustainability*, 1703–4; Light, *The Law of the Corporation as Environmental Law*, 203.
12. Khan et al., *Corporate Sustainability*, 1699, 1706, 1722.
13. Light, *The Law of the Corporation as Environmental Law*, 208; Lee, "Modernizing" Regulation S-K.

14. Richardson & Sjåfjell, *Capitalism, the Sustainability Crisis, and the Limitations of Current Business Governance*, 2–3, 4 (quoting World Business Council for Sustainable Development).

15. Fink, *Letter to CEOs*.

16. Fink, *Letter to CEOs*.

Chapter 6. Measuring and Reporting Sustainability

1. Global Reporting Initiative, *Sustainability and Reporting Trends in 2025*, 4, 7; Alliance for Corporate Transparency, *2018 Research Report*, 28.

2. Global Reporting Initiative & SustainAbility, *Insights from the GRI Corporate Leadership Group on Reporting 2025*, 6, 7, 24.

3. Global Reporting Initiative & SustainAbility, *Insights from the GRI Corporate Leadership Group on Reporting 2025*, 9; RE100, *Companies*.

4. Directive (EU) 2014/95/EU; Williams, *The Cooperative Movement*, 28 (advocating for corporate charters to require non-financial accounting of externalities); Ioannou & Serafeim, *The Consequences of Mandatory Corporate Sustainability Reporting*.

5. Ioannou & Serafeim, *The Consequences of Mandatory Corporate Sustainability Reporting*; Frank Bold Law Firm, *Responsible Companies*.

6. Alliance for Corporate Transparency, *2018 Research Report*, 12 [citing EU Directive 2014/95/EU].

7. Morrow, *Non-Financial Reporting*.

8. Amadeo, *Largest Economies in the World*; Alliance for Corporate Transparency, *2018 Research Report*, 6.

9. Alliance for Corporate Transparency, *2018 Research Report*, 26, 68.

10. Alliance for Corporate Transparency, *2018 Research Report*, 7, 32.

11. Alliance for Corporate Transparency, *2018 Research Report*, 7, 8, 27.

12. Alliance for Corporate Transparency, *2018 Research Report*, 9.

13. International Co-operative Alliance & Sustainability Solutions Group, *Sustainability Reporting for Co-operatives*, 4; International Co-operative Alliance, *Guidance Notes to the Co-operative Principles*, 94.

14. International Co-operative Alliance & Sustainability Solutions Group, *Sustainability Reporting for Co-operatives*, 11.

15. International Co-operative Alliance & Sustainability Solutions Group, *Sustainability Reporting for Co-operatives*, 12, 15–18.

16. International Co-operative Alliance & Sustainability Solutions Group, *Sustainability Reporting for Co-operatives*, 20.

17. International Co-operative Alliance & Sustainability Solutions Group, *Sustainability Reporting for Co-operatives*, 20, 27.

18. International Co-operative Alliance & Sustainability Solutions Group, *Sustainability Reporting for Co-operatives*, 24.

19. Duguid, *Non-Financial Tools and Indicators for Measuring the Impact of Co-Operatives*, 50.

20. Henrÿ, *Sustainable Development and Cooperative Law*, 9; Fici, *Italy*, 493, 495 [citing Art. 4, Legislative Decree 220/2002].

21. Williams, *The Cooperative Movement*, 27–28, 29 [citing Pretty 2003, 1913].

22. International Co-operative Alliance, *Blueprint for a Co-operative Decade*, 15.

23. International Co-operative Alliance & Sustainability Solutions Group, *Sustainability Reporting for Co-operatives*, 8.

24. International Co-operative Alliance & Sustainability Solutions Group, *Sustainability Reporting for Co-operatives*, 3; International Co-operative Alliance & Sustainability Solutions Group, *Co-operatives and Sustainability*; Duguid & Balkan, *Talking the Talk*, 8 [citing ICA research Co-operatives and Sustainability (Dale et al., 2013)].

25. Duguid & Balkan, *Talking the Talk*, 1, 8, 11.

26. Duguid & Balkan, *Talking the Talk*, 4, 6. 7 [citing Duguid, 2015a and b].

27. Duguid & Balkan, *Talking the Talk*, 13, 14.

28. Duguid & Balkan, *Talking the Talk*, 15–16, 17, 20.

29. Duguid & Balkan, *Talking the Talk*, 20, 21.

30. Bollas-Araya et al., *Sustainability Reports in European Cooperative Banks*, 32, 37 [citing Rodríguez-Gutiérrez et al., 2013, and Fernández and Souto, 2009, at 32, 36, 46–48, 50]; Duguid & Balkan, *Talking the Talk*, 7 [citing Bollas-Araya et al., 46].

31. Duguid & Balkan, *Talking the Talk*, 22.

32. Research interview with Maria Vicente; research interview with Milagros Perez.

33. Global Reporting Initiative, *Sustainability and Reporting Trends in 2025*, 4, 5.

34. Global Reporting Initiative, *Sustainability and Reporting Trends in 2025*, 5.

35. Economía Circular, *Circular Economy*; Burger, *Materials Consumption and Solid Waste*, 184.

36. Global Reporting Initiative, *Sustainability and Reporting Trends in 2025*, 8, 9.

37. Liao, *Limits to Corporate Reform and Alternative Legal Structures*, 274–75.

38. Interface, *The Interface Story*; Esty & Winston, *Green to Gold*.

39. Richardson & Sjåfjell, *Capitalism, the Sustainability Crisis, and the Limitations of Current Business Governance*, 6, 17.

40. Global Reporting Initiative, *Sustainability and Reporting Trends in 2025*, 27.

41. International Co-operative Alliance, *86th ICA International Co-operative Day 14th UN International Day of Cooperative 5 July 2008*, 1, 2; International Co-operative Alliance & Sustainability Solutions Group, *Sustainability Reporting for Co-operatives*, 28.

42. International Co-operative Alliance & Sustainability Solutions Group, *Sustainability Reporting for Co-operatives*, 3, 94.

43. International Co-operative Alliance, *Guidance Notes to the Co-operative Principles*, 85–86, 87–88, 93.

44. International Co-operative Alliance, *Guidance Notes to the Co-operative Principles*, 92, 93.

45. International Co-operative Alliance, *Guidance Notes to the Co-operative Principles*, 94–95; Henrÿ, *Sustainable Development and Cooperative Law*, 4.

46. Henrÿ, *Sustainable Development and Cooperative Law*, 9.

47. International Co-operative Alliance, *Guidance Notes to the Co-operative Principles*, 93.

Part Two. Case Studies

1. Intergovernmental Panel on Climate Change, *Special Report on Global Warming of 1.5°C*, A.1.2.
2. Intergovernmental Panel on Climate Change, *Special Report on Global Warming of 1.5°C*, A.2.2; Intergovernmental Panel on Climate Change, *Characteristics of Four Illustrative Model Pathways*.
3. Intergovernmental Panel on Climate Change, *Characteristics of Four Illustrative Model Pathways*.
4. U.S. Envtl. Prot. Agency, *Inventory of U.S. Greenhouse Gas Emissions and Sinks: 1990–2018* ; U.N. Env't Programme, *The Emissions Gap Report 2014*, Table 4.1; Gilbert, *One-Third of Our Green Gas Emissions Come from Agriculture* [citing S. J. Vermeulen, B. M. Campbell, & J. S. I. Ingram, 37 *Ann. Rev. Envtl. Resource* 195 (2012)].
5. Scanlan, *Sustainable Sewage*.
6. Rainforest Action Network, *Banking on Climate Change*, 8.
7. International Energy Agency, *Key CO2 Emissions Trends*, xii, Table 1; International Maritime Organization, *Greenhouse Gas Emissions*.
8. Intergovernmental Panel on Climate Change, *Special Report on Global Warming of 1.5°C*, C.2.2.
9. Intergovernmental Panel on Climate Change, *Special Report on Global Warming of 1.5°C*, C.2.5, C.3.4, C.3.5.
10. Pope Francis, Encyclical Letter, *Laudato Si'*; Intergovernmental Panel on Climate Change, *Special Report on Global Warning of 1.5°C*, D.6.3, D.7.

Chapter 7. Supportive Cooperative Ecosystems in Spain and the United States

1. Birchall, *A Comparative Analysis of Co-operative Sectors in Scotland, Finland, Sweden and Switzerland*, at 16, Table 1.2.
2. Diamantopoulos, *The Developmental Movement Model*, 48–49; Diamantopoulos, *Breaking Out of Co-Operation's 'Iron Cage,'* 208–10.
3. Diamantopoulos, *Breaking Out of Co-Operation's 'Iron Cage,'* 204.
4. Dayen, *Amazon Is Thriving Thanks to Taxpayer Dollars;* Williams, *The Cooperative Movement*, 20 [telling the story of Middlefield, Ohio's, city council voting to subsidize Walmart and refusing to do so for their existing H & H Hunt's Ace Hardware cooperative, which then went out of business].
5. García, *Spain*, 702.
6. Williams, *The Cooperative Movement*, 139, 143; Mondragón, *Home*.
7. Dubb, *How Much Outside Help Do Worker Co-Ops Need to Get to Scale?*

8. Williams, *The Cooperative Movement*, 137–38 [The steel mill, named ULGOR, preceded today's FAGOR].

9. Williams, *The Cooperative Movement*, 138; Eroski, *Annual Report 2017*, 6; U.S. Dep't of Agric., *Spain Retail Foods Report*, 2.

10. Williams, *The Cooperative Movement*, 138, 145.

11. Williams, *The Cooperative Movement*, 142.

12. Agència Catalana de Notícies, *Catalan Company Cata Buys Bankrupt Domestic Appliance Business Fagor*. [Fagor Electrodomésticos declared bankruptcy in 2013, and another corporation, Cata, purchased it.]

13. Cheney, *Values at Work*.

14. Williams, *The Cooperative Movement*, 3, 13, 24, 145.

15. Bamburg, *Mondragón Through a Critical Lens*. [In a short reflection article on a visit to Mondragón, Prof. Jill Bamburg notes she "did not see much traction in Mondragón in either practice or articulated values," focused on environmental sustainability.]

16. Mondragón Team Academy, *A Model That Supports Environmental Sustainability, Social Justice and Economic Democracy Through Radical Innovation on Education*, 18–19.

17. Mondragón, *Mondragón Green Community*; Mondragón, *Mondragón Eko*; Mondragón, *Corporate Management Model*, 51, 61.

18. García, *Spain*, 702.

19. COCETA, *What Is Coceta?*; CECOP, *What Is CECOP?*

20. COCETA, *What Is Coceta?*

21. FEVECTA, *Who Are We?*; Law 11/1985, of Oct. 25, on the Cooperatives of the Valencian Community; FEVECTA, *What Do We Do?*; FEVECTA, *Info Fevecta*.

22. FEVECTA, *Valencian Work Cooperatives Already Exceed Pre-Crisis Levels of Business and Employment*.

23. FEVECTA, *Who Are We?*

24. Impulso Cooperativo, *Nothing is what it was . . .* ; Serlicoop, *Serlicoop's Mission and Vision*; Grupo Sorolla Educación, *Know Us*.

25. Cooperativas Agro-alimentarias España, *What Are Agro-Food Cooperatives?*

26. Cooperatives Agro-alimentarias Comunitat Valenciana, *Larger Producer Organizations Offer Containment and Consumption Promotion*; Cooperatives Agro-alimentarias Comunitat Valenciana, *Our Cooperatives*.

27. Email with Rodrigo Gouveia; Hispacoop, *What Is Hispacoop?*; Hispacoop, *Do You Know the Difference Between Ecological, Biological and Organic?*

28. CECOP, *Our Work*; COCETA, *Business Innovation Conference*.

29. FEVECTA, *Publications*; FEVECTA, *Committed to Equality and CSR*.

30. Generalitat de Catalunya, *A Little History*.

31. Law 12/2015, of July 9, on cooperatives; Confederació de Cooperatives de Catalunya, *Mission, Vision and Values*; Confederació de Cooperatives de Catalunya, *What Are Cooperative Federations?*

32. García, *Spain*, 702, 703, note 4 [citing Constitución Española].

33. García, *Spain*, 703–4, 707.
34. García, *Spain*, 705 [citing Law 27/1999].
35. García, *Spain*, 706.
36. García, *Spain*, 708, 712, 715.
37. García, *Spain*, 712.
38. Law 5/2011, of March 29, on Social Economy, Article 4 [Spain amended this in 2015 with Law 31/2015, of Sept. 9, which updated regulations to foster and promote self-employment in the social economy]; Article 4 of the Spanish Social Economy Law states the following principles: "a) Primacy of people and the social purpose over capital, which is embodied in autonomous and transparent, democratic and participatory management, which leads to prioritizing decision-making more in terms of people and their contributions of work and services rendered to the entity or for a social purpose, that contributes to social capital. b) Application of the results from the economic activity mainly in terms of the work provided and service or activity carried out by the members and partners or by its members and, where appropriate, the social purpose of the entity. c) Promotion of internal solidarity and with society that favors the commitment to local development, equal opportunities for men and women, social cohesion, the insertion of people at risk of social exclusion, the generation of stable employment and quality, the reconciliation of personal, family and work life, and sustainability. d) Independence from public authorities"; CEPES, *What Is the Social Economy?*
39. Law 5/2011, of March 29, on social economy, Article 5.
40. Law 5/2011, of March 29, on social economy, Article 8, Article 9.
41. CEPES, *CEPES in Europe*.
42. CEPES, *Social Economy Companies in Spain*; CEPES, *State Statistics*.
43. FEVECTA, *In Figures*.
44. Research interview with María José Ortolá Sastre.
45. Generalitat Valenciana, *Making Cooperatives*, 4.
46. Generalitat Valenciana, *Procedure Detail*.
47. Email with María José Ortolá Sastre.
48. Burns, *Worker-Owners Cheer Creation of $1.2 Million Co-op Development Fund in NYC*, 5, 7.
49. Wikipedia, *Valencian Community*; NYC Planning, *Population*.
50. Generalitat Valenciana, *Making Cooperatives*, 3. [The employment number here is much higher than that provided by CEPES, above, and it is unclear why.]
51. Ajuntament de Barcelona, *The Commissioner's Office*; Ajuntament de Barcelona, *Resources*.
52. Ajuntament de Barcelona, *Budget*.
53. Ajuntament de Barcelona, *What Is Social and Solidarity Economy?*
54. Ajuntament de Barcelona, *Towards a New Socio-Economic Policy*.
55. H.R. 561 (United States).
56. U.S. Federation of Worker Cooperatives, *About the USFWC*; U.S. Federation of Worker Cooperatives, *Worker Ownership*.

57. National Council of Farmer Cooperatives, *About NCFC*; National Council of Farmer Cooperatives, *Farmer Co-ops Applaud Repeal of 2015 Water of the United States Rule*; National Council of Farmer Cooperatives, *Priorities & Policy Resolutions*.

58. National Cooperative Business Association CLUSA, *Education & Learning*.

59. Consumer Federation of America, *Issues*; Consumer Federation of America, *Reports*.

60. Harvard Transactional Law Clinics et al., *Tackling the Law, Together*, 1 [citing 26 U.S.C. §§ 1381–88 and *Puget Sound Plywood, Inc. v. Commissioner*, 44 T.C. 305 (1965), acq., 1966-1 C.B. 3]; Czarchorska-Jones et al., *United States*, 774; 26 U.S.C. § 501 (a) [Section 501(a) exempts organizations described in § 501(c) from the federal income tax, including organizations described in § 501(c)(12) (cooperative telephone and electric companies)]; 26 U.S.C. § 501(c)(14) [certain cooperative banks]; 26 U.S.C. § 501(c)(16) [famers' cooperative crop financing organizations].

61. Williams, *The Cooperative Movement*, 12; Czarchorska-Jones et al., *United States*, 766; 7 U.S.C. § 291.

62. 7 U.S.C. § 291.

63. Federal Farm Loan Act, Pub. L. No. 64-158, 39 Stat. 360; National Cooperative Bank, *Community Impact*.

64. National Cooperative Business Association CLUSA, *Our History*.

65. H.R. 561 (United States).

66. U.S. Dep't of Agric., *Rural Cooperative Development Grant Program*.

67. National Cooperative Business Association CLUSA, *Congressional Caucus*; Pro-Publica, *Bills Sponsored by Jared Polis (D-Colo.)*; National Cooperative Business Association CLUSA, *Cooperative for a Better Tomorrow*, 2.

68. Duda, *Elements of the Democratic Economy*; Shepherd, *2.3 Million Small Businesses Nationwide Owned by Aging Boomers Preparing to Retire Puts 1 in 6 Employees' Jobs at Risk, Based on a Project Equity Study*; Burns, *Worker-Owners Cheer Creation of $1.2 Million Co-op Development Fund in NYC*; 26 U.S.C. § 1042 (c)(2)(A-E).

69. Lechleitner, *Landmark Employee Ownership Act, Signed into Law Yesterday, Will Amend Lending Landscape for Worker Co-Ops*.

70. National Cooperative Business Association CLUSA, *State Cooperative Statute Library*; Sustainable Economies Law Center, *State-by-State Co-op Law Info*; Czarchorska-Jones et al., *United States*, 760.

71. Czarchorska-Jones et al., *United States*, 763, 767, 771; Dean & Geu, *The Uniform Limited Cooperative Association Act*, 66.

72. Czarchorska-Jones et al., *United States*, 771–72.

73. Dean & Geu, *The Uniform Limited Cooperative Association Act*, 72.

74. Dean & Geu, *The Uniform Limited Cooperative Association Act*, 75.

75. Uniform Law Commission, *Uniform Limited Cooperative Association Act*, 1.

76. Uniform Law Commission, *Uniform Limited Cooperative Association Act*, 1, 3.

77. Uniform Law Commission, *Uniform Limited Cooperative Association Act*, 5.

78. Levinson et al., *Alleviating Food Insecurity via Cooperative By-laws*, 248–50.

79. Oatfield, *Governor Brown Signs California Worker Cooperative Act, AB 816*; Assembly Bill 816, *Cooperative Corporations*, § 1(a) (California); García, *Spain*, 705 [citing Law 27/1999].

80. Burns, *Worker-Owners Cheer Creation of $1.2 Million Co-op Development Fund in NYC*; NYC Small Business Services, *Working Together*, 5, 7; Duda, *Elements of the Democratic Economy*.

81. Research interview with Ruth Rohlich; Madison Cooperative Development Coalition, *About*.

82. Ifateyo, *$5 Million for Co-op Development in Madison*; research interview with Ruth Rohlich.

83. Ifateyo, *$5 Million for Co-op Development in Madison*; Willy Street Co-op, *About Our Co-op*.

84. Madison Cooperative Development Coalition, *Home*; research interview with Ruth Rohlich [the first and to date only loan was to Union Cab, an existing worker-cooperative that used the loan to buy two accessible vehicles to refocus their business on serving people with special needs]; Madison Cooperative Development Coalition, *Year-End Report*.

85. Research interview with Ruth Rohlich.

86. Research interview with Ruth Rohlich.

87. Sustainable Economies Law Center, *Choice of Entity*.

88. Harvard Transactional Law Clinics et al., *Tackling the Law, Together*, 2, 4–6; Sustainable Economies Law Center, *Co-op Bylaws and Other Governance Documents*.

89. Author notes from conference call with cooperative attorneys, held by the U.S. Dep't of Agric. on Dec. 6, 2018 (on file with author).

90. Real Pickles Cooperative, Inc., *Bylaws*, Article I, §§ 4–6.

91. Levinson et al., *Alleviating Food Insecurity via Cooperative By-laws*, 269.

92. Real Pickles Cooperative, Inc., *Bylaws*, Article IV, § 7.

93. Real Pickles Cooperative, Inc., *Bylaws*, Article II, § 7.

94. Real Pickles Cooperative, Inc., *Articles of Organization*, Article VI.

Chapter 8. Renewable Energy

1. U.S. Envtl. Prot. Agency, *Global Emissions by Economic Sector*.

2. Mackie & Murray, Risky Business, 18–19; U.S. Energy Info. Admin., U.S. Renewable Energy Consumption Surpasses Coal for the First Time in Over 130 Years; Watts & Ambrose, Coal Industry Will Never Recover After Coronavirus Pandemic, Say Experts.

3. U.S. Energy Info. Admin., *Electricity Explained*; RED Eléctrica de España, *Renewable Energy in the Spanish Electricity System*, 10.

4. Sustainable Development Solutions Network, *Mapping the Renewable Energy Sector to the Sustainable Development Goals*, 12.

5. European Commission, *Clean Energy for All Europeans Package*.

6. United Nations, *About the Sustainable Development Goals*; European Commission, *Clean Energy for All Europeans Package*.

7. Som Energia, *Home*; Cobb EMC, *Solar Initiatives*; National Rural Electric Cooperative Association, *Solar*.

8. Som Energia, *About Us*; Som Energia. *Bylaws of Som Energia, SCCL*, Article; research interview with Irene Machuca.

9. Research interview with Irene Machuca; Som Energia, *Home*.

10. Som Energia, *About Us*; Som Energia, *Bylaws of Som Energia, SCCL*, Article 5, Article 7, Article 21.

11. Research interview with Irene Machuca [For this research, I interviewed one of the elected members of the Governing Board, Irene Machuca. Based in Seville, Spain, she has been involved in the cooperative since 2013, just three years after it originated].

12. Som Energia, *What Do We Do?*

13. Som Energia, *The Electrical System and Som Energia*.

14. Research interview with Prof. Germán Valencia Martin, April 8, 2019 (citing "Asi funciona el Mercado electrico en España," https://lucera.es.blog/asi-funciona-mercado-electrico-espana); Som Energia, *The Electrical System and Som Energia*.

15. RED Eléctrica de España, *Renewable Energy in the Spanish Electricity System*, 8.

16. Binnie & Rodríguez, *Spain Scraps "Sun Tax" in Measures to Cool Electricity Prices*.

17. Royal Decree-Law 15/2018, of Oct. 5, on urgent measures for the energy transition and consumer protection; Binnie & Rodríguez, *Spain Scraps "Sun Tax" in Measures to Cool Electricity Prices;* Gobierno de España, Council of Ministers, *Government Approves Plan to Combat Long-Term Unemployment*.

18. Neslen, *Spain Plans Switch to 100% Renewable Electricity by 2050*; Farand, *Spain Unveils Climate Law to Cut Emissions to Net Zero by 2050*.

19. Instituto Internacional de Derecho y Medio Ambiente, *The Six Priorities of the New Spanish Ministry for Ecological Transition, According to Environmental Lawyers;* Beam, *Italy to Phase Out Coal by 2025*; Watts & Ambrose, *Coal Industry Will Never Recover After Coronavirus Pandemic, Say Experts*.

20. Neslen, *Spain Plans Switch to 100% Renewable Electricity by 2050*.

21. Government of Spain, Council of Ministers, *Government Approves Plan to Combat Long-Term Unemployment;* Fernández, *Want to Use Clean Energy? This Cooperative from Madrid May Have the Answer*.

22. Research interview with Irene Machuca; Instituto Internacional de Derecho y Medio Ambiente, *The Spanish Strategic Framework for Climate & Energy Lacks Ambition and It Is Not Enough to Tackle Climate Change*.

23. Research interview with Irene Machuca.

24. Research interview with Irene Machuca.

25. Research interview with Irene Machuca; Som Energia, *Produces Renewable Energy*.

26. Som Energia, *Payment of Interest on Voluntary Contributions to Share Capital*; Som Energia, *Differences Between Contributing Capital Stock and Participating in the Generation kWh Project*.

27. Generation kWh, *Home*; Som Energia, *Payment of Interest on Voluntary Contributions to Share Capital*; Som Energia, *Differences Between Contributing Capital Stock and Participating in the Generation kWh Project*; email with Marco Sanchez.

28. Som Energia, *Produces Renewable Energy*.

29. Research interview with Irene Machuca.

30. Som Energia, *Bells*; Planter Solar, *We Open the Engineering Competition to Install 100 Photovoltaic Self-Production Roofs in Central Catalonia*.

31. Som Energia, *Bylaws of Som Energia, SCCL*, Article 2, Article 28.

32. Som Energia, *Governing Council*; Som Energia, *Bylaws of Som Energia, SCCL*, Article 34, Article 37, Article 40, Article 44; research interview with Irene Machuca.

33. Som Energia, *Processes*.

34. Som Energia, *Processes*.

35. Som Energia, *Som Energia School*.

36. Research interview with Irene Machuca.

37. Som Energia, *Home*.

38. Enercoop Grupo, *90th Anniversary of Enercoop*.

39. Enercoop Grupo, *Nine Decades of Fair, Clean and Efficient Distribution*.

40. Enercoop Grupo, *Installations*.

41. Enercoop Grupo, *90th Anniversary of Enercoop*; Enercoop Grupo, *Renewable Energy*; Enercoop Grupo, *Introduction*.

42. Enercoop Grupo, *The Enercoop Group Achieves Access to the Electricity Mega-Contract of the Public Universities of the Valencian Community*; Enercoop Grupo, *Cooperative Purchasing Group*; Enercoop Grupo, *The Enercoop Group Presents to Minister Clement Its Strategy to Empower the Consumer and Reduce the Cost of Electricity* [Enercoop is identified as Unión Electro Industrial, S.L.].

43. Enercoop Grupo, *The Enercoop Group Achieves Access to the Electricity Mega-Contract of the Public Universities of the Valencian Community*; Enercoop, *Join Us*.

44. Enercoop Grupo, *Today More Than Ever, Electric Cooperative*; research interview with Maria Vicente; SENEO, *About Us*, https://www.seneo.org/cont/2-quienes-somos; Enercoop Grupo, *Home*; Enercoop Grupo, *The Enercoop Group Achieves Access to the Electricity Mega-Contract of the Public Universities of the Valencian Community*.

45. Research interview with Prof. Germán Valencia Martin, April 8, 2019.

46. National Rural Electric Cooperative Association, *Electric Co-op Facts & Figures*; research interview with Prof. Germán Valencia Martin, April 8, 2019.

47. Research interview with Prof. Germán Valencia Martin, April 8, 2019; Directive (EU) 96/92, of the European Parliament and of the Council of Dec. 19, 1996, Concerning Common Rules for the Internal Market in Electricity; Law 54/1997, of Nov. 27, on the Electricity Sector; Law 17/2007, of July 4, amending Law 54/1997, of Nov. 27, on the Electricity Sector, to adapt it to the provisions of Directive 2003/54/EC, of the European Parliament and of the Council of June 26, 2003, on common rules for the internal electricity market.

48. Santos, *Spain*.

49. Order ITC/1522/2007, of May 24, which establishes the regulation of the guarantee of the origin of electricity from renewable energy sources and high-efficiency cogeneration [this was modified by Order ITC/2914/2011, of 27 of October and by the Order IET/931/2015, of 20 of May]; Circular 1/2018, of April 18, from the National Commission of Markets and Competition, which manages the guarantee system of origin of electricity from highly efficient renewable and cogeneration energy sources; Directive (EU) 2018/2001, of the European Parliament and of the Council of Dec. 11, 2018, on the Promotion of the Use of Energy from Renewable Sources, Article 2.12; research interview with Prof. Germán Valencia Martin, June 29, 2019.

50. Enercoop Grupo, *Today More Than Ever, Electric Cooperative*.

51. Enercoop Grupo, *Renewable Energy*; Enercoop Grupo, *The Enercoop Group Presents to Minister Clement Its Strategy to Empower the Consumer and Reduce the Cost of Electricity*.

52. Enercoop Grupo, *Cooperative Will Lead on Lowering Consumer's Electric Bill*; Enercoop Grupo, *90th Anniversary of Enercoop*.

53. Enercoop Grupo, *Cooperative Will Lead on Lowering Consumer's Electric Bill*.

54. Enercoop Grupo, *The Enercoop Group Presents to Minister Clement Its Strategy to Empower the Consumer and Reduce the Cost of Electricity*.

55. Enercoop Grupo, *90th Anniversary of Enercoop*; Enercoop Grupo, *Cooperativism*; Enercoop Grupo, *Nine Decades of Fair, Clean and Efficient Distribution*.

56. Enercoop Grupo, *Nine Decades of Fair, Clean and Efficient Distribution*.

57. Cobb EMC, *Our Mission*; research interview with Tim Jarrell and Kristen Delaney; email with Tim Jarrell and Kristen Delaney.

58. Cobb EMC, *Our Mission*; research interview with Tim Jarrell and Kristen Delaney; email with Tim Jarrell and Kristen Delaney; National Rural Electric Cooperative Association, *Case Study: Cobb EMC*, 1–2.

59. National Rural Electric Cooperative Association, *Solar*; research interview with Tim Jarrell and Kristen Delaney; Southern Power, *Sandhills Solar Facility*; Georgetown Climate Center, *Georgia Climate and Energy Profile*; National Rural Electric Cooperative Association, *Case Study: Cobb EMC*, 2–3. [To clarify, other cooperatives obtain a greater percentage of their energy supply from the broad category of renewables, but not as much as Cobb EMC from solar specifically (e.g., Vermont Electric Cooperative's power supply (not considering the RECs) includes gas/oil/coal for only 0.85% and the rest is from renewables and nuclear, but solar supplies only 1.51%. https://www.vermontelectric.coop/keeping-the-lights-on/power-supply.]

60. U.S. Energy Info. Admin., *How Much of U.S. Energy Consumption and Electricity Generation Comes from Renewable Energy Sources?*; U.S. Energy Info. Admin., *Renewable Energy Explained*; Smart Electric Power Alliance, *Community Solar Program Design Models*, 7; Smart Electric Power Alliance, *2019 Utility Solar Market Snapshot*, 6.

61. Smart Electric Power Alliance, *2019 Utility Solar Market Snapshot*, 18.

62. National Rural Electric Cooperative Association, *We Are America's Electric Cooperatives*.

63. Pacyniak, *Greening the Old New Deal: Reforming Rural Electric Cooperative Governance*, 10; Jeter et al., *Democracy and Dysfunction*, 376; U.S. Energy Info. Admin., *Investor-Owned Utilities Served 72% of U.S. Electricity Customers in 2017*. [The Public Utilities Holding Company Act of 1935 (PUHCA), also known as the Wheeler Rayburn Act, was designed to break up the holding companies, among other things. Yet, 80 years after PUHCA aimed to reduce such highly concentrated power, in 2017, only 168 investor-owned utilities control vast amounts of U.S. electric generation and provide it to de jure or de facto monopoly service territories for 110 million Americans.]

64. Rural Electrification Act of 1936 (United States); Exec. Order No. 7037 (United States) [establishing the Rural Electrification Administration (May 11, 1935); the REA was renamed the RUS in 1994 in a reorganization of the Department of Agriculture]; Pacyniak, *Greening the Old New Deal* [providing an in-depth historical overview of the development of electric cooperatives in the U.S.]; Brown, *Electricity for Rural America*.

65. National Rural Electric Cooperative Association, *We Are America's Electric Cooperatives*.

66. Jeter et al., *Democracy and Dysfunction*, 365.

67. National Rural Electric Cooperative Association, *Electric Co-op Facts & Figures*; Pacyniak, *Greening the Old New Deal*, 40–41; National Rural Electric Cooperative Association, Comments on Emission Guidelines for Greenhouse Gas Emissions from Existing Electric Utility Generating Units.

68. National Rural Electric Cooperative Association, *We Are America's Electric Cooperatives*; Van Atten et al., *Benchmarking Air Emissions of the 100 Largest Electric Power Producers in the United States*; Pacyniak, *Greening the Old New Deal*, 8.

69. Van Atten et al., *Benchmarking Air Emissions of the 100 Largest Electric Power Producers in the United States*.

70. Pacyniak, *Greening the Old New Deal*, 5–8, 42–43, 62; National Rural Electric Cooperative Association, *Flexible Approach Important to Energy Sector Transition, Matheson Tells Congress*.

71. Pacyniak, *Greening the Old New Deal*, 18.

72. Tri-State Generation and Transmission Association, *Tri-State Announces Transformative Responsible Energy Plan Actions to Advance Cooperative Clean Energy*; Law360, *Electric Co-Ops Ditch Plans for $2.8B Coal Plant in Kansas*.

73. Blockstein & Gibson, *Electric Cooperatives*; Hsu & Kelly, *How Georgia Became a Surprising Bright Spot in the U.S. Solar Industry*; Smart Electric Power Alliance, *2019 Utility Solar Market Snapshot*; Frangoul, *From California to Texas, These Are the US States Leading the Way in Solar*; RE100, *Companies*; Sullivan, *Apple Now Runs on 100% Green Energy, and Here's How It Got There*.

74. Blockstein & Gibson, *Electric Cooperatives*.

75. Cobb EMC, *Time Lapse of Sandhills Solar Facility Construction*; Cobb EMC, *Corporate Environmental Responsibility*; research interview with Tim Jarrell and Kristen Delaney.

76. Cobb EMC, *Energy Portfolio*; research interview with Tim Jarrell and Kristen Delaney; email with Tim Jarrell and Kristen Delaney; National Rural Electric Cooperative Association, *Case Study: Cobb EMC*, 2–3.

77. Research interview with Tim Jarrell and Kristen Delaney; Cobb EMC, *Policy No. 614*; Cobb EMC, *Distributed Generation Service Schedule DG-1*.

78. Gulley, *Rural Co-Ops and Public Utilities Have Voluntarily Built Nearly 100MW of Community Solar. Here's Why*; research interview with Tim Jarrell and Kristen Delaney.

79. Smart Electric Power Alliance, *Community Solar Program Design Models*, 6–8.

80. Smart Electric Power Alliance, *Community Solar Program Design Models*, 6, 9, 14.

81. Research interview with Tim Jarrell and Kristen Delaney.

82. Research interview with Tim Jarrell and Kristen Delaney; email with Tim Jarrell and Kristen Delaney.

83. Cobb EMC, *EV Charging for Business*.

84. Cobb EMC, *NiteFlex*; National Rural Electric Cooperative Association, *Case Study: Cobb EMC*, 2–6 [describing how *Cobb EMC* has had such a strong member response to its variable rates].

85. Research interview with Tim Jarrell and Kristen Delaney.

86. Green Power EMC, *About Green Power EMC*; research interview with Tim Jarrell and Kristen Delaney; email with Tim Jarrell; Oglethorpe Power, *Our Company*.

87. Green Power EMC, *About Green Power EMC*; Green Power EMC, *Frequently Asked Questions*.

88. Green Power EMC, *Home*; Green Power EMC, *Green Power EMC to Significantly Expand Its Solar Energy Portfolio with Construction Across Four Georgia Locations*.

89. Cobb EMC, *Corporate Environmental Responsibility*; research interview with Tim Jarrell and Kristen Delaney; email with Kristen Delaney.

90. Cobb EMC, *Corporate Environmental Responsibility*.

91. Email with Tim Jarrell and Kristen Delaney.

92. Research interview with Tim Jarrell and Kristen Delaney; email with Tim Jarrell and Kristen Delaney; Cobb EMC, *Strategic Plan 2020–2022*.

93. Cobb EMC, *Amended and Restated Bylaws of Cobb Electric Membership Corporation*.

94. Cobb EMC, *Annual Meeting*; research interview with Tim Jarrell and Kristen Delaney; email with Kristen Delaney; email with Tim Jarrell and Kristen Delaney.

95. Cobb EMC, *Board of Directors*; Cobb EMC, Amended and Restated Bylaws of Cobb Electric Membership Corporation, § 3.02, § 3.04, § 3.05; Cobb EMC, *Policy No. 611*.

96. Cobb EMC, *Policy No. 611*.

97. Cobb EMC, *Our Mission*; email with Tim Jarrell and Kristen Delaney.

Chapter 9. Food and Agriculture

1. International Panel of Experts on Sustainable Food Systems, *Breaking Away from Industrial Food and Farming Systems*, 8; Vermeulen et al., *Climate Change and Food Systems*, 195–222.

2. Vermeulen et al., *Climate Change and Food Systems*, 195–222.

3. Scarborough et al., *Dietary Greenhouse Gas Emissions of Meat-Eaters, Fish-Eaters, Vegetarians and Vegans in the UK*, 179–92.

4. Lehner & Rosenberg, *Legal Pathways to Carbon-Neutral Agriculture*, 10845.

5. Lehner & Rosenberg, *Legal Pathways to Carbon-Neutral Agriculture*, 10858–75.

6. Author observation at Consum, El Campello, Spain, 2019. [Internationally, coop is added to the name of cooperatives to indicate their distinct business form. Consum's headquarters have a sign with the "coop" designation on it, but none of the stores I saw used it.]

7. Consum, *The Commitment of Consum Is to Be Cooperative*; Consum, *History* [Consum has grown by acquiring large distribution companies and consolidating with smaller consumer cooperatives. For 14 years, Consum was part of the Eroski Group, which is part of Mondragon. Consum left this alliance in 2004 due to "divergences in the organizational model . . ."]; Consum, *Together as a Cooperative: Sustainability Report 2019.*

8. Research interview with Carman Picot and Ana Mffi Garcia, April 12, 2019.

9. Climent & Sanchis-Palacio, *The Consum Model*, 19–22, 26–27, 96–97 (2014); research interview with Carman Picot and Ana Mffi Garcia, April 12, 2019.

10. Research interview with Carman Picot and Ana Mffi Garcia, April 12, 2019.

11. Consum, *Mission, Vision & Core Values*; research interview with Carman Picot and Ana Mffi Garcia, April 12, 2019.

12. Research interview with Carman Picot and Ana Mffi Garcia, April 12, 2019; Consum, *We Have Memory.*

13. U.S. Dep't of Agric., *Spain Retail Foods Report.*

14. U.S. Dep't of Agric., *Spain Retail Foods Report*; Consum, *Sustainability Report 2017*; Consum, *Growth Works When It Is Sustainable.*

15. National Coop Grocers, *About Us.*

16. EuroCoop, *Who We Are*; EuroCoop, *Co-op Distinctiveness* (Euro Coop represents 7,000 European consumer retail cooperatives).

17. Research interview with Carman Picot and Ana Mffi Garcia, April 12, 2019.

18. Research interview with Carman Picot and Ana Mffi Garcia, April 12, 2019.

19. Consum, *Sustainability Report 2017*, 80; research interview with Carman Picot and Ana Mffi Garcia, April 15, 2019.

20. Consum, *Environmental Management to Not Leave a Footprint*; Consum, *Sustainability Report 2017*, 82–83; research interview with Sheila Ongie.

21. Consum, *Growth Works When It Is Sustainable*; Consum, *Sustainability Report 2017*, 31.

22. Consum, *Consum Opens Its 9th Eco-Efficient Supermarket of the Year in Baza;* Consum, *Environmental Management to Not Leave a Footprint*; Consum, *Consum Promotes 25 Initiatives to Achieve the SDGs on Its 5th Anniversary.*

23. European Commission, *EU Circular Economy Action Plan*; Rankin, *European Parliament Votes to Ban Single-Use Plastics*; European Commission, *European Parliament Votes for Single-Use Plastics Ban*; Directive (EU) 2019/904, of the European Parliament and of the Council of June 5, 2019, on the Reduction of the Impact of

Certain Plastic Products on the Environment [Article 5 (Restrictions on placing on the market) states: "Member States shall prohibit the placing on the market of the single-use plastic products listed in Part B of the Annex and of products made from oxo-degradable plastic." And single-use plastic cutlery, cotton buds, straws, and stirrers are mentioned in Part B of the Annex]; Consum, *Consum Removes the Plastic Bags from Its Online Shop*; Consum, *Environmental Management to Not Leave a Footprint*; research interview with Carman Picot and Ana Mffi Garcia, April 12, 2019; author observation at Consum, El Campello, Spain, 2019.

24. Research interview with Carman Picot and Ana Mffi Garcia, April 15, 2019; Garrote, *Quarterly Reports on Spain's Organic Sector*.

25. Consum, *The Commitment of Consum Is to Be Cooperative*; Consum, *Consum Increases up to 7 Weeks' Paid Paternity Leave, Two Weeks More Than Required by Law*; Consum, *Consum and Red Cross Renew Their Partnership in Employment and Social Action for Vulnerable Groups*; Consum, *Consum Increases the Salary of Its Workers by 1.5%*; Consum, *Consum Promotes 25 Initiatives to Achieve the SDGs on Its 5th Anniversary*.

26. Consum, *Sustainability Report 2017*, 89, 90–92; Consum, *A Real Commitment to Society*.

27. Research interview with Carman Picot and Ana Mffi Garcia, April 12, 2019.

28. Consum, *Management, Administration and Control Bodies*; research interview with Carman Picot and Ana Mffi Garcia, April 12, 2019.

29. Research interview with Carman Picot and Ana Mffi Garcia, April 12, 2019 (citing Art. 4, 2, 3 Consumer Code of Good Corporate Governance); Consum, *Regulation of the Management and Audit Control Committee*, 11–12.

30. Garrigues Law Firm, *Publication of the Law on Non-Financial Information and Diversity in Spain*; research interview with Carman Picot and Ana Mffi Garcia, April 30, 2019.

31. Research interview with Carman Picot and Ana Mffi Garcia, April 30, 2019.

32. Consum, *Ethical Code and Complaints Channel*; research interview with Carman Picot and Ana Mffi Garcia, Apr. 30, 2019.

33. BioTrèmol, *Stores*; Research Interview with Carmen Llinares and Jesus Arnaiz; bioTrèmol, *About Us*; bioTrèmol, *Mares Madrid*.

34. Research interview with Carmen Llinares and Jesus Arnaiz.

35. Research interview with Carmen Llinares and Jesus Arnaiz.

36. Reid, *Are Food Co-Ops Still Relevant?*; bioTrèmol, *The Whys of bioTrèmol*.

37. Garrote, *Quarterly Reports on Spain's Organic Sector*; research interview with Carmen Llinares and Jesus Arnaiz.

38. Garrote, *Quarterly Reports on Spain's Organic Sector*.

39. Supermarket News, *Maturing U.S. Organic Sector Grows 6.4% in 2017*; Reid, *Are Food Co-Ops Still Relevant?*; Parker, *Can Food Co-Ops Survive the New Retail Reality?*

40. Parker, *Can Food Co-Ops Survive the New Retail Reality?*; author observation at Hanover Coop, Hanover, New Hampshire, United States, 2019.

41. Research interview with Carmen Llinares and Jesus Arnaiz.

42. Research interview with Carmen Llinares and Jesus Arnaiz.

43. Research interview with Carmen Llinares and Jesus Arnaiz.

44. Navarro, *A Common Destiny*.

45. BioTrèmol, *About Us*; research interview with Carmen Llinares and Jesus Arnaiz.

46. Navarro, *A Common Destiny*; research interview with Carmen Llinares and Jesus Arnaiz.

47. Research interview with Carmen Llinares and Jesus Arnaiz.

48. Research interview with Carmen Llinares and Jesus Arnaiz.

49. Research interview with Carmen Llinares and Jesus Arnaiz.

50. BioTrèmol, *The Whys of bioTrèmol*; bioTrèmol, *Corporate Bylaws*, Article 1.

51. BioTrèmol, *Corporate Bylaws*, Article 1.

52. BioTrèmol, *Corporate Bylaws*, Article 3, Article 4.

53. Research interview with Carmen Llinares and Jesus Arnaiz; bioTrèmol, *Corporate Bylaws*, Article 23, Article 29, Article 30; bioTrèmol, *Mares Madrid*.

54. Research interview with Carmen Llinares and Jesus Arnaiz.

55. Research interview with Milagros Perez.

56. Bodegas Pinoso, *Winery*.

57. Bodegas Pinoso, *Bylaws*, Article 1, Article 3; research interview with Milagros Perez. [In addition to wine, they also joined together to produce nuts, primarily almonds, and purchase farm supplies.]

58. Research interview with Milagros Perez.

59. Research interview with Milagros Perez.

60. MX2 Global, *Spanish Wine History*; Food & Wines from Spain, *Spanish Wine in Figures*; Smith, *Spain Dominates Organic Wine Market*.

61. Food & Wines from Spain, *Spanish Wine in Figures*; Smith, *Spain Dominates Organic Wine Market*; see also Gilinsky Jr. et al., *Sustainability in the Global Wine Industry*, 42; Smith, *Spanish Organic Wines*.

62. Research interview with Milagros Perez; Maida, *Top 5 Vendors in the Organic Wine Market from 2017 to 2021* [Bodegas Pinoso is not far behind the world's largest organic wine producer, Emiliana Organic Vineyards with 800 hectares of organic grapes in Chile]; Bodegas Pinoso, *Winery* [Bodegas Pinoso's total annual production in 2018 was 5.5 million liters of wine annually from 1,200 hectares of vines. Half of their production is organic].

63. Bodegas Pinoso, *Winery*; Food & Wines from Spain, *Spain's Devotion to Organic and Sustainable Winemaking*; Gilinsky Jr. et al., *Sustainability in the Global Wine Industry*, 39; Zucca et al., *Sustainable Viticulture and Winery Practices in California*, 189–90, 192. [By 2006, almost 1,000 California vineyards had self-assessed using these tools.]

64. Krajnc, *Impact of Wine Production Life Cycle on the Environment*; Federación Española del Vino, *Why WFCP?* [Wineries for Climate Protection created a sustainability certification for Spanish wineries, but it is in the early stages and does not have many certified wineries.]

65. Krajnc, *Impact of Wine Production Life Cycle on the Environment*; Federación Española del Vino, *Why WFCP?*
66. Research interview with Milagros Perez.
67. Bodegas Pinoso, *Winery*; Council Regulation (EC) 834/2007 of June 28, 2007, on Organic Production and Labeling of Organic Products and Repealing Regulation (EEC) 2092/91; Commission Implementing Regulation (EU) 203/2012 of March 8, 2012, amending Regulation (EC) 889/2008, Laying Down Detailed Rules for the Implementation of Council Regulation (EC) 834/2007, as Regards Detailed Rules on Organic; Pages, *Albet I Noya* [However, Albet I Noya pioneered organic winemaking in Spain in 1979 when a Danish wine expert came to Spain and instigated organic production for export to Denmark. Albet I Noya is still producing organic wines in 2019, with eighty hectares of vineyards.]; Food & Wines from Spain, *Spain's Devotion to Organic and Sustainable Winemaking*; Bodegas Pinoso, *Ecological Wines*.
68. Krajnc, *Viticulture*.
69. Bodegas Pinoso, *Ecological Wines*; Bodegas Pinoso, *Bodega* ["la calidad, la sostenibilidad y la innovación . . ."].
70. Research interview with Milagros Perez.
71. Research interview with Milagros Perez.
72. Research interview with Milagros Perez.
73. Research interview with Milagros Perez.
74. Research interview with Milagros Perez.
75. Research interview with Milagros Perez.
76. Bodegas Pinoso, *Ecological Wines*; research interview with Milagros Perez.
77. Research interview with Milagros Perez.
78. Research interview with Milagros Perez.
79. Research interview with Milagros Perez.
80. Research interview with Milagros Perez.
81. Research interview with Milagros Perez; Trading Economics, *Spain Youth Unemployment Rate*.
82. Bodegas Pinoso, *Bylaws*, Articles 2 and 4 [translated from direct quote: "fomentar practicas de cultivo y tecnicas de produccion y de gestion de los residuos respetuosas con el medio ambiente, en especial para proteger la calidad de las aguas, del suelo y del paisaje y para preservar y/o potenciar la biodiversidad"].
83. Research interview with Milagros Perez.
84. Research interview with Milagros Perez.
85. Bodegas Pinoso, *Bylaws*, Article 1, Article 2, Article 28.
86. Cerveses Lluna Bio, *Home*; research interview with Maria Vicente; JustinTOOLs, *Convert 40000 Liters to US Beer Barrels* [Converting liters to U.S. barrels of beer, this is approximately 340 barrels. 1 US beer barrel = 117 liters].
87. Research interview with Maria Vicente.
88. Research interview with Maria Vicente.
89. Research interview with Maria Vicente.

90. Research interview with Maria Vicente.
91. Research interview with Maria Vicente.
92. Brewers Association, *National Beer Sales & Production Data*; Hede & Watne, *Leveraging the Human Side of the Brand Using a Sense of Place*, 207–24; Levay et al., *The Demographic and Political Composition of Mechanical Turk Samples*.
93. Acitelli, *What Is Organic Beer?*
94. Moyle, *10 Best Organic Beers*.
95. Furnari, *Wolaver's Organic Ceases Production*; Acitelli, *What Is Organic Beer?*
96. Research interview with Maria Vicente [there are some other coops producing organic beers in Spain: Gabarera outside of Madrid; Bandolera in Cordova produces one or two organic beers, but the rest are conventional; and Asgaya in Asturias]; Full Barrel Cooperative Brewery & Taproom, *Announcing Full Barrel Co-op*; Flying Bike Cooperative Brewery, *Home*; 4th Tap Brewing Co-Op, *Home*; Black Star Co-op, *Home* [however, none of these co-ops use their websites to promote environmental sustainability practices. Black Star simply mentions supporting local farms, with no details].
97. Brewers Association, *Sustainability Manuals*; Brewers Association, *Sustainability Benchmarking Tool*; Brewers Association, *Sustainability Benchmarking Report*, 4, 6.
98. Brewers Association, *Sustainability Benchmarking*.
99. Brewers Association, *Sustainability Benchmarking Report*, 2.
100. Brewers Association, *Sustainability Benchmarking Report*, 7.
101. Brewers Association, *Sustainability Benchmarking Report*, 14; research interview with Maria Vicente.
102. Research interview with Maria Vicente.
103. Research interview with Maria Vicente.
104. Research interview with Maria Vicente.
105. Cerveses Lluna Bio, *Environment*; research interview with Maria Vicente; SENEO, *About Us*; Enercoop Grupo, *Home*.
106. Research interview with Maria Vicente.
107. Research interview with Maria Vicente.
108. International Co-operative Alliance, *Cooperative Identity, Values & Principles*.
109. Research interview with Maria Vicente.
110. Research interview with George Siemon, May 31, 2019.
111. Research interview with George Siemon, May 31, 2019.
112. Research interview with George Siemon, May 31, 2019.
113. Research interview with George Siemon, May 31, 2019.
114. Research interview with George Siemon, Nov. 27, 2019; Schultz, *Organic Valley's Longtime CEO Steps Down*; research interview with George Siemon, May 31, 2019.
115. Research interview with George Siemon, May 31, 2019.
116. Research interview with George Siemon, May 31, 2019; Schultz, *Organic Valley's Longtime CEO Steps Down*.

117. Barrett, *'Struggling to Tread Water'*; Flynn, *Group Ranks 160 Organic Milk Brands*; B Corps, *Danone North America*.
118. Schultz, *Organic Valley's Longtime CEO Steps Down*; Lu, *Organic Valley Lays Off 39 Employees*.
119. Lu, *Organic Valley Lays Off 39 Employees*; research interview with George Siemon, Nov. 27, 2019.
120. Research interview with Jonathon Reinbold.
121. Research interview with George Siemon, May 31, 2019.
122. Research interview with Jonathon Reinbold.
123. Research interview with George Siemon, May 31, 2019.
124. Research interview with George Siemon, May 31, 2019.
125. Research interview with George Siemon, May 31, 2019.
126. Research interview with George Siemon, May 31, 2019.
127. U.S. Envtl. Prot. Agency, Inventory of U.S. Greenhouse Gas Emissions and Sinks: 1990–2015, 5-2, table 5-15; Lehner & Rosenberg, *Legal Pathways to Carbon-Neutral Agriculture*, 10848, 10852, 10855 [citing Ranjith P. Udawatta & Shibu Jose, Agroforestry Strategies to Sequester Carbon in Temperate North America, 86 *Agroforestry Sys.* 225, 239 (2012); citing Tiziano Gomiero et al., Environmental Impact of Different Agricultural Management Practices: Conventional vs. Organic Agriculture, 30 *Critical Rev. Plant Sci.* 95, 101–4, 109–11 (2011)].
128. Research interview with George Siemon, May 31, 2019; research interview with Jonathan Reinbold.
129. Research interview with George Siemon, May 31, 2019; research interview with Jonathan Reinbold.
130. Research interview with George Siemon, May 31, 2019; Schultz, *Organic Valley's Longtime CEO Steps Down*.
131. Research interview with Jonathan Reinbold.
132. Research interview with George Siemon, Nov. 27, 2019; research interview with George Siemon, May 31, 2019.
133. CROPP Cooperative, *Farmer Owned*; research interview with George Siemon, May 31, 2019.
134. Research interview with George Siemon, May 31, 2019.
135. CROPP Cooperative, *Farmer Owned*.
136. Research interview with George Siemon, May 31, 2019.
137. Research interview with George Siemon, May 31, 2019.

Chapter 10. Water

1. Milwaukee Metropolitan Sewage District, *Sustainable Water Reclamation*, 40.
2. Scanlan, *Sustainable Sewage*.
3. U.N.G.A. Res. 64/292, Human Right to Water and Sanitation; U.N. Department of Economic and Social Affairs, *Human Right to Water*; U.N. Committee on Economic, Social and Cultural Rights, *General Comment No. 15: The Right to Water*.

4. Scanlan et al., *Realizing the Promise of the Great Lakes Compact*; World Health Organization, *Drinking-Water*.
5. Knowledge@Wharton, *America's Neglected Water Systems Face a Reckoning*.
6. Clarke, *Inside the Bottle*, 9.
7. Barlow & Clarke, *Blue Gold*, 85; Water & Wastes Digest, *Report Ranks World's 50 Largest Private Water Utilities*.
8. Scanlan et al., *Realizing the Promise of the Great Lakes Compact*, 44.
9. Research interview with Joan Arévalo i Vilà.
10. Aigua.Coop, *The Exhibition "Water and Social Economy" of the Olesana Mining Community Is Inaugurated*; Aigua.Coop, *Thanks for the 150 Years of Loyalty*.
11. Research interview with Joan Arévalo i Vilà; Aigua.Coop, *For a New Management Model* [SCCL stands for Sociedad Cooperativa Catalana Limitada]; Aigua.Coop, *Thanks for the 150 Years of Loyalty*; Comunitat Minera Olesana, *Presentation and Welcome*; Aigua.Coop, *The Exhibition "Water and Social Economy" of the Olesana Mining Community Is Inaugurated*; Aigua.Coop, *25 Years of Cooperativism*.
12. Research interview with Joan Arévalo i Vilà.
13. Research interview with Joan Arévalo i Vilà; Comunitat Minera Olesana, *Report 2018*; Steinfort & Kishimoto, *From Terrassa to Barcelona*.
14. Comunitat Minera Olesana, *Report 2018*.
15. Generalitat de Catalunya, *What Is a Cooperative?*; Aigua.Coop, *For a New Management Model*; research interview with Joan Arévalo i Vilà; Comunitat Minera Olesana, *Report 2018*; Aigua.Coop, *Bylaws*, Article 2.
16. Research interview with Joan Arévalo i Vilà; Aigua.Coop, *You Write Water, You Read Democracy*; Law 7/1985, of April 2, Regulating the Bases of the Local Regime, Article 26.1.a [makes water service a public obligatory municipal service], Article 85.2, Article 86 [Article 86 makes water service an essential public municipal service which can be provided in monopoly, and Article 85.2 allows municipal water to be managed directly by the municipality or indirectly by a private person by means of license or concession.].
17. Food & Water Watch, *Water Privatization*.
18. Aigua.Coop, *For a New Management Model*.
19. Aigua.Coop, *For a New Management Model*; Comunitat Minera Olesana, *Report 2018*; research interview with Joan Arévalo i Vilà.
20. Albuquerque & Roaf, *On the Right Track*, 162.
21. Albuquerque & Roaf, *On the Right Track*, 162.
22. Aigua.Coop, *For a New Management Model*; Getzner et al., *Comparison of European Water Supply and Sanitation Systems*, 11–15.
23. Arana, *Water and Territory in Latin America*, 146–47; Social y Solidaria, *Federation of Drinking Water Cooperatives of the Province of Buenos Aires (FEDECAP)*.
24. Birchall, *A Comparative Analysis of Co-operative Sectors in Scotland, Finland, Sweden and Switzerland*, 33; P2P Foundation, *Water Cooperatives*.
25. University of Wisconsin-Madison Center for Cooperatives, *Water Cooperatives*; EJ Water Cooperative, Inc., *Our Story*.

26. Research interview with Joan Arévalo i Vilà.

27. Research interview with Joan Arévalo i Vilà.

28. Aigua.Coop, *25 Years of Cooperativism*; research interview with Joan Arévalo i Vilà; Forte, *Per Capita Daily Water Consumption in Spain 2000–2018* [however, the World Health Organization recommends an even lower usage or 80 liters per day]; U.S. Geological Survey, *Water Q&A*; *Gallons to Liters Conversion*.

29. Research interview with Joan Arévalo i Vilà.

30. Factor Energia, *Finally There Is Another Light*; research interview with Joan Arévalo i Vilà. [In 2019, 432 members were using this deal for renewable electricity for their households.]

31. Aigua.Coop, *Services Aigua.Coop.*

32. Research interview with Joan Arévalo i Vilà.

33. Ajuntament de Saint Hilari Sacalm, *Home*; research interview with Joan Arévalo i Vilà.

34. Comunitat Minera Olesana, *Bylaws*, Preamble 5, Article 2, Article 30.

35. Comunitat Minera Olesana, *Bylaws*, Article 63.

36. Aigua.Coop, *Bylaws*, Articles 2, Article 22(a)(2).

37. Comunitat Minera Olesana, *Bylaws*, Preamble 4, Article 39 and Article 48; Aigua. Coop, *You Write Water, You Read Democracy.*

38. Aigua.Coop, *You Write Water, You Read Democracy.*

39. Aigua.Coop, *Bylaws*, Article 15, Article 16, Article 18, Article 19, Article 26, Article 27, Article 28, Article 30, Article 32.

40. Research interview with Joan Arévalo i Vilà.

Chapter 11. Finance

1. Harvey, *Climate Crisis*; U.S. Commodity Futures Trading Commission, *Managing Climate Risk in the U.S. Financial System*, ii-iii.

2. Rainforest Action Network, *Banking on Climate Change*, 10; World Resources Institute, *Green Targets.*

3. Climent & Sanchis-Palacio, *Caixa Popular* [describing the making of a regional cooperative bank and its social model; this book is only available in Spanish and Valenciano]; Caixa Popular, *Origin and History*; research interview with Francesc "Paco Alós" Alabajos.

4. Research interview with Francesc "Paco Alós" Alabajos.

5. Research interview with Francesc "Paco Alós" Alabajos.

6. Caixa Popular, *Member Cooperatives*; research interview with Francesc "Paco Alós" Alabajos.

7. ABC Valencian Community, *Caixa Popular Increases its Profits by 20% and Reaches 12 Million Euros*; Research interview with Francesc "Paco Alós" Alabajos; Caixa Popular, *Bylaws*, Article 9, Article 10.

8. Research interview with Francesc "Paco Alós" Alabajos.

9. Caixa Popular, *Bylaws*, Article 3, Article 4; Climent & Sanchis-Palacio, *Caixa Popular*; ABC Valencian Community, *Caixa Popular Increases Its Profits by 20% and Reaches 12 Million Euros*.

10. Research interview with Francesc "Paco Alós" Alabajos.

11. FEVECTA, *Valencian Work Cooperatives Already Exceed Pre-crisis Levels of Business and Employment*; H.R. 561 (United States).

12. ABC Valencian Community, *Caixa Popular Increases Its Profits by 20% and Reaches 12 Million Euros*; Bollas-Araya et al., *Sustainability Reports in European Cooperative Banks*, 32, 36 [citing Rodríguez-Gutiérrez et al., 2013, and Fernández & Souto, 2009].

13. Martín & Sevillano, *Cooperative and Saving Banks in Europe*, 147; New Economics Foundation, *Stakeholder Banks*, 4; research interview with Francesc "Paco Alós" Alabajos.

14. Anzilotti, *More U.S. Businesses Are Becoming Worker Co-Ops*.

15. Shared Capital Cooperative, *About*.

16. Climent & Sanchis-Palacio, *Caixa Popular*; European Association of Co-Operative Banks, *Characteristics of the Co-Operative Banking Model*; Bollas-Araya et al., *Sustainability Reports in European Cooperative Banks*, 32, 40 [citing Rodríguez-Gutiérrez et al., 2013, and Fernández & Souto, 2009; citing Bank of Spain, 2012].

17. International Co-operative Alliance, *Guidance Notes to the Co-operative Principles*, 91.

18. Research interview with Francesc "Paco Alós" Alabajos.

19. Caixa Popular, *Bylaws*, Article 9, §§ 3, 4; research interview with Francesc "Paco Alós" Alabajos.

20. Pilcher, *Closing the Gender Pay Gap in Banking*; research interview with Francesc "Paco Alós" Alabajos.

21. Research interview with Francesc "Paco Alós" Alabajos; Anderson et al., *CEO–Worker Pay Ratios in the Banking Industry*, 1–2.

22. Anderson et al., *CEO–Worker Pay Ratios in the Banking Industry*, 1–3; research interview with Francesc "Paco Alós" Alabajos.

23. Equator Principles, *About*.

24. Caixa Popular, *Cooperative Principles*; Caixa Popular, *Committed to the Environment*; research interview with Francesc "Paco Alós" Alabajos.

25. Research interview with Francesc "Paco Alós" Alabajos; National Centre for Business and Sustainability for Co-operatives UK, *Key Social and Co-operative Performance Indicators*, 22.

26. Caixa Popular, *Committed to the Environment*.

27. Research interview with Francesc "Paco Alós" Alabajos.

28. Mazzucato & Semieniuk, *Financing Renewable Energy*, 8–22.

29. Economist Intelligence Unit, *Crédit Agricole's 2017 Net Income Rises by 3%*; Corporate Finance Institute, *Top Banks in Spain*; Rainforest Action Network, *Explore the Data*; Rainforest Action Network, *Banking on Climate Change*, 5.

30. Crédit Agricole Group, *Climate Finance*; Rainforest Action Network, *Banking on Climate Change*, 6, 44.

31. Santander, *Sustainability Report*, 5, 46–49.

32. Santander, *Sustainability Report*, 46–49; Mazzucato & Semieniuk, *Financing Renewable Energy*, 8–22 [Unfortunately, Mazzucato and Semieniuk (2004–14) analysis does not identify cooperative banks as a category of financial actors.]

33. Caixa Popular, *What Is a Work Cooperative?*; Caixa Popular, *Cooperative Principles*.

34. Caixa Popular, *Bylaws*, Article 2, Article 7; Caixa Popular, *What Is a Work Cooperative?*

35. Global Alliance for Banking on Values, *Principles*; Global Alliance for Banking on Values, *Updated Research*.

36. Research interview with Francesc "Paco Alós" Alabajos; Caixa Popular, *Bylaws*, Article 9, Article 10.

Chapter 12. Trade

1. World Trade Organization, *Climate Change and the Potential Relevance of WTO Rules.*

2. Groom, *U.S. Solar Group Says Trump Tariffs Killings Jobs.*

3. U.N. Conference on Trade & Development, *Climate Change and Trade.*

4. Kyoto Protocol to the UNFCCC, 1, 2; Hemmings, *Bunker Fuels and the Kyoto Protocol*, 1, 6; Subsidiary Body for Scientific and Technological Advice, Guidelines, Schedule and Process for Consideration; BBC News, *Reality Check*; National Public Radio, *Getting to Zero Carbon.*

5. International Energy Agency, *Key CO2 Emissions Trends*, xii, Table 1; International Maritime Organization, *Greenhouse Gas Emissions.*

6. Earle et al., eds., *Beyond NAFTA 2.0*, at 63–64 (citing Ben Beachy, "Discussion Paper: A New, Climate-Friendly Approach to Trade," Sierra Club, 2016).

7. European Parliament, *The European Parliament Declares Climate Emergency*; European Commission, *The European Green Deal*, 21.

8. Jaffee & Howard, *Who's the Fairest of Them All?*; World Fair Trade Organization Europe, *The 10 Principles of Fair Trade.*

9. World Fair Trade Organization, *The Big Idea of Fair Trade Is to Challenge the Purpose of Business.*

10. Fairtrade International, *What Is Fairtrade?*; Cooperative Coffees, *History of Fair Trade*; Jaffee & Howard, *Who's the Fairest of Them All?*, 816, 820.

11. International Co-operative Alliance, *Guidance Notes to the Co-operative Principles*, 92–93.

12. Research interview with Matthew Earley, May 21, 2019.

13. Research interview with Matthew Earley, May 21, 2019.

14. Just Coffee Cooperative, *About Just Coffee*; research interview with Matthew Earley, May 21, 2019.

15. Just Coffee Cooperative, *Coffee Tracker*; Just Coffee Cooperative, *Carbon Footprint Report*, 3 Cooperative Coffees, *Mission & Vision*.

16. Coffee & Cocoa International, *Carbon Tax Is New Phase in Roaster Co-operatives' Sustainable Agenda*, 40–41; research interview with Matthew Earley, May 21, 2019.

17. Research interview with Matthew Earley, May 21, 2019; Coffee & Cocoa International, *Carbon Tax Is New Phase in Roaster Co-operatives' Sustainable Agenda*, 40–41.

18. Research interview with Matthew Earley, May 21, 2019.

19. Jaffee & Howard, *Who's the Fairest of Them All?*, 816; Coffee & Cocoa International, *Carbon Tax Is New Phase in Roaster Co-operatives' Sustainable Agenda*, 40–41.

20. Research interview with Matthew Earley, July 8, 2019.

21. Research interview with Matthew Earley, May 21, 2019, July 17, 2019; Just Coffee Cooperative, *Sustainability*; Just Coffee Cooperative, *Coffee Tracker*.

22. Research interview with Matthew Earley, May 21, 2019.

23. Just Coffee Cooperative, *Doing Justice to Every Bag of Coffee We Roast*; Just Coffee Cooperative, *Sustainability*.

24. Research Interview with Matthew Earley, May 21, 2019.

25. Research interview with Matthew Earley, May 21, 2019.

26. Just Coffee Cooperative, *Sustainability*; research interview with Matthew Earley, May 21, 2019.

27. Just Coffee Cooperative, *Carbon Footprint Report*, 3.

28. Just Coffee Cooperative, *Carbon Footprint Report*, 5.

29. Research interview with Matthew Earley, May 31, 2019.

30. Cooperative Coffees, Proposal to Board of Directors, *Carbon, Climate, Coffee*, 1.

31. Research interview with Matthew Earley, May 30, 2019 [Just Coffee's total footprint in 2017 was 235.7 metric tons of CO_2]; Coffee & Cocoa International, *Carbon Tax Is New Phase in Roaster Co-operatives' Sustainable Agenda*, 40–41.

32. Research interview with Matthew Earley, May 21, 2019; Just Coffee Cooperative, *Carbon Footprint Report*, 15; Coffee & Cocoa International, *Carbon Tax Is New Phase in Roaster Co-operatives' Sustainable Agenda*, 40–41.

33. Mackie & Murray, *Risky Business*, 16; U.S. Commodity Futures Trading Commission, *Managing Climate Risk in the U.S. Financial System*, ii.

34. Hsu, *Carbon Pricing*, 82; Trujillo, *International Trade*, at 201–3, 216–24.

35. Hsu, *Carbon Pricing*, 75–77, 83–84.

36. Hsu, *Carbon Pricing*, 86–89; Lehner & Rosenberg, *Legal Pathways to Carbon-Neutral Agriculture* (providing more information about agriculture and carbon sequestration).

37. Research interview with Matthew Earley, May 21, 2019.

38. Research interview with Matthew Earley, May 21, 2019.

39. Research interview with Matthew Earley, May 21, 2019.

40. Just Coffee Cooperative, *Bylaws*, Article 2, Article 2.D; research interview with Matthew Earley, May 21, 2019.
41. Just Coffee Cooperative, *Second Restated Articles of Incorporation*, Article 5-8; research interview with Matthew Earley, May 21, 2019.
42. Research interview with Matthew Earley, May 21, 2019.

Chapter 13. Findings and Lessons for a Livable Planet

1. Montt et al., *Does Climate Action Destroy Jobs? An Assessment of the Employment Implications of the 2-Degree Goal*, 522, 531.

Appendix

1. Kvale, *InterViews*, 54.
2. Esty & Winston, *Green to Gold*, 24.
3. Research email with Pablo Sanchez.

BIBLIOGRAPHY

Acitelli, Tom, *What Is Organic Beer?*, Food Republic (June 2, 2016), https://www
.foodrepublic.com/2016/06/02/what-is-organic-beer/.

Agència Catalana de Notícies, *Catalan Company Cata Buys Bankrupt Domestic Appliance Business Fagor*, CatalanNews (July 29, 2014), https://www.catalannews.com/
business/item/catalan-company-cata-buys-bankrupt-domestic-appliance-business
-fagor.

Agresta, Pablo Ascasíbar, *The Work of Cooperatives Before the Challenge of Business Innovation* [*Las cooperativas de trabajo ante el reto de la innovación empresarial*] (2013),
COCETA, https://www.coceta.coop/publicaciones/jornada-innovacion-2013-agresta
.pdf.

Aigua.Coop, *Aigua.Coop: For a New Management Model* [*Aigua.Coop: per un nou model
de gestió*], http://aigua.coop/benvinguda-del-president/.

Aigua.Coop, Bylaws [Estatutos] (2018) (on file with author).

Aigua.Coop, *Services Aigua.Coop* [*Serveis Aigua.Coop*], http://aigua.coop/serveis-aigua
-coop/.

Aigua.Coop, *Thanks for the 150 Years of Loyalty* [*Un agraïment per 150 anys de fidelitat*],
http://aigua.coop/opinions/un-agraiment-per-150-anys-de-fidelitat/.

Aigua.Coop, *The Exhibition "Water and Social Economy" of the Olesana Mining Community Is Inaugurated* [*S'inaugura l'exposició "Aigua i Economia Social" de la Comunitat
Minera Olesana*] (2019), http://aigua.coop/noticies/sinaugura-lexposicio-aigua-i
-economia-social-emmarcada-dins-la-commemoracio-del-150e-aniversari/.

Aigua.Coop, *You Write Water, You Read Democracy* [*S'escriu aigua, es llegeix democrà-
cia*], http://aigua.coop/opinions/sescriu-aigua-es-llegeix-democracia/.

Aigua.Coop, *25 Years of Cooperativism* [*25 Anys de cooperativisme*], http://aigua.coop/
opinions/25-anys-de-cooperativisme/.

Ajuntament de Barcelona, *Budget*, https://ajuntament.barcelona.cat/economia-social
-solidaria/en/budget.

Ajuntament de Barcelona, *The Commissioner's Office*, https://ajuntament.barcelona
.cat/economia-social-solidaria/en/the-comissioners-office.

Ajuntament de Barcelona, *Resources*, https://ajuntament.barcelona.cat/omic/en/
responsible-consumption/resources.

Ajuntament de Barcelona, *Towards a New Socio-Economic Policy*, https://ajuntament
.barcelona.cat/economia-social-solidaria/en/towards-new-socio-economic-policy.

Ajuntament de Barcelona, *What Is Social and Solidarity Economy?*, https://ajuntament
.barcelona.cat/economia-social-solidaria/en/what-Social-and-Solidarity-Economy.

Ajuntament de Saint Hilari Sacalm, *Home*, https://www.santhilari.cat/.

Albuquerque, Catarina de, & Virginia Roaf, *On the Right Track: Good Practices in Re-
alising the Rights to Water and Sanitation* (UNESCO World Water Assessment Programme 2012), https://www.worldwatercouncil.org/sites/default/files/Thematics/
On_The_Right_Track_Book.pdf.

Alliance for Corporate Transparency, *2018 Research Report: The State of Corporate
Sustainability Disclosure Under the EU Non-Financial Reporting Directive* (2019),
https://www.allianceforcorporatetransparency.org/assets/2018_Research_Report
_Alliance_Corporate_Transparency-66d0af6a05f153119e7cffe6df2f11b094affe9aaf
4b13ae14db04e395c54a84.pdf.

Amadeo, Kimberly, *Largest Economies in the World: Why China Is the Largest, Even though
Some Say It's the U.S.*, The Balance (March 18, 2020), https://www.thebalance
.com/world-s-largest-economy-3306044.

Anderson, Sarah, et al., Institute for Policy Studies & Public Citizen, *CEO–Worker Pay
Ratios in the Banking Industry* (2018), https://inequality.org/wp-content/uploads/
2018/04/Bank-Pay-Ratios.pdf.

Anzilotti, Eillie, *More U.S. Businesses Are Becoming Worker Co-Ops: Here's Why*, Fast
Company (May 21, 2018), https://www.fastcompany.com/40572926/more-u-s
-businesses-are-becoming-worker-co-ops-heres-why.

Arana, Vladimir, *Water and Territory in Latin America: Trends, Challenges and Opportunities* (Springer 2016).

Assembly Bill 816, Cooperative Corporations: Worker Cooperatives, 2015–16 Sess.
(California 2015).

Bamburg, Jill, Fifty by Fifty, *Mondragón through a Critical Lens*, Medium (Oct. 3, 2017),
https://medium.com/fifty-by-fifty/mondragon-through-a-critical-lens-b29de
8c6049.

Barlow, Maude, & Tony Clarke, *Blue Gold: The Fight to Stop the Corporate Theft of the
World's Water* (New House 2005).

Barrett, Rick, *'Struggling to Tread Water': Dairy Farmers Are Caught in an Economic System with No Winning Formula*, Milwaukee J. Sentinel (Feb. 18, 2020), https://www
.jsonline.com/in-depth/news/special-reports/dairy-crisis/2019/05/16/wisconsin
-dairy-farms-closing-milk-prices-drop-economics-get-tough/3508060002/.

B Corps, *About B Lab*, https://bcorporation.net/about-b-lab.

B Corps, *B Impact Report: Danone North America*, https://bcorporation.net/directory/
danone-north-america.

B Corps, *Home*, https://bcorporation.net.

Beam, *Italy to Phase Out Coal by 2025*, CleanTechnica (Oct. 30, 2017), https://clean technica.com/2017/10/30/italy-phase-coal-2025/.

Benefit Corporation, *International Legislation*, https://benefitcorp.net/international -legislation.

Benefit Corporation, *State by State Status of Legislation*, https://benefitcorp.net/ policymakers/state-by-state-status.

Benefit Corporation, *State by State Status of Legislation: Delaware*, https://benefitcorp .net/policymakers/state-by-state-status?state=Delaware.

Binnie, Isla, & Jose Elías Rodríguez, *Spain Scraps "Sun Tax" in Measures to Cool Electricity Prices*, Reuters (Oct. 5, 2018), https://www.reuters.com/article/us-spain -politics-electricity/spain-scraps-sun-tax-in-measures-to-cool-electricity-prices -idUSKCN1MF1T0.

BioTrèmol, Amador Navarro, *A Common Destiny* [*Un destino común*] (June 10, 2018), https://biotremol.com/un-destino-comun/.

BioTrèmol, Bylaws of the "bioTrèmol" Cooperative [Estatutos sociales de la cooperativa "bioTrèmol"] (2012), https://biotremol.com/wp-content/uploads/2019/04/ 130312-estatutos-bioTrèmol-3.pdf.

BioTrèmol, *Home*, https://biotremol.com/.

BioTrèmol, *Mares Madrid*, https://biotremol.com/en-mares-madrid/.

BioTrèmol, *The Whys of bioTrèmol* [*Los porqués de bioTrèmol*] (2018), https://biotremol .com/los-porques-de-biotremol/.

BioTrèmol, *About Us* [*Sobre Nosotros*], https://biotremol.com/inicio/.

BioTrèmol, *Stores* [*Tiendas*], https://biotremol.com/tiendas/.

Birchall, Johnston, Co-operative Development Scotland, *A Comparative Analysis of Co-operative Sectors in Scotland, Finland, Sweden, and Switzerland* (2009), https://institute .coop/sites/default/files/resources/094%202009_Birchall_A%20comparative %20analysis%20of%20cooperative%20sectors%20in%20Scotland,%20Finland, %20Sweden,%20and%20Switzerland.pdf.

Black Star Co-op, *Home*, https://blackstar.coop.

Blockstein, Joshua, & Trevor Gibson, *Electric Cooperatives: The Key to Widespread-Solar Development in the Dakotas?*, Smart Electric Power Alliance (July 18, 2019), https://sepapower.org/knowledge/electric-cooperatives-the-key-to-widespread -solar-deployment-in-the-dakotas/.

Bodegas Pinoso, *Winery* [*Bodega*], https://www.bodegaspinoso.com/es/bodega/.

Bodegas Pinoso, *Ecological Wines*, https://www.bodegaspinoso.com/en/ecological -soul/.

Bodegas Pinoso, Bylaws of the Agrarian Cooperative "The Bodega of Pinoso, Coop. V" [Estatutos Sociales de la Cooperativa Agraria, "La Bodega de Pinoso, Coop. V"] (2005) (on file with author).

Bodegas Pinoso, *Ecological Wines* [*Vinos Ecológicos*], https://www.bodegaspinoso .com/es/alma-ecologica/.

Bodegas Pinoso, *Winery*, https://www.bodegaspinoso.com/en/winery/.

Bollas-Araya, Helena María, et al., *Sustainability Reports in European Cooperative Banks: An Exploratory Analysis*, 115 REVESCO. Cooperative Stud. Mag. 30 (2014).

Brewers Association, *National Beer Sales & Production Data* (2019), https://www.brewersassociation.org/statistics-and-data/national-beer-stats/.

Brewers Association, *Sustainability Benchmarking*, https://s3-us-west-2.amazonaws.com/brewersassoc/wp-content/uploads/2017/07/SustainabilityBenchmarking_1page.pdf.

Brewers Association, *Sustainability Benchmarking Report* (2017), https://www.brewersassociation.org/wp-content/uploads/2019/01/2017-Sustainability-Benchmarking-Report.pdf.

Brewers Association, *Sustainability Benchmarking Tool*, https://www.brewersassociation.org/educational-publications/sustainability-benchmarking-tool/.

Brewers Association, *Sustainability Manuals* (2016), https://www.brewersassociation.org/brewing-industry-updates/sustainability-manuals/.

British Academy, *Future of the Corporation* (2017), https://www.thebritishacademy.ac.uk/programmes/future-of-the-corporation.

British Academy, *Principles for Purposeful Business: How to Deliver the Framework for the Future of the Corporation* (2019), https://www.thebritishacademy.ac.uk/sites/default/files/future-of-the-corporation-principles-purposeful-business.pdf.

Brown, D. Clayton, *Electricity for Rural America: The Fight for the REA* (Robert Sobel ed., Greenwood Press 1941).

Burger, Michael, *Materials Consumption and Solid Waste*, in *Legal Pathways to Deep Decarbonization in the United States* 183 (Michael B. Gerrard & John C. Dernbach, eds., Envtl. L. Inst. Press 2019).

Burns, Rebecca, *Worker-Owners Cheer Creation of $1.2 Million Co-op Development Fund in NYC*, In These Times (July 2, 2014), https://inthesetimes.com/working/entry/16901/new_york_co_ops.

Business Roundtable, *Statement on the Purpose of a Corporation* (Aug. 19, 2019), https://opportunity.businessroundtable.org/wp-content/uploads/2020/03/BRT-Statement-on-the-Purpose-of-a-Corporation-with-Signatures.pdf.

Caixa Popular, *Bylaws* [*Estatutos sociales*], https://www.caixapopular.es/cms/estatico/rvia/caixapopular/ruralvia/es/particulares/informacion_institucional/gobierno/docs/Estatutos-Caixa-Popular.pdf?exp=TRUE.

Caixa Popular, *Caixa Popular Increases Its Profits by 20% and Reaches 12 Million Euros* [*Incrementa un 20% sus beneficios y alcanza los 12 millones de euros*], ABC Valencian Community (Jan. 17, 2019), https://www.abc.es/espana/comunidad-valenciana/abci-caixa-popular-incrementa-20-por-ciento-beneficios-y-alcanza-12-millones-euros-resultados-201901171446_noticia.html.

Caixa Popular, *Committed to the Environment* [*Comprometidos con el medio ambiente*], https://www.caixapopular.es/cms/estatico/rvia/caixapopular/ruralvia/es/particulares/segmentos/responsabilidad_social/medio_ambiente/medio ambiente/medioambiente/medioambiente.html?exp=TRUE.

Caixa Popular, *Cooperative Principles* [*Principios cooperativos*], https://www.caixapopular .es/cms/estatico/rvia/caixapopular/ruralvia/es/empresas/segmentos/somos _cooperativa/principios_cooperativos/index.html?exp=TRUE.

Caixa Popular, *Member Cooperatives* [*Cooperativas socias*], https://www.caixapopular.es/ cms/dinamico/caixapopular/ruralvia/es/particulares/informacion_institucional/ cooperativas_socias/productos/cooperativas_socias/cooperativas_socias.html.

Caixa Popular, *Origin and History* [*Origen e historia*], https://www.caixapopular .es/cms/estatico/rvia/caixapopular/ruralvia/es/particulares/informacion _institucional/la_entidad/la_entidad/index.html?exp=TRUE#.

Caixa Popular, *What Is a Work Cooperative?* [*¿Qué es una Cooperativa de Trabajo?*], https://www.caixapopular.es/cms/estatico/rvia/caixapopular/ruralvia/es/ particulares/segmentos/somos_cooperativa/cooperativa_de_trabajo/index.html.

Carbon Tax Is New Phase in Roaster Co-operatives' Sustainable Agenda, 44 Coffee & Cocoa International 40 (2017), https://online.fliphtml5.com/nzim/fmkx/#p=40.

Cartwright, Mark, *Feudalism*, Ancient History Encyclopedia (Nov. 22, 2018), https:// www.ancient.eu/Feudalism/.

Cassidy, John, *Piketty's Inequality Story in Six Charts*, New Yorker (March 26, 2014), https://www.newyorker.com/news/john-cassidy/pikettys-inequality-story-in-six -charts.

CECOP, *Our Work*, https://cecop.coop/works/0/0.

CECOP, *What Is CECOP?*, https://cecop.coop/aboutCecop.

CEPES, *CEPES in Europe*, https://www.cepes.es/principal/cepes_europe&lng=en#.

CEPES, *Social Economy Companies in Spain* [*Empresas de economía social en España*], https://www.cepes.es/cifras&lng=en.

CEPES, *Home*, https://www.cepes.es/.

CEPES, *State Statistics: Cooperatives Q4 2018*, https://www.cepes.es/social/statistics&. t=cooperatives.

CEPES, *What Is the Social Economy?* [*¿Qué es la economía social?*], https://www.cepes .es/social/econ_social_whats_it.

Cerveses Lluna Bio, *Environment* [*Medi ambient*] (2015), http://cerveseslluna.com/es/ cat/medi-ambient/.

Cerveses Lluna Bio, *Home*, http://cerveseslluna.com/en/.

Cheney, George, *Values at Work: Employee Participation Meets Market Pressure at Mondragón* (Cornell Univ. Press 1999).

Circular 1/2018, de 18 de abril, de la Comisión Nacional de los Mercados y la Competencia, por la que se regula la gestión del sistema de garantía de origen de la electricidad procedente de fuentes de energía renovables y de cogeneración de alta eficiencia [Circular 1/2018, of April 18, from the National Commission of Markets and Competition, which manages the guarantee system of origin of electricity from highly efficient renewable and cogeneration energy sources], Boletín Oficial del Estado [B.O.E.], Apr. 27, 2018 (Spain), https://www.boe.es/buscar/doc.php?id =BOE-A-2018–5717.

Clark, Bill, B Lab, *Model Benefit Corporation Legislation* (2017), https://benefitcorp.net/sites/default/files/Model%20benefit%20corp%20legislation%20_4_17_17.pdf.

Clarke, Tony, *Inside the Bottle: An Exposé of the Bottled Water Industry* (Polaris Inst. 2005).

Climent, Vanessa Campos, & Joan Ramon Sanchis-Palacio, *Caixa Popular: A Different Social Cooperative Banking Model* [*Caixa popular: Un model de banca cooperativa, social i diferent*] (Universitat de València ed., Vincle 2015).

Climent, Vanessa Campos, & Joan Ramon Sanchis-Palacio, *The Consum Model: A Responsible and Sustainable Cooperative* [*El modelo Consum: una cooperativa responsible y sostenible*] (Vincle 2014).

Cobb EMC, Amended and Restated Bylaws of Cobb Electric Membership Corporation (2016), https://cobbemc.com/sites/cobbemc/files/Current%20Site%20PDFs/Bylaws%20and%20Service%20Rules/2016/Master-Bylaws%20Amended%20May%2026%202016.pdf.

Cobb EMC, *Annual Meeting*, https://cobbemc.com/annual-meeting.

Cobb EMC, *Board of Directors*, https://cobbemc.com/board.

Cobb EMC, *Corporate Environmental Responsibility*, https://cobbemc.com/content/corporate-environmental-responsibility.

Cobb EMC, *Distributed Generation Service Schedule DG-1* (2020), https://cobbemc.com/sites/cobbemc/files/Current%20Site%20PDFs/Rates-Residential/Rate-Schedule-DG-1.pdf.

Cobb EMC, *Energy Portfolio*, https://cobbemc.com/content/energy-portfolio (June 16, 2020).

Cobb EMC, *EV Charging for Business*, https://cobbemc.com/content/ev-charging-business.

Cobb EMC, *NiteFlex: Charge Your EV and Smart Appliances for FREE*, https://cobbemc.com/content/niteflex.

Cobb EMC, *Our Mission*, https://cobbemc.com/ourmission.

Cobb EMC, *Policy No. 611: Open Meetings* (2017), https://cobbemc.com/sites/cobbemc/files/Current%20Site%20PDFs/Bylaws%20and%20Service%20Rules/Policy-611-1-24-2017.pdf.

Cobb EMC, *Policy No. 614: Distributed Generation Policy* (2019), https://cobbemc.com/sites/cobbemc/files/Current%20Site%20PDFs/Bylaws%20and%20Service%20Rules/Cobb-EMC-Distributed-Generation-Policy-Final.pdf.

Cobb EMC, *Solar Initiatives*, https://www.cobbemc.com/content/solar-initiatives.

Cobb EMC, *Strategic Plan 2020–2022* (2020), https://cobbemc.com/sites/cobbemc/files/Current%20Site%20PDFs/Our%20Mission/3YStrategic%20Plan.pdf.

Cobb EMC, *Time Lapse of Sandhills Solar Facility Construction*, YouTube (Dec. 20, 2016), https://www.youtube.com/watch?v=XMM0DqxxUqc.

COCETA, *Business Innovation Conference: The Cooperative for the Production and Consumption of Green Energy* [*Jornada Innovación Empresarial: La cooperativa de producción y consumo de energía verde*] (2013), https://www.coceta.coop/publicaciones/jornada-innovacion-2013-som-energia.pdf.

COCETA, *What Is Coceta?* [*¿Qué es Coceta?*], https://www.coceta.coop/coceta.asp.

Commission Implementing Regulation (EU) 203/2012 of 8 March 2012, amending Regulation (EC) 889/2008 Laying Down Detailed Rules for the Implementation of Council Regulation (EC) 834/2007, as Regards Detailed Rules on Organic Wine, 2012 O.J. (L 71) 42.

Comunitat Minera Olesana, *Annual Report 2018—Writing Board of Directors* [*Memòria 2018—escrit consell rector*] (2018), https://www.cmineraolesana.cat/estatus-i-memoria .html.

Comunitat Minera Olesana, Bylaws [Estatutos de la sociedad] (Adopted 2005, Amended 2011), https://www.cmineraolesana.cat/uploads/static/documents/ estatutos-cmo-2014.pdf.

Comunitat Minera Olesana, *Presentation and Welcome* [*Presentació i benvinguda*], https://www.cmineraolesana.cat/presentacio-i-benvinguda.html.

Confederació de Cooperatives de Catalunya, *Mission, Vision and Values* [*Missió, visió i valors*] (2017), https://www.coopcrativescatalunya.coop/index.php/ca/homepage/ missio-visio-i-valors.

Confederació de Cooperatives de Catalunya, *What Are Cooperative Federations?* [*¿Què són les federacians de cooperatives?*] (2019), https://www.cooperativescatalunya.coop/ index.php/ca/homepage/les-federacions.

Constitución Española [C.E], art. 192, B.O.E. n. 311, Dec. 29, 1978 (Spain).

Consum, *A Real Commitment to Society* [*Un compromiso real con la sociedad*], https:// decirhaciendo.consum.es/compromiso-social/.

Consum, *Consum and Red Cross Renew Their Partnership in Employment and Social Action for Vulnerable Groups* (2018), https://www.consum.es/en/consum-red-cross -employment.

Consum, *Consum Increases the Salary of Its Workers by 1.5%* (2018), https://www .consum.es/en/consum-increases-salary.

Consum, *Consum Increases up to 7 Weeks' Paid Paternity Leave, Two Weeks More Than Required by Law* (2018), https://www.consum.es/en/consum-increases-7-weeks -paternity.

Consum, *Consum Opens Its 9th Eco-Efficient Supermarket of the Year in Baza* (2018), https://www.consum.es/en/consum-opens-its-9th-eco-efficient-supermarket-of -the-year-in-baza.

Consum, *Consum Promotes 25 Initiatives to Achieve the SDGs on Its 5th Anniversary* [*Consum impulsa 25 iniciativas para la consecución de los ODS en su 5 aniversario*], Sep. 17, 2020, https://www.consum.es/consum-impulsa-25-iniciativas-apoyo-ods -5-aniversario.

Consum, *Consum Removes the Plastic Bags from Its Online Shop* (2018), https://www .consum.es/en/consum-removes-plastics-bags.

Consum, *Environmental Management to Not Leave a Footprint* [*Una gestión ambiental para no dejar huella*], https://decirhaciendo.consum.es/politicas-ambientales/.

Consum, *Ethical Code and Complaints Channel*, https://www.consum.es/en/ethical -code-and-complaints-channel.

Consum, *Growth Works When It Is Sustainable* [*El crecimiento funciona cuando es sostenible*], https://decirhaciendo.consum.es/desarrollo-sostenible/.

Consum, *History*, https://www.consum.es/en/history.

Consum, *Management, Administration and Control Bodies*, https://www.consum.es/en/organigram.

Consum, *Mission, Vision & Core Values*, https://www.consum.es/en/mission-vision-and-core-values.

Consum, *Regulation of the Management and Audit Control Committee* (2018), https://www.consum.es/wp-content/uploads/2018/04/Regulation_Control_Committee_EN.pdf.

Consum, *Report for the Future: Sustainability Report 2017* [*Hacer memoria para hacer futuro: memoria de sostenibilidad 2017*] (2018), https://decirhaciendo.consum.es/wp-content/uploads/memorias/2017/mobile/index.html#p=1.

Consum, *The Commitment of Consum Is to Be Cooperative* [*El compromiso de la gente consum es hacer cooperativa*], https://decirhaciendo.consum.es/trabajadores-comprometidos/.

Consum, *Together as a Cooperative: Sustainability Report 2019* [*Juntos es cooperativa: memoria de sostenibilidad 2019*] (2020), https://decirhaciendo.consum.es/wp-content/uploads/memorias/2019/mobile/index.html#p=1.

Consum, *We Have Memory* [*Tenemos memoria*], https://decirhaciendo.consum.es/tenemos-memoria/.

Consumer Federation of America, *Issues*, https://consumerfed.org/issues/.

Consumer Federation of America, *Reports*, https://consumerfed.org/reports/.

Cooperative Coffees, *History of Fair Trade*, https://coopcoffees.coop/fair-trade/history-of-fair-trade/.

Cooperative Coffees, *Mission & Vision*, https://coopcoffees.coop/about/mission-vision/.

Cooperative Coffees, Proposal to Board of Directors, *Carbon, Climate, Coffee* (2017) (on file with author).

Cooperatives Agro-alimentarias Comunitat Valenciana, *Larger Producer Organizations, Offer Containment and Consumption Promotion* [*Organizaciones de productores más grandes, contención de la oferta y promoción del consumo*] (2018), http://www.cooperativesagroalimentariescv.com/organizaciones-de-productores-mas-grandes-contencion-de-la-oferta-y-promocion-del-consumo/.

Cooperatives Agro-alimentarias Comunitat Valenciana, *Our Cooperatives* [*Nuestras cooperativas*], http://www.cooperativesagroalimentariescv.com/nuestras-cooperativas/.

Cooperativas Agro-alimentarias España, *What Are Agro-Food Cooperatives?* [*¿Qué es cooperativas agro-alimentarias?*], http://www.agro-alimentarias.coop/informacion_corporativa.

Corporate Finance Institute, *Top Banks in Spain: An Overview of Spain's Leading Financial Institutions*, https://corporatefinanceinstitute.com/resources/careers/companies/top-banks-in-spain/.

Council of Ministers, *Government Approves Plan to Combat Long-Term Unemployment*, Gobierno de España (April 5, 2019) https://www.lamoncloa.gob.es/lang/en/gobierno/councilministers/Paginas/2019/20190405council.aspx.

Council Regulation (EC) 834/2007 of 28 June 2007 on Organic Production and Labeling of Organic Products and Repealing Regulation (EEC) 2092/91, 2007 O.J. (L 189) 1.

Co-operative Heritage Trust, *Our History*, https://www.co-operativeheritage.coop/our -history-1.

Co-operative Heritage Trust, *Who We Are*, https://www.co-operativeheritage.coop/who-we-are.

Crédit Agricole Group, *Climate Finance*, https://www.credit-agricole.com/en/responsible-and-committed/our-csr-strategy-partnering-a-sustainable-economy/climate-finance.

CROPP Cooperative, *Farmer Owned: Governance*, https://www.farmers.coop/farmer -owned.

Czarchorska-Jones, Barbara, et al., *United States*, in International Handbook of Cooperative Law 759 (Dante Cracogna et al. eds., Springer 2013).

Dayen, David, *Amazon Is Thriving Thanks to Taxpayer Dollars*, New Republic (Jan. 9, 2018), https://newrepublic.com/article/146540/amazon-thriving-thanks-taxpayer -dollars.

Dean, James B., & Thomas Earl Geu, *The Uniform Limited Cooperative Association Act: An Introduction*, 13 Drake J. Agric. L. 63 (2008).

Dernbach, John C., *Introduction*, in Legal Pathways to Deep Decarbonization in the United States 1 (Michael B. Gerrard & John C. Dernbach eds., Envtl. L. Inst. Press 2019).

Dernbach, John C., *Sustaining America*, 29 Envtl. F. 30 (2012).

Diamantopoulos, Mitch, *Breaking Out of Co-Operation's 'Iron Cage': From Movement Degeneration to Building a Developmental Movement*, 83 Annals Pub. & Cooperative Econ. 199 (2012).

Diamantopoulos, Mitch, *The Developmental Movement Model: A Contribution to the Social Movement Approach to Co-operative Development*, 45 J. Co-operative Stud. 42 (2012).

Directive (EU) 96/92, of the European Parliament and of the Council of 1 Dec. 19, 1996, Concerning Common Rules for the Internal Market in Electricity, 1996 O.J. (L 27) 1.

Directive (EU) 2014/95/EU of the European Parliament and of the Council of Oct. 22, 2014, Amending Directive 2013/34/EU as Regards Disclosure of Non-Financial and Diversity Information by Certain Large Undertakings and Groups, 2014 O.J. (L 330) 1.

Directive (EU) 2018/2001, of the European Parliament and of the Council of Dec. 11, 2018, on the Promotion of the Use of Energy from Renewable Sources, 2018 O.J. (L 328) 82.

Directive (EU) 2019/904, of the European Parliament and of the Council of June 5, 2019, on the Reduction of the Impact of Certain Plastic Products on the Environment, 2019 O.J. (L 155) 1.

Dodge v. Ford Motor Co. 170 N.W. 668 (Mich. 1919).

Dubb, Steve, *How Much Outside Help Do Worker Co-Ops Need to Get to Scale?*, Shelterforce (July 10, 2014), https://shelterforce.org/2014/07/10/how_much_outside _help_do_worker_co-ops_need_to_get_to_scale/.

Duda, John, *Elements of the Democratic Economy: Worker Cooperatives*, Next System Project (Jan. 30, 2019), https://thenextsystem.org/learn/stories/worker-cooperatives ?mc_cid=793c5aa0ef&mc_eid=70bc7f58ca.

Duguid, Fiona, *Non-Financial Tools and Indicators for Measuring the Impact of Co-Operatives*, 5 J. Co-Operative Acct. & Reporting 40 (2017).

Duguid, Fiona & Donna Balkan, *Talking the Talk: Canadian Co-operatives and Sustainability Reporting*, 4 J. Co-Operative Acct. & Reporting 1 (2016).

Earle, Ethan, et al. eds., *Beyond NAFTA 2.0: Toward a Progressive Trade Agenda for People and Planet* (Rosa Luxemburg Stiftung 2019).

Eccles, Robert G., et al., *The Impact of Corporate Sustainability on Organizational Processes and Performance*, 60 Mgmt. Sci. 2835 (2014).

Economía Circular, *Circular Economy: Towards an Eco-Efficient Economy in the Use of Resources*, https://economiacircular.org/EN/?page_id=62.

Economist Intelligence Unit, *Crédit Agricole's 2017 Net Income Rises by 3%* (Feb. 19, 2018), https://www.eiu.com/industry/article/1256442509/credit-agricoles-2017 -net-income-rises-by-3/2018–02–19.

EJ Water Cooperative, Inc., *Our Story*, https://www.ejwatercoop.com/about-us/our -story.

Electric Co-Ops Ditch Plans for $2.8B Coal Plant in Kansas, Law360 (Jan. 15, 2020), https:// www.law360.com/energy/articles/1234795/electric-co-ops-ditch-plans-for-2–8b- coal-plant-in-kansas?nl_pk=b242149f-7156–4186-a61e-b3b4a2d71d0e&utm_ source=newsletter&utm_medium=email&utm_campaign=energy?copied=1.

Enercoop Grupo, *Cooperative Purchasing Group* [*Grupo de compras cooperativo*], https:// www.enercoop.es/grupo-de-compras-cooperativo/.

Enercoop Grupo, *Cooperative Will Lead on Lowering Consumer's Electric Bill* [*Cooperativa dará protagonismo al consumidor para rebajar su factura eléctrica*], https://www .enercoop.es/facturaelectrica/.

Enercoop Grupo, *Cooperativism* [*Cooperativismo*], https://www.enercoop.es/ cooperativismo/.

Enercoop Grupo, *Home*, https://www.enercoop.es.

Enercoop Grupo, *Installations* [*Instalaciones*], https://www.enercoop.es/instalaciones -tecnicas/.

Enercoop Grupo, *Introduction* [*Introduccion*], https://www.enercoop.es/introduccion/.

Enercoop Grupo, *Join Us* [*Ùnete a nosotros*], https://www.enercoop.es/unete-a -nosotros/.

Enercoop Grupo, *Nine Decades of Fair, Clean and Efficient Distribution* [*Nueve décadas de distribución justa, limpia y eficiente*], https://www.enercoop.es/historia/.

Enercoop Grupo, *Renewable Energy* [*Energias renovables*], https://www.enercoop.es/energias-renovables/.

Enercoop Grupo, *The Enercoop Group Achieves Access to the Electricity Mega-Contract of the Public Universities of the Valencian Community* [*El grupo enercoop logra acceder al megacontrato de electricidad de las universidades públicas de la comunitat valenciana*], https://www.enercoop.es/contratouniversidades/.

Enercoop Grupo, *The Enercoop Group Presents to Minister Clement Its Strategy to Empower the Consumer and Reduce the Cost of Electricity* [*El grupo enercoop expone ante el conseller climent su estrategia para empoderar al consumidor y abaratar el coste de la electricidad*], https://www.enercoop.es/visitaconsellereconomia/.

Enercoop Grupo, *Today More Than Ever, Electric Cooperative* [*Hoy más que nunca, cooperativa eléctrica*], https://www.enercoop.es/filosofia-y-futuro/.

Enercoop Grupo, *90th Anniversary of Enercoop* [*90 Aniversario de Enercoop*], https://www.enercoop.es/90-aniversario-enercoop/.

Equator Principles, *About*, https://equator-principles.com/about/.

Eroski, *Annual Report 2017* [*Memoria 2017*], https://corporativo.eroski.es/wp-content/uploads/2018/06/Memoria_Eroski_2017_.pdf.

Esty, Daniel C., & Andrew S. Winston, *Green to Gold: How Smart Companies Use Environmental Strategy to Innovate, Create Value, and Build Competitive Advantage* (Yale Univ. Press 2006).

EuroCoop, *Co-op Distinctiveness*, https://www.eurocoop.coop/our-priorities/co-op-distinctiveness.html.

EuroCoop, *Who We Are*, https://www.eurocoop.coop/about-us/Who-We-Are/.

European Association of Co-Operative Banks, *Characteristics of the Co-Operative Banking Model*, https://www.eacb.coop/en/cooperative-banks/definition-and-characteristics.html.

European Commission, *Clean Energy for All Europeans Package* (2020), https://ec.europa.eu/energy/topics/energy-strategy/clean-energy-all-europeans_en.

European Commission, *EU Circular Economy Action Plan* (2020), https://ec.europa.eu/environment/circular-economy/index_en.htm.

European Commission, *European Parliaments Vote for Single-Use Plastics Ban* (2019), https://ec.europa.eu/environment/efe/news/european-parliament-votes-single-use-plastics-ban-2019–01–18_en.

European Commission, *The European Green Deal*, COM (2019) 640 final (Nov. 12, 2019), https://eur-lex.europa.eu/legal-content/EN/TXT/?qid=1588580774040&uri=CELEX:52019DC0640.

European Parliament, Press Release, The European Parliament Declares Climate Emergency (Nov. 29, 2019), https://www.europarl.europa.eu/news/en/press-room/20191121IPR67110/the-european-parliament-declares-climate-emergency.

Exec. Order No. 7037, 7 C.F.R. § 1700.1 (1935) (United States).

Factor Energia, *Finally There Is Another Light* [*Por fin hay otra luz*], https://www .factorenergia.com/es/.

Fairtrade International, *What is Fairtrade?*, https://www.fairtrade.net/about/what-is -fairtrade.

Farand, Chloé, *Spain Unveils Climate Law to Cut Emissions to Net Zero by 2050*, Climate Home News (May 18, 2020), https://www.climatechangenews.com/2020/05/18/ spain-unveils-climate-law-cut-emissions-net-zero-2050/.

Federación Española del Vino, *Why WFCP?*, https://www.fev.es/cambio-climatico/ english/why-wfcp_228_1_ap.html.

Federal Farm Loan Act, Pub. L. No. 64-158, 39 Stat. 360 (United States).

Fernández, Miguel Ezquiaga, *Want to Use Clean Energy? This Cooperative from Madrid May Have the Answer*, El País (April 24, 2019), https://english.elpais.com/elpais/ 2019/04/22/inenglish/1555918325_976914.html.

FEVECTA, *Committed to Equality and CSR* [*Compromesos amb la igualtat i la RSE*], https://www.fevecta.coop/info-fevecta/compromesos-amb-la-igualtat/.

FEVECTA, *In Figures* [*En xifres*], https://www.fevecta.coop/cooperatives-de-treball/en -xifres/.

FEVECTA, *Info Fevecta*, https://www.fevecta.coop/info-fevecta-1/.

FEVECTA, *Publications* [*Publicacions*], https://www.fevecta.coop/publicacions/.

FEVECTA, *Valencian Work Cooperatives Already Exceed Pre-crisis Levels of Business and Employment* [*Las cooperativas de trabajo valencianas superan ya los niveles de empresas y empleo anteriores a la crisis*] (2017), https://www.fevecta.coop/index.asp?ra_id=7& no_id=2079#.XpXLBS2ZNQP.

FEVECTA, *What Do We Do?* [*¿Qué fem?*], https://www.fevecta.coop/info-fevecta/que -fem/.

FEVECTA, *What Is a Worker Cooperative?* [*¿Qué es una cooperativa de treball?*], https:// www.fevecta.coop/cooperatives-de-treball/que-es-una-cooperativa-de-treball/.

FEVECTA, *Who Are we?* [*¿Qui som?*], https://www.fevecta.coop/info-fevecta/qui -som/.

Fici, Antonio, *A European Statute for Social and Solidarity-Based Enterprise* (2017), https://www.europarl.europa.eu/RegData/etudes/STUD/2017/583123/IPOL _STU(2017)583123_EN.pdf.

Fici, Antonio, *An Introduction to Cooperative Law*, in International Handbook of Cooperative Law 3 (Dante Cracogna et al. eds., Springer 2013).

Fici, Antonio, *Italian Co-operative Law Reform and Co-operative Principles* (Euricse, Working Paper N. 002|10, 2010), https://www.euricse.eu/wp-content/uploads/ 2015/03/WP_002_FICI.pdf.

Fici, Antonio, *Italy*, in International Handbook of Cooperative Law 479 (Dante Cracogna et al., eds., Springer 2013).

Fink, Larry, BlackRock, *Letter to CEOs: A Fundamental Reshaping of Finance* (2020), https://www.blackrock.com/corporate/investor-relations/larry-fink-ceo-letter?cid =ppc:CEOLetter:PMS:US:NA&gclid=CjoKCQiAo4XxBRD5ARIsAGFygj_BTTxyp

_DLRHd6kmLfWREmqqcWsTnb4ZMWrLXyXlhctNlC3KNLHKUaAppDEALw
_wcB&gclsrc=aw.ds.

Flying Bike Cooperative Brewery, *Home*, https://flyingbike.coop.

Flynn, Dan, *Group Ranks 160 Organic Milk Brands; Says USDA Has Failed Consumers*,
Food Safety News (Aug. 29, 2018), https://www.foodsafetynews.com/2018/08/
group-ranks-160-organic-milk-brands-says-usda-has-failed-consumers/.

Food & Water Watch, *Water Privatization: Facts and Figures* (2015), https://www
.foodandwaterwatch.org/insight/water-privatization-facts-and-figures.

Food & Wines from Spain, *Spain's Devotion to Organic and Sustainable Winemaking*,
https://www.foodswinesfromspain.com/spanishfoodwine/global/wine/features/
feature-detail/REG2020847620.html.

Food & Wines from Spain, *Spanish Wine in Figures* (2017), https://www
.foodswinesfromspain.com/spanishfoodwine/wcm/idc/groups/public/
documents/documento_anexo/mde3/nzcw/~edisp/dax2017770176.pdf.

Foreign Agric. Serv., U.S. Dep't of Agric., SP1813, Spain Retail Foods Report (2018),
https://apps.fas.usda.gov/newgainapi/api/report/downloadreportbyfilename
?filename=Retail%20Foods_Madrid_Spain_5–29–2018.pdf.

Forte, Fernando, *Per Capita Daily Water Consumption in Spain 2000–2018*, Statista
(Sept. 23, 2019), https://www.statista.com/statistics/801686/per-capita-daily
-water-consumption-in-spain/.

4th Tap Brewing Coop, *Home*, https://www.4thtap.coop.

Fox, Ed, Letter to Members, *A Rose by Any Other Name . . .* , Co-op News (Aug. 14,
2018), http://coopnews.coop/a-rose-by-any-other-name/.

Frangoul, Anmar, *From California to Texas, These Are the US States Leading the Way
in Solar*, CNBC (Sept. 19, 2018), https://www.cnbc.com/2018/09/19/the-us-states
-leading-the-way-in-solar.html.

Frank Bold Law Firm, *Responsible Companies*, https://en.frankbold.org/our-work/
programme/responsible-companies.

Frank Bold Law Firm, *The Purpose of the Corporation Project: Concept Note*, http://www
.purposeofcorporation.org/documents/project_outputs/purpose_project_concept
_note.pdf.

Frank Bold Law Firm, *What Is the Purpose of the Corporation?*, YouTube (Oct. 6, 2016),
https://www.youtube.com/watch?v=mwLhoR04Qe8&feature=emb_logo.

Friedman, Thomas L., *The World Is Flat: A Brief History of the Twenty-First Century*
(Farrar, Straus and Giroux 2006).

Full Barrel Cooperative Brewery & Taproom (@FullBarrelCoop), *Announcing Full
Barrel Co-op—Vermont's First Co-operative Brewery*, BeerAdvocate (Sept. 27, 2014),
https://www.beeradvocate.com/community/threads/announcing-full-barrel-co
-op-vermonts-first-co-operative-brewery.215549/.

Furnari, Chris, *Wolaver's Organic Ceases Production*, Brewbound (Dec. 30, 2015),
https://www.brewbound.com/news/wolavers-organic-ceases-production.

Gallons to Liters Conversion, http://www.gallonstoliters.com.

García, Gemma Fajardo, *Cooperative Finance and Cooperative Identity* (Euricse, Working Paper N. 045|12, 2012), https://www.euricse.eu/wp-content/uploads/2015/03/1358347206_n2283.pdf.

García, Gemma Fajardo, *Spain*, in International Handbook of Cooperative Law 701 (Dante Cracogna et al. eds., Springer 2013).

García, Gemma Fajardo, et al., *New Study Group on European Cooperative Law: "Principles" Project* (Euricse, Working Paper N. 024|12, 2012), https://www.euricse.eu/wp-content/uploads/2015/03/1329215779_n1963.pdf.

García, Gemma Fajardo, et al., *Principles of European Cooperative Law: Principles, Commentaries and National Reports* (Intersentia 2017).

Garrote, Conchi, *New: Quarterly Reports on Spain's Organic Sector*, Organic Food Iberia (Nov. 22, 2018), https://www.organicfoodiberia.com/news/el-lanzamiento-del-sector-ecologico-de-espana-2/.

Garrigues Law Firm, *Publication of the Law on Non-Financial Information and Diversity in Spain* (2018), https://www.garrigues.com/en_GB/new/publication-law-non-financial-information-and-diversity-spain.

Generalitat de Catalunya, *A Little History* [*Una mica d'història*] (2018), https://treball.gencat.cat/ca/ambits/economia_social/que_es_l_economia_social/coneixer_l_economia_social/una_mica_d_historia/.

Generalitat de Catalunya, *Classes of Cooperatives* [*Classes de cooperatives*] (2015), https://treball.gencat.cat/ca/ambits/economia_social/que_es_l_economia_social/que_son_les_cooperatives_i_les_s/que_es_una_cooperativa/classes_de_cooperatives/.

Generalitat de Catalunya, *Conditions that Cooperatives May Have* [*Condicions que poden tenir les cooperatives*] (2015), https://treball.gencat.cat/ca/ambits/economia_social/que_es_l_economia_social/que_son_les_cooperatives_i_les_s/que_es_una_cooperativa/condicions_de_les_cooperatives/.

Generalitat de Catalunya, *Cooperative Regulations* [*Normativa de cooperatives*] (2019), https://treball.gencat.cat/ca/ambits/economia_social/recursos/normativa/normativa_cooperatives/.

Generalitat de Catalunya, *What Is a Cooperative?* [¿*Què és una cooperative?*], https://treball.gencat.cat/ca/ambits/economia_social/que_es_l_economia_social/que_son_les_cooperatives_i_les_s/que_es_una_cooperativa/.

Generalitat Valenciana, *Making Cooperatives: Support Plan and Promotion of Cooperativism of the Valencian Community I* [*Fent Cooperatives: I Plan de apoyo y fomento del cooperativismo de la comunitat valenciana*] (2018/2019), http://www.gvaoberta.gva.es/documents/7843050/162503991/Plan+de+apoyo+y+fomento+del+cooperativismo+de+la+Comunitat+Valenciana+2018–2019.pdf/b161a7b2-c487–4587–9912-bf73bfcb4fee.

Generalitat Valenciana, *Procedure Detail* [*Detalle de Procedimientos*], https://www.gva.es/es/inicio/procedimientos?id_proc=18081.

Generation kWh, *Home*, https://www.generationkwh.org/.

Georgetown Climate Center, *Georgia Climate and Energy Profile: Regulation and Policy*, https://www.georgetownclimate.org/clean-energy/clean-energy-and-climate-data.html?state=GA#panel2–5.

Getzner, Michael, et al., *Comparison of European Water Supply and Sanitation Systems: Final Report* (Abridged Version) (Vienna Chamber of Lab. 2018), https://www.akeuropa.eu/sites/default/files/2019–01/Studie%20197b%20-%20Comparison%20of%20European%20water%20supply%20and%20sanitation%20systems. . . . pdf.

Gilbert, Natasha, *One-Third of Our Green Gas Emissions Come from Agriculture*, Nature (Oct. 31, 2012), https://www.nature.com/news/one-third-of-our-greenhouse-gas-emissions-come-from-agriculture-1.11708.

Gilinsky Jr., Armand, et al., *Sustainability in the Global Wine Industry*, 8 Agric. & Agric. Sci. Procedia 37 (2016).

Global Alliance for Banking on Values, *Principles* (2020), http://www.gabv.org/about-us/our-principles.

Global Alliance for Banking on Values, *Updated Research: Real Economy—Real Returns: The Business Case for Values-Based Banking* (2018), http://www.gabv.org/news/updated-research-real-economy-real-returns-the-business-case-for-values-based-banking-2018.

Global Justice Now, *69 of the Richest 100 Entities on the Planet Are Corporations, not Governments, Figures Show* (Oct. 17, 2018), https://www.globaljustice.org.uk/news/2018/oct/17/69-richest-100-entities-planet-are-corporations-not-governments-figures-show.

Global Reporting Initiative, *Sustainability and Reporting Trends in 2025: Preparing for the Future* (May 2015), https://www.globalreporting.org/resourcelibrary/Sustainability-and-Reporting-Trends-in-2025–1.pdf.

Global Reporting Initiative & SustainAbility, *Insights from the GRI Corporate Leadership Group on Reporting 2025: Future Trends in Sustainability Reporting* (Jan. 2017), https://sustainability.com/wp-content/uploads/2017/01/GRI-CLG_Report-FutureTrends2025-v12.pdf.

Gonick, Larry, & Tim Kasser, *Hypercapitalism: The Modern Economy, Its Values, and How to Change Them* (New Press 2018).

Greenfield, Kent, *A Skeptic's View of Benefit Corporations*, 1 Emory Corp. Governance & Accountability Rev. 17 (2014).

Greenfield, Kent, *The Failure of Corporate Law: Fundamental Flaws and Progressive Possibilities* (Univ. of Chicago Press 2006).

Green Power EMC, *About Green Power EMC*, https://greenpoweremc.com/content/about-green-power-emc-0.

Green Power EMC, *Frequently Asked Questions*, https://greenpoweremc.com/content/frequently-asked-questions.

Green Power EMC, *Green Power EMC to Significantly Expand Its Solar Energy Portfolio with Construction Across Four Georgia Locations* (2018), https://greenpoweremc

.com/content/green-power-emc-significantly-expand-its-solar-energy-portfolio -construction-across-four.

Green Power EMC, *Home*, https://greenpoweremc.com.

Groom, Nichola, *U.S. Solar Group Says Trump Tariffs Killings Jobs; White House Says 'Fake News,'* Reuters (Dec. 3, 2019), https://www.reuters.com/article/uk-usa-solar-tariffs/ u-s-solar-industry-to-lose-62000-jobs-due-to-trump-tariffs-study-idUSKBN 1Y71V8.

Grupo Sorolla Educación, *Know Us [Conócenos]*, https://www.gruposorolla.es/saber -mas.

Gulley, Thomas, *Rural Co-Ops and Public Utilities Have Voluntarily Built Nearly 100MW of Community Solar. Here's Why*, Greentech Media (Aug. 14, 2017), https:// www.greentechmedia.com/articles/read/rural-coops-and-public-utilities-have -built-nearly-100mw-of-community-solar.

Harvard Transactional Law Clinics et al., *Tackling the Law, Together: A Legal Guide to Worker Cooperatives Generally and in Massachusetts* (2015), http://clinics.law.harvard .edu/tlc/files/2015/12/Tackling-the-Law-Together-Legal-Guide-to-Worker-Co-Ops -FINAL1.pdf.

Harvey, Fiona *Climate Crisis: What Is COP and Can It Save the World?*, The Guardian (Dec. 2, 2019), https://www.theguardian.com/news/2019/dec/02/climate-crisis -what-is-cop-and-can-it-save-the-world.

Hede, Anne-Marie & Torgeir Watne, *Leveraging the Human Side of the Brand Using a Sense of Place: Case Studies of Craft Breweries*, 29 J. Marketing Mgmt. 207 (2013).

Heede, Richard, *Tracing Anthropogenic Carbon Dioxide and Methane Emissions to Fossil Fuel and Cement Producers, 1854–2010*, 122 Climatic Change 229 (2014).

Hemmings, Bill, Eur. Fed. for Transp. & Env't, *Bunker Fuels and the Kyoto Proto-col: How ICAO and the IMO Failed the Climate Change Test* (2009), https://www .transportenvironment.org/sites/te/files/media/2009_06_aviation_shipping _icao_imo_history.pdf.

Henrÿ, Hagen, *Basics and New Features of Cooperative Law: The Case of Public Interna-tional Cooperative Law and the Harmonisation of Cooperative Laws*, 17 Unif. L. Rev. 197 (2012).

Henrÿ, Hagen, *Sustainable Development and Cooperative Law: Corporate Social Respon-sibility or Cooperative Social Responsibility?* (Univ. of Oslo Faculty of Law, Research Paper No. 2012–23, 2012).

Hispacoop, *Do You Know the Difference Between Ecological, Biological, and Organic? [¿Conoces la diferencia entre ecológico, biológico y orgánico?]*, https://www.hispacoop .com/home/index.php/informacion-consumidor/478-conoces-la-diferencia-entre -ecologico-biologico-y-organico.

Hispacoop, *What Is Hispacoop? [¿Qué es Hispacoop?]*, https://www.hispacoop.com/ home/index.php/features/template-framework.

H.R. 561, 115th Cong. (2017) (United States).

Hsu, Andrea & Mary Louise Kelly, *How Georgia Became a Surprising Bright Spot in the U.S. Solar Industry*, NPR (June 14, 2019), https://www.npr.org/2019/06/

24/733795962/how-georgia-became-a-surprising-bright-spot-in-the-u-s-solar
-industry.

Hsu, Shi-Ling, *Carbon Pricing*, in Legal Pathways to Deep Decarbonization in the United States 70 (Michael B. Gerrard & John C. Dernbach eds., Envtl. L. Inst. Press 2019).

Ifateyo, Ajowa Nzinga, *$5 Million for Co-op Development in Madison*, Grassroots Economic Organizing (Jan. 26, 2015), https://geo.coop/story/5-million-co-op -development-madison.

Impulso Cooperativo, *Nothing Is What It Was . . . [Nada es lo que era . . .]* (2020), https://www.impulsocooperativo.com/.

Instituto Internacional de Derecho y Medio Ambiente, *The Six Priorities of the New Spanish Ministry for Ecological Transition, According to Environmental Lawyers*, https://iidma.org/index.php/en/the-six-priorities-of-the-new-spanish-ministry-for -ecological-transition-according-to-environmental-lawyers/.

Instituto Internacional de Derecho y Medio Ambiente, *The Spanish Strategic Framework for Climate & Energy Lacks Ambition and It Is not Enough to Tackle Climate Change*, https://iidma.org/index.php/en/strategicframework/.

Interface, *The Interface Story*, https://www.interface.com/US/en-US/sustainability/ our-history-en_US.

Intergovernmental Panel on Climate Change, *Characteristics of Four Illustrative Model Pathways* (2018), https://www.ipcc.ch/site/assets/uploads/sites/2/2019/02/SPM3b .png.

Intergovernmental Panel on Climate Change, *Special Report on Global Warning of 1.5°C: Summary for Policymakers* (2018), https://report.ipcc.ch/sr15/pdf/sr15_spm _final.pdf.

International Co-operative Alliance, *About Us*, https://www.ica.coop/en/about-us.

International Co-operative Alliance, *Blueprint for a Co-operative Decade* (2017), https:// www.ica.coop/sites/default/files/publication-files/blueprint-for-a-co-operative -decade-english-1707281677.blueprint-for-a-co-operative-decade-english?_ga=2 .35003554.2036395128.1586453075–1477488065.1585585042.

International Co-operative Alliance, *Cooperative Identity, Values & Principles*, https:// www.ica.coop/en/cooperatives/cooperative-identity.

International Co-operative Alliance, *Guidance Notes to the Co-operative Principles* (2015), https://www.ica.coop/sites/default/files/publication-files/ica-guidance -notes-en-310629900.pdf.

International Co-operative Alliance, *What Is a Cooperative?*, https://www.ica.coop/ en/cooperatives/what-is-a-cooperative.

International Co-operative Alliance, *86th ICA International Co-operative Day 14th UN International Day of Cooperative 5 July 2008: Confronting Climate Change through Co-operative Enterprise* (2008), https://www.aciamericas.coop/IMG/pdf/2008-idc -en-2.pdf.

International Co-operative Alliance & Sustainability Solutions Group, *Co-operatives and Sustainability: An Investigation into the Relationship* (2013), https://www

.ica.coop/sites/default/files/attachments/Sustainability%20Scan%202013–
12–17%20EN_0.pdf.

International Co-operative Alliance & Sustainability Solutions Group, *Sustainability Reporting for Co-operatives: A Guidebook* (2016), https://www.ica.coop/sites/
default/files/publication-files/ica-sustainability-reporting-guidebook-1575997496
.pdf.

International Energy Agency, *Global Energy Review 2020: The Impacts of the COVID-19
Crisis on Global Energy Demand and CO_2 Emissions* (2020), https://www.iea.org/
reports/global-energy-review-2020.

International Energy Agency, *Key CO2 Emissions Trends, Excerpt from: CO2 Emissions
from Fuel Combustion* (2016), https://safety4sea.com/wp-content/uploads/2017/
02/IEA-Key-CO2-Emissions-Trends-2016.pdf.

International Maritime Organization, *Greenhouse Gas Emissions*, http://www.imo
.org/en/OurWork/Environment/PollutionPrevention/AirPollution/Pages/GHG
-Emissions.aspx.

International Panel of Experts on Sustainable Food Systems, *Breaking Away from Industrial Food and Farming Systems: Seven Case Studies of Agroecological Transition*
(2018), https://www.ipes-food.org/_img/upload/files/CS2_web.pdf.

Ioannou, Ioannis, & George Serafeim, *The Consequences of Mandatory Corporate Sustainability Reporting* (Harv. Bus. Sch. Res., Working Paper No. 11-100, 2017).

Jaffee, Daniel, & Philip H. Howard, *Who's the Fairest of Them All? The Fractured Landscape of U.S. Fair Trade Certification*, 33 Agric. & Hum. Values 813 (2016).

Jeter, Debra C., et al., *Democracy and Dysfunction: Rural Electric Cooperatives and the
Surprising Persistence of the Separation of Ownership and Control*, 70 Ala. L. Rev. 361
(2018).

Just Coffee Cooperative, *About Just Coffee: Why We Do What We Do*, https://justcoffee
.coop/about/.

Just Coffee Cooperative, Bylaws (2016) (on file with author).

Just Coffee Cooperative, *Carbon Footprint Report* (2018) (on file with author).

Just Coffee Cooperative, *Coffee Tracker: Transparency in Effect*, https://justcoffee.coop/
coffee-tracker/.

Just Coffee Cooperative, *Doing Justice to Every Bag of Coffee We Roast*, https://justcoffee
.coop/just-great-coffee/.

Just Coffee Cooperative, Second Restated Articles of Incorporation (2016) (on file
with author).

Just Coffee Cooperative, *Sustainability: Using Business for Good*, https://justcoffee
.coop/sustainability-coffee/.

JustinTOOLs, *Convert 40000 Liters to US Beer Barrels*, https://www.justintools.com/
unit-conversion/volume.php?k1=liters&k2=us-beer-barrels&q=40000.

J. W. Hampton, Jr. & Co. v. United States, 276 U.S. 394 (1928).

Kelly, Marjorie, & Shanna Ratner, Wealth Creation in Rural Communities, Keeping Wealth Local: Shared Ownership and Wealth Control for Rural Communities

(2009), https://www.marjoriekelly.com/wp-content/uploads/2009/09/Keeping
-Wealth-Local.pdf.

Kelly, Marjorie, & Sarah Stranahan, Fifty by Fifty, *Employee Ownership and Ecological Sustainability*, Medium (May 25, 2018), https://medium.com/fifty-by-fifty/
employee-ownership-and-ecological-sustainability-2137e90de21d.

Kelly, Marjorie, & Tim Howard, *The Making of a Democratic Economy: Building Prosperity for the Many, not Just the Few* (Berrett-Koehler Pub. 2019).

Khan, Mozaffar, et al., *Corporate Sustainability: First Evidence on Materiality*, 91 Acct.
Rev. 1697 (2016).

Knowledge@Wharton, *America's Neglected Water Systems Face a Reckoning*, Wharton
Sch. of the Univ. of Pa. (June 10, 2015), https://knowledge.wharton.upenn.edu/
article/americas-neglected-water-systems-face-a-reckoning/.

Krajnc, Urška, *Impact of Wine Production Life Cycle on the Environment*, eVineyard
(Oct. 25, 2016), https://www.evineyardapp.com/blog/2016/10/25/impact-the
-wine-production-life-cycle-on-the-environment/.

Krajnc, Urška, *Viticulture: Why Are You Going Organic?*, eVineyard (Oct. 13, 2015),
https://www.evineyardapp.com/blog/2015/10/13/viticulture-why-going-organic/.

Kubiszewski, Ida, et al., *Beyond GDP: Measuring and Achieving Global Genuine Progress*, 93 Ecological Econ. 57 (2013).

Kurland, Nancy B., *Accountability and the Public Benefit Corporation*, 60 Bus. Horizons
519 (2017).

Kvale, Steinar, *InterViews: An Introduction to Qualitative Research Interviewing* (Astrid
Virding ed., Sage Pub. 1996).

Kyoto Protocol to the UNFCCC, UNFCCC Third Conference of the Parties, 3d Sess.,
U.N. Doc. FCCC/CP/1997/7/Add.2.

Lechleitner, Elizabeth, *Landmark Employee Ownership Act, Signed into Law Yesterday,
Will Amend Lending Landscape for Worker Co-Ops*, NCBA CLUSA (Aug. 14, 2018),
https://ncbaclusa.coop/blog/landmark-employee-ownership-act-signed-into-law
-yesterday-will-amend-lending-landscape-for-worker-co-ops/.

Lee, Allison Herren, *"Modernizing" Regulation S-K: Ignoring the Elephant in the Room*,
U.S. Securities & Exchange Commission (Jan. 30, 2020), https://www.sec.gov/
news/public-statement/lee-mda-2020–01–30.

Lehner, Peter, & Nathan A. Rosenberg, *Legal Pathways to Carbon-Neutral Agriculture*,
47 Envtl L. Rep. 10845 (2017).

Levay, Kevin E., et al., *The Demographic and Political Composition of Mechanical Turk
Samples*, 6 SAGE Open 1 (2016).

Levinson, Ariana R., et al., *Alleviating Food Insecurity via Cooperative By-laws*, 26 Geo.
J. Poverty L. Pol'y 227 (2019).

Ley 7/1985, de 2 de abril, Reguladora de las Bases del Régimen Local [Law 7/1985,
of April 2, Regulating the Bases of the Local Regime], Boletín Oficial del Estado
[B.O.E.], April 3, 1985 (Spain), https://www.boe.es/buscar/act.php?id=BOE-A
-1985–5392.

Ley 11/1985, de 25 de octubre, de Cooperativas de la Comunidad Valenciana [Law 11/1985, of Oct. 25, on the Cooperatives of the Valencian Community], Diari Oficial de la Generalitat Valenciana [D.O.] [Official Gazette of the Generalitat Valenciana], Oct. 25, 1985 (Valencia), http://www.dogv.gva.es/portal/ficha_disposicion_pc.jsp ?sig=1387/1985&L=1.

Ley 54/1997, de 27 de noviembre, del Sector Eléctrico [Law 54/1997, of Nov. 27, on the Electricity Sector], Boletín Oficial del Estado [B.O.E.], Nov. 27, 1997 (Spain), https://www.boe.es/buscar/doc.php?id=BOE-A-1997–25340.

Ley 27/1999, de 16 de julio, de Cooperativas [Law 27/1999, of July 16, on Cooperatives], Boletín Oficial del Estado [B.O.E.], July 16, 1999 (Spain), https://boe.es/ buscar/doc.php?id=BOE-A-1999–15681.

Ley 8/2003, de 24 de marzo, de Cooperativas de la Comunidad Valenciana [Law 8/2003, of March 24, on the Cooperatives of the Valencian Community], Diari Oficial de la Generalitat Valenciana [D.O.] [Official Gazette of the Generalitat Valenciana], March 24, 2003 (Valencia), http://www.dogv.gva.es/portal/ficha_disposicion _pc.jsp?sig=1404/2003&L=1.

Ley 17/2007, de 4 de julio, por la que se modifica la Ley 54/1997, de 27 de noviembre, del Sector Eléctrico, para adaptarla a lo dispuesto en la Directiva 2003/54/CE, del Parlamento Europeo y del Consejo, de 26 de junio de 2003, sobre normas comunes para el mercado interior de la electricidad [Law 17/2007, of July 4, amending Law 54/1997, of Nov. 27, on the Electricity Sector, to adapt it to the provisions of Directive 2003/54/EC, of the European Parliament and of the Council of June 26, 2003 on common rules for the internal electricity market], Boletín Oficial del Estado [B.O.E.], July 4, 2007 (Spain), https://www.boe.es/buscar/act.php?id=BOE -A-2007–13024.

Ley 5/2011, de 29 de marzo, de Economía Social [Law 5/2011, of March 29, on Social Economy], Boletín Oficial del Estado [B.O.E.], March 29, 2011 (Spain), https:// www.boe.es/diario_boe/txt.php?id=BOE-A-2011–5708.

Liao, Carol, Limits to Corporate Reform and Alternative Legal Structures, in Company Law and Sustainability: Legal Barriers and Opportunities 274 (Beate Sjåfjell & Benjamin J. Richardson eds., Cambridge Univ. Press 2015).

Light, Sarah E., The Law of the Corporation as Environmental Law, 71 Stan. L. Rev. 137 (2019).

Light, Sarah E., & Eric W. Orts, Parallels in Public and Private Environmental Governance, 5 Mich. J. Envtl. & Admin. L. 1 (2015).

Light, Sarah E., & Michael P. Vandenbergh, Private Environmental Governance, in Elgar Encyclopedia of Environmental Law: Environmental Decision Making (Lee Paddock et al. eds., Edward Elgar Pub. 2016).

Lipton, Martin, The Future of the Corporation, Harv. L. Sch. F. on Corp. Governance (Nov. 7, 2018), https://corpgov.law.harvard.edu/2018/11/07/the-future-of-the -corporation/.

Llei 12/2015, del 9 de juliol, de cooperatives [Law 12/2015, of July 9, on Cooperatives], Diari Oficial de la Generalitat de Catalunya [D.O.] [Official Gazette of the

Generalitat de Catalunya], July 9, 2015 (Catalonia), http://www.icab.cat/files/242–492978-DOCUMENTO/1434848.pdf.

Lu, Jennifer, *Organic Valley Lays Off 39 Employees*, La Crosse Trib. (June 28, 2019), https://lacrossetribune.com/business/local/organic-valley-lays-off-employees/article_4b0f8796–5478–51ce-9fa4–14fb30f362a2.html.

Mackie, David, & Jessica Murray, J.P. Morgan, Risky Business: The Climate and the Macroeconomy (2020).

Madison Cooperative Development Coalition, *About*, https://www.mcdcmadison.org/about/.

Madison Cooperative Development Coalition, *Home*, https://www.mcdcmadison.org.

Madison Cooperative Development Coalition, *Year-End Report* (2018) (on file with author).

Maida, Jesse, *Top 5 Vendors in the Organic Wine Market from 2017 to 2021: Technavio*, BusinessWire (May 11, 2017), https://www.businesswire.com/news/home/20170511005161/en/Top-5-Vendors-Organic-Wine-Market-2017.

Mar Pages, *Albet I Noya—The Pioneer in Organic Winemaking in Spain*, Once in a Lifetime Journey (Aug. 2, 2016), https://www.onceinalifetimejourney.com/wine-tourism/spain/pioneer-organic-winemaking-in-spain-at-albet-i-noya/.

Martín, Rebeca Anguren, & José Manuel Marqués Sevillano, *Cooperative and Savings Banks in Europe: Nature, Challenges and Perspectives*, 6 Banks & Bank Systems 121 (2011).

Mazzucato, Mariana, & Gregor Semieniuk, *Financing Renewable Energy: Who Is Financing What and Why It Matters*, 127 Technological Forecasting & Soc. Change 8 (2018).

Maturing U.S. Organic Sector Grows 6.4% in 2017, Supermarket News (May 21, 2018), https://www.supermarketnews.com/organic-natural/maturing-us-organic-sector-grows-64–2017.

Millstone, Carina, *Frugal Value: Designing Business for a Crowded Planet* (Routledge 2017).

Milwaukee Metropolitan Sewage District, *Sustainable Water Reclamation* (2012), https://www.mmsd.com/application/files/9314/8416/1452/Sustainability_Plan.pdf.

Modern Corporation, *Statement on Company Law*, https://themoderncorporation.wordpress.com/company-law-memo/.

Mondragón, *Corporate Management Model* (2012), https://www.mondragon-corporation.com/wp-content/uploads/2017/Corporate-Management-Model.pdf.

Mondragón, *Home*, https://www.mondragon-corporation.com/en/.

Mondragón, *Mondragón Eko*, https://www.mondragon-corporation.com/en/our-businesses/corporate-projects/mondragon-eko/.

Mondragón, *Mondragón Green Community*, https://www.mondragon-corporation.com/en/our-businesses/corporate-projects/mondragon-green-comunity/.

Mondragón Team Academy, *A Model That Supports Environmental Sustainability, Social Justice and Economic Democracy through Radical Innovation on Education* (2014) (on file with author).

Montt, Guillermo, et al., *Does Climate Action Destroy Jobs? An Assessment of the Employment Implications of the 2-Degree Goal*, 157 Int'l Lab. Rev. 519 (2018).

Morgan, Bronwen, *Legal Models Beyond the Corporation in Australia: Plugging a Gap or Weaving a Tapestry?*, 14 Soc. Enter. J. 180 (2018).

Morrow, Paige, *Non-Financial Reporting: How to Comply? What Does It Mean for the Climate?*, Responsible Investor (Jan. 5, 2017), https://www.responsible-investor .com/articles/non-financial-reporting-fb.

Möslein, Florian, *Certifying 'Good' Companies: Regulatory Schemes for Corporate Sustainability in a Comparative Perspective*, in The Cambridge Handbook of Corporate Law, Corporate Governance and Sustainability 669 (Beate Sjåfjell & Christopher M. Bruner eds., Cambridge Univ. Press 2019).

Moyle, Nick, *10 Best Organic Beers*, Independent (Nov. 23, 2017), https://www .independent.co.uk/extras/indybest/food-drink/beer-cider-perry/best-organic -beer-light-tasting-samuel-smith-gluten-free-a8072226.html.

MX2 Global, *Spanish Wine History* (2019), https://www.mx2global.com/spanish -wine-history/.

National Centre for Business and Sustainability for Co-operatives UK, *Key Social and Co-operative Performance Indicators* (2004), https://library.uniteddiversity.coop/ Measuring_Progress_and_Eco_Footprinting/Key%20Social%20and%20Co -operative%20Performance%20Indicators%3A%20Guidance%20document.pdf.

National Cooperative Bank, *Community Impact*, https://www.ncb.coop/about-us/ community-impact.

National Cooperative Bank, *The 2017 NCB Co-op 100* (2017), https://impact.ncb.coop/ hubfs/Co-ops/NCB-Co-op-100-2017-Web.pdf.

National Cooperative Business Association CLUSA, *About Us: Our History*, https:// ncbaclusa.coop/about-us/our-history/.

National Cooperative Business Association CLUSA, *Congressional Caucus*, https:// ncbaclusa.coop/advocacy/congressional-caucus/.

National Cooperative Business Association CLUSA, *Cooperative for a Better Tomorrow: Creating Economic Opportunity for Americans and People Around the World* (2016), https://ncbaclusa.coop/content/uploads/2016/12/transition-paper-FINAL-EDITS .compressed.pdf.

National Cooperative Business Association CLUSA, *Education & Learning*, https:// ncbaclusa.coop/membership/education-and-learning/.

National Cooperative Business Association CLUSA, *State Cooperative Statute Library*, https://ncbaclusa.coop/resources/state-cooperative-statute-library/.

National Council of Farmer Cooperatives, *About NCFC*, https://ncfc.org/about-ncfc/.

National Council of Farmer Cooperatives, *Farmer Co-ops Applaud Repeal of 2015 Water of the United States Rule* (2019), https://ncfc.org/press-release/farmer-co-ops -applaud-repeal-2015-waters-united-states-rule/.

National Council of Farmer Cooperatives, *Priorities & Policy Resolutions* (2020), https://ncfc.org/priorities-policy-resolutions/.

National Coop Grocers, *About Us*, https://www.ncg.coop/about-us.

National Public Radio, *Getting to Zero Carbon: The Climate Challenge*, https://www.npr
.org/series/735519565/getting-to-zero-carbon-the-climate-challenge.

National Rural Electric Cooperative Association, *Advancing Energy Access for All, Case
Study: Cobb EMC* (2019) (on file with author).

National Rural Electric Cooperative Association, *America's Electric Cooperatives: Electric Co-op Facts & Figures* (2019), https://www.electric.coop/wp-content/uploads/
2020/05/NRECA-Fact-Sheet-5–2020–1.pdf.

National Rural Electric Cooperative Association, Comments on Emission Guidelines
for Greenhouse Gas Emissions from Existing Electric Utility Generating Units
(Oct. 31, 2018), https://www.electric.coop/wp-content/uploads/2018/10/NRECA
-Comm-proposed-ACE.pdf.

National Rural Electric Cooperative Association, *Flexible Approach Important to Energy
Sector Transition, Matheson Tells Congress* (2019), https://www.electric.coop/flexible
-approach-important-to-energy-sector-transition-matheson-tells-congress/.

National Rural Electric Cooperative Association, *Solar*, https://www.electric.coop/wp
-content/Renewables/solar.html.

National Rural Electric Cooperative Association, *We Are America's Electric Cooperatives*, https://www.electric.coop/our-mission/americas-electric-cooperatives/.

Neslen, Arthur, *Spain Plans Switch to 100% Renewable Electricity by 2050*, The Guardian (Nov. 12, 2018), https://www.theguardian.com/environment/2018/nov/13/
spain-plans-switch-100-renewable-electricity-2050.

New Economics Foundation, *Stakeholder Banks: Benefits of Banking Diversity* (2013),
https://neweconomics.org/uploads/files/e0b3bd2b9423abfec8_pem6i6six.pdf.

NYC Planning, *Population—Current and Projected Population* (2018), https://www1
.nyc.gov/site/planning/planning-level/nyc-population/current-future-populations
.page.

NYC Small Business Services, *Working Together: A Report on the Third Year of the
Worker Cooperative Business Development Initiative* (2017), https://www1.nyc.gov/
assets/sbs/downloads/pdf/about/reports/worker_coop_report_fy17.pdf.

Oatfield, Christina, *Governor Brown Signs California Worker Cooperative Act, AB 816*,
Sustainable Economies Law Center (Aug. 12, 2015), https://www.theselc.org/
governor_brown_signs_california_worker_cooperative_act.

Oglethorpe Power, *Our Company*, https://opc.com/about/our-company/.

Oliver, Melvin L., & Thomas M Shapiro, Black Wealth/White Wealth: A New Perspective on Racial Inequality (Routledge, 2nd ed. 2006).

Orden ITC/1522/2007, de 24 de mayo, por la que se establece la regulación de la garantía del origen de la electricidad procedente de fuentes de energía renovables y de
cogeneración de alta eficiencia [Order ITC/1522/2007, of May 24, which establishes
the regulation of the guarantee of the origin of electricity from renewable energy
sources and high-efficiency cogeneration], Boletín Oficial del Estado [B.O.E.], June
1, 2007 (Spain), https://www.boe.es/buscar/act.php?id=BOE-A-2007–10868.

Orr, David W., *Dangerous Years: Climate Change, the Long Emergency, and the Way Forward* (Yale Univ. Press 2016).

Packel, Israel, *The Law of Cooperatives* (3d ed., M. Bender 1956).

Pacyniak, Gabriel, *Greening the Old New Deal: Reforming Rural Electric Cooperative Governance*, 85 Mo. L. Rev. (2020).

Paris Agreement (Dec. 13, 2015), in UNFCCC, COP Report No. 21, Addendum, at 21, U.N. Doc. FCCC/CP/2015/10/Add, 1 (Jan. 29, 2016).

Parker, Stephanie, *Can Food Co-Ops Survive the New Retail Reality?*, Civil Eats (Feb. 28, 2018), https://civileats.com/2018/02/28/can-food-coops-survive-the-new-retail -reality/.

Piketty, Thomas, *Capital and Ideology* (Arthur Goldhammer trans., Harvard Univ. Press 2020).

Pilcher, Jeffry, *Closing the Gender Pay Gap in Banking*, The Financial Brand (Feb. 13, 2018), https://thefinancialbrand.com/70404/closing-gender-pay-gap-banking/.

Pope Francis, Encyclical Letter, *Laudato Si': On Care for Our Common Home* (Vatican Press 2015), http://w2.vatican.va/content/dam/francesco/pdf/encyclicals/ documents/papa-francesco_20150524_enciclica-laudato-si_en.pdf.

ProPublica, *Bills Sponsored by Jared Polis (D-Colo.)*, https://projects.propublica.org/ represent/members/P000598-jared-polis/bills-sponsored/115.

P2P Foundation, *Water Cooperatives* (2014), https://wiki.p2pfoundation.net/Water _Cooperatives.

Rankin, Jennifer, *European Parliament Votes to Ban Single-Use Plastics*, The Guardian (March 27, 2019), https://www.theguardian.com/environment/2019/mar/27/the -last-straw-european-parliament-votes-to-ban-single-use-plastics.

Rainforest Action Network, *Banking on Climate Change: Fossil Fuel Finance Report* (2020), https://www.ran.org/wp-content/uploads/2020/03/Banking_on_Climate _Change__2020_vF.pdf.

Rainforest Action Network, *Explore the Data* (2019), https://www.ran.org/banking onclimatechange2019/#data-panel.

Real Decreto-ley 15/2018, de 5 de octubre, de medidas urgentes para la transición energética y la protección de los consumidores [Royal Decree-Law 15/2018, of Oct. 5, on urgent measures for the energy transition and consumer protection], Boletín Oficial del Estado [B.O.E.], Oct. 5, 2018 (Spain), https://www.boe.es/buscar/doc .php?id=BOE-A-2018–13593.

Real Pickles Cooperative, Inc., Articles of Organization (2012) (on file with author).

Real Pickles Cooperative, Inc., Bylaws (Adopted 2012, Amended 2019) (on fire with author).

Reality Check: Are Ships More Polluting than Germany?, BBC News (April 12, 2018), https://www.bbc.com/news/world-43714029.

RED Eléctrica de España, *Renewable Energy in the Spanish Electricity System* (2017), https://www.ree.es/sites/default/files/11_PUBLICACIONES/Documentos/ Renewable-2017.pdf.

Reeves, Richard V., & Eleanor Krause, *Raj Chetty in 14 Charts: Big Findings on Opportunity and Mobility We Should All Know*, Brookings (Jan. 11, 2018), https://www .brookings.edu/blog/social-mobility-memos/2018/01/11/raj-chetty-in-14-charts -big-findings-on-opportunity-and-mobility-we-should-know/.

Reich, Robert B., *Higher Wages Can Save America's Economy—and Its Democracy*, Salon (Sept. 3, 2013, 10:12 PM), https://www.salon.com/test/2013/09/03/higher _wages_can_save_americas_economy_and_its_democracy_partner/.

Reich, Robert B., *Saving Capitalism: For the Many, Not the Few* (Vintage 2016).

Reid, Stuart, *Are Food Co-Ops Still Relevant?*, Cooperative Grocer Network (2017), https://www.grocer.coop/articles/are-food-co-ops-still-relevant.

Reiser, Dana Brakman, *Benefit Corporations—A Sustainable Form of Organization?*, 46 Wake Forest L. Rev. 591 (2011).

Research email with Kristen Delaney, Jan. 5, 2020 (on file with author).

Research email with Marco Sanchez, consulting translator, April 10, 2019 (on file with author).

Research email with María José Ortolá Sastre, Feb. 12, 2019 (on file with author).

Research email with Pablo Sanchez, Executive Director of B Lab Spain, Jan. 16, 2019 (on file with author).

Research email with Rodrigo Gouveia, Dec. 3, 2018 (on file with author).

Research email with Tim Jarrell, Jan. 23, 2020 (on file with author).

Research email with Tim Jarrell and Kristen Delaney, Jan. 15, 2020 (on file with author).

Research interview with Carmen Llinares, Board Vice President, and Jesus Arnaiz, Board Secretary, bioTrèmol, April 12, 2019 (on file with author).

Research interviews with Carman Picot, Executive of Institutional Relations and CSR, and Ana Mffi Garcia, Environmental Technician, Consum, April 12, 2019 and April 30, 2019 (on file with author).

Research interview with Francesc "Paco Alós" Alabajos, Director of Commercial Department and Director of Social Responsibility, Caixa Popular, March 22, 2019 (on file with author).

Research interview with George Siemon, Founder and Former CEO, CROPP Cooperative/Organic Valley, May 31, 2019, and Nov. 27, 2019 (on file with author).

Research interview with Irene Machuca, Governing Board Member, Som Energia, March 19, 2019 (on file with author).

Research interview with Joan Arévalo i Vilà, President, Mining Community SCCL Olesana, and President, La Cooperativa de L'Aigua or Aigua.Coop, Apr. 16, 2019 (on file with author).

Research interview with Jonathon Reinbold, former Senior Manager for Sustainability, CROPP Cooperative/Organic Valley, Nov. 14, 2019 (on file with author).

Research interview with María José Ortolá Sastre, Subdirector General of Social Economy and Entrepreneurship, Valencia, Feb. 11, 2019 (on file with author).

Research interview with Maria Vicente, Co-founder, Cerveses Lluna, Feb. 13, 2019 (on file with author).

Research interview with Matthew Earley, Co-founder and President of Board of Directors of Just Coffee, May 21, May 30, May 31, July 8, July 17, 2019 (on file with author).

Research interview with Milagros Perez, Communication Director, Bodegas Pinoso, Jan. 31, 2019 (on file with author).

Research interview with Prof. Germán Valencia Martin, Law School, University of Alacant, April 8, April 14, June 29, 2019 (on file with author).

Research interview with Prof. Tom Webb, St. Mary's University, Nova Scotia, Canada, Nov. 28, 2018 (on file with author).

Research interview with Ruth Rohlich, Business Development Specialist, City of Madison, May 21, 2019 (on file with author).

Research interview with Sheila Ongie, Sustainability Manager, National Coop Grocers, Dec. 4, 2019 (on file with author).

Research interview with Tim Jarrell, Vice President of Power Supply, and Kristen Delaney, Vice President, Marketing and Corporate Communications, Cobb EMC, Dec. 13, 2019 (on file with author).

RE100, *Companies*, https://there100.org/companies.

RE100, *Progress and Insights Annual Report, November 2018* (2018), http://media.virbcdn.com/files/c1/36cff88178ac5f22-AppendixTable-newIKEAchange.pdf.

Richardson, Benjamin J., & Beate Sjåfjell, *Capitalism, the Sustainability Crisis, and the Limitations of Current Business Governance*, in Company Law and Sustainability: Legal Barriers and Opportunities 1 (Beate Sjåfjell & Benjamin J. Richardson eds., Cambridge Univ. Press 2015).

Ritchie, Hannah, & Max Roser, *CO_2 and Greenhouse Gas Emissions*, Our World in Data (Dec. 2019), https://ourworldindata.org/co2-and-other-greenhouse-gas-emissions.

Rural Development, U.S. Dep't of Agric., Rural Cooperative Development Grant Program, https://www.rd.usda.gov/programs-services/rural-cooperative-development-grant-program.

Rural Electrification Act of 1936, 7 U.S.C. § 901 *et seq.* (2018) (United States).

Santander, *Sustainability Report* (2017), https://www.santander.com/content/dam/santander-com/en/documentos/informe-anual-de-sostenibilidad/2017/IAS-2017-Sustainability%20report%202017–10-en.pdf.

Santos, Javier, *Spain: The Regulation of Renewables* (2008), Int'l Fin. L. Rev., https://www.iflr.com/Article/2025527/Spain-The-regulation-of-renewables.html.

Scanlan, Melissa K., *Climate Change, System Change, and the Path Forward*, in Law and Policy for a New Economy: Sustainable, Just, and Democratic 1 (Melissa K. Scanlan ed., Edward Elgar Pub. 2017).

Scanlan, Melissa K., *Sustainable Sewage*, in Energy, Governance and Sustainability 243 (Jordi Jaria i Manzano et al. eds., Edward Elgar Pub. 2016).

Scanlan, Melissa K., et al., *Realizing the Promise of the Great Lakes Compact: A Policy Analysis for State Implementation*, 8 Vt. J. Envtl. L. 39 (2006).

Scarborough, Peter, et al., *Dietary Greenhouse Gas Emissions of Meat-Eaters, Fish-Eaters, Vegetarians and Vegans in the UK*, 125 Climatic Change 179 (2014).

Schultz, Rob, *Organic Valley's Longtime CEO Steps Down*, Wis. St. J. (March 15, 2019), https://madison.com/wsj/business/organic-valleys-longtime-ceo-steps-down/article_b276698a-b1b5–5fed-862e-0c7e19145ccd.html.

SENEO, *About Us [Quiénes somos]*, https://www.seneo.org/cont/2-quienes-somos.

Serlicoop, *Serlicoop's Mission and Vision* [*La misión y visión de Serlicoop*], https://www
    .serlicoop.com/sobre-nosotros-serlicoop/.

Shaik, Anwar, *Capitalism: Competition, Conflict, Crises* (Oxford Univ. Press 2016).

Shared Capital Cooperative, *About*, https://sharedcapital.coop/about/.

Shepherd, Lori, *2.3 Million Small Businesses Nationwide Owned by Aging Boomers
    Preparing to Retire Puts 1 in 6 Employees' Jobs at Risk, Based on a Project Equity
    Study*, Cision PRWeb (May 8, 2017), https://www.prweb.com/releases/2017/05/
    prweb14310931.htm.

Sjåfjell, Beate, et al., *Shareholder Primacy: The Main Barrier to Sustainable Companies*,
    in Company Law and Sustainability: Legal Barriers and Opportunities 79 (Beate
    Sjåfjell & Benjamin J. Richardson eds., Cambridge Univ. Press 2015).

Smart Electric Power Alliance, *Community Solar Program Design Models* (2018), http://
    go.sepapower.org/l/124671/2018–04–19/2dqsrp/124671/49916/Community
    _Solar_Program_Design_Models.pdf.

Smart Electric Power Alliance, *2019 Utility Solar Market Snapshot* (2019), https://
    sepapower.org/resource/2019-utility-solar-market-snapshot/.

Smith, Adrienne, *Spain Dominates Organic Wine Market*, Food & Wines from Spain
    (Nov. 28, 2017), https://www.foodswinesfromspain.com/spanishfoodwine/global/
    whats-new/news/new-detail/spain-organic-wine-market.html.

Smith, Adrienne, *Spanish Organic Wines*, Food & Wines from Spain (June 20, 2016),
    https://www.foodswinesfromspain.com/spanishfoodwine/global/wine/features/
    feature-detail/REG2017734204.html.

Social y Solidaria, *Federation of Drinking Water Cooperatives of the Province of Buenos
    Aires (FEDECAP)* [*Federación de cooperativas de aqua potable de la provincia de Buenos
    Aires (FEDECAP)*] (2019), https://socialysolidaria.com/federacion-de-cooperativas
    -dc-agua-potable-de-la-provincia-de-buenos-aires-fedecap/.

Som Energia, *About Us* [*¿Quiénes somos?*], https://www.somenergia.coop/es/quienes
    -somos/.

Som Energia, Bylaws of Som Energia, SCCL [Estatutos de Som Energia, SCCL] (2019),
    https://www.somenergia.coop/estatuts/Estatutos_Som_Energia_2019.pdf.

Som Energia, *Campaigns* [*Campañas*], https://www.somenergia.coop/es/participa/
    campanas/#ilumina.

Som Energia, *Differences Between Contributing Capital Stock and Participating in the
    Generation kWh Project* [*Diferencias entre aportar en el capital social y participar en el
    Proyecto Generation kWh*] (2020), https://es.support.somenergia.coop/article/633
    -diferencias-entre-invertir-en-el-capital-social-y-el-proyecto-generation-kwh.

Som Energia, *Governing Council* [*Consell rector*], https://www.somenergia.coop/ca/qui
    -som/#consellrector.

Som Energia, *Home*, https://www.somenergia.coop/.

Som Energia, *Payment of Interest on Voluntary Contributions to Share Capital* [*Pago
    de los intereses de las aportaciones voluntarias al capital social*] (2019), https://
    blog.somenergia.coop/som-energia/2019/01/pago-de-los-intereses-de-las
    -aportaciones-voluntarias-al-capital-social/.

Som Energia, *Processes* [*Procesos*] (2019), https://participa.somenergia.coop/processes/SomAG2019.

Som Energia, *Produce Renewable Energy* [*Produce energía renovable*], https://www.somenergia.coop/es/produce-energia-renovable/.

Som Energia, *Som Energia School* [*Escuela de Som Energia*], https://www.somenergia.coop/es/participa/escuela-som-energia/.

Som Energia, *The Electrical System and Som Energía* [*El sistema eléctrico y Som Energía*], https://es.support.somenergia.coop/article/652-el-sistema-electrico-y-som-energia.

Som Energia, *We Open an Engineering Competition to Install 100 Photovoltaic Self-Production Roofs in Central Catalonia* [*Obrim el concurs d'enginyeries per install-lar 100 cobertes d'autoproducció fotovoltaica a la Catalunya Central*], Som Energia (April 1, 2019), https://blog.somenergia.coop/grupos-locales/catalunya/ripolles/2019/04/obrim-el-concurs-denginyeries-per-instal%C2%B7lar-100-cobertes-dauto produccio-fotovoltaica-a-la-catalunya-central/.

Som Energia, *What Do We Do?* [*¿Què fem?*], https://www.somenergia.coop/ca/qui-som/#quefem.

Southern Power, *Sandhills Solar Facility* (2020), https://www.southerncompany.com/content/dam/southern-company/pdf/southernpower/Sandhills_Solar_Facility _factsheet.pdf.

Steinfort, Lavinia, & Satoko Kishimoto, Dennis Burke ed., *From Terrassa to Barcelona: Cities and Citizens Reclaim Public Water and Other Essential Services*, Transnat'l Inst. (April 4, 2017), https://www.tni.org/en/article/from-terrassa-to-barcelona-cities -and-citizens-reclaim-public-water-and-other-essential.

Stranahan, Sarah, & Marjorie Kelly, Fifty by Fifty, *Mission-Led Employee-Owned Firms: The Best of the Best* (2019), https://www.fiftybyfifty.org/wp-content/uploads/2019/05/missionledemployeeownedfirms-1.pdf.

Study Group on European Cooperative Law, *Draft Principles of European Cooperative Law*, Euricse (2015), https://www.euricse.eu/wp-content/uploads/2015/04/PECOL-May-2015.pdf.

Subsidiary Body for Scientific and Technological Advice, Communications from Parties Included in Annex I to the Convention: Guidelines, Schedule and Process for Consideration, U.N. Doc. FCCC/SBSTA/1996/9/Add.1 (June 25, 1996), https://unfccc.int/cop3/resource/docs/1996/sbsta/09a01.htm#N_5_.

Sullivan, Mark, *Apple Now Runs on 100% Green Energy, and Here's How It Got There*, Fast Company (April 9, 2018), https://www.fastcompany.com/40554151/how -apple-got-to-100-renewable-energy-the-right-way.

Sustainable Development Solutions Network, *Mapping the Renewable Energy Sector to the Sustainable Development Goals: An Atlas* (2019), https://irp-cdn.multiscreensite .com/be6d1d56/files/uploaded/190603-mapping-renewables-report-interactive .pdf.

Sustainable Economies Law Center, *Choice of Entity*, Co-Op Law, https://www.co -oplaw.org/governance-operations/entity/.

Sustainable Economies Law Center, *Cooperatives*, https://www.theselc.org/cooperatives.

Sustainable Economies Law Center, *Co-op Bylaws and Other Governance Documents*, Co-Op Law, https://www.co-oplaw.org/legal-tools/cooperative-bylaws/.

Sustainable Economies Law Center, *State-by-State Co-op Law Info*, Co-Op Law, https://www.co-oplaw.org/statebystate/.

Trading Economics, *Spain Youth Unemployment Rate*, https://tradingeconomics.com/spain/youth-unemployment-rate.

Tri-State Generation and Transmission Association, *Tri-State Announces Transformative Responsible Energy Plan Actions to Advance Cooperative Clean Energy* (2020), https://www.tristategt.org/tri-state-announces-transformative-responsible-energy-plan-actions-advance-cooperative-clean-energy.

Trujillo, Elizabeth, *International Trade*, in Legal Pathways to Deep Decarbonization in the United States 197 (Michael B. Gerrard & John C. Dernbach eds., Envtl. L. Inst. Press 2019).

Uniform Law Commission, *Uniform Limited Cooperative Association Act* (2007), https://www.uniformlaws.org/HigherLogic/System/DownloadDocumentFile.ashx?DocumentFileKey=56dcb635-374c-f1a9-cb11-d701041c939f&forceDialog=0.

United Nations, *About the Sustainable Development Goals*, https://www.un.org/sustainabledevelopment/sustainable-development-goals/.

University of Wisconsin-Madison Center for Cooperatives, *Bylaws for Cooperatives, Including a Sample Outline* (2019), https://resources.uwcc.wisc.edu/Legal/SampleBylaws.pdf.

University of Wisconsin-Madison Center for Cooperatives, *History*, https://uwcc.wisc.edu/about-uwcc/history/.

University of Wisconsin-Madison Center for Cooperatives, *Types of Co-ops*, https://uwcc.wisc.edu/about-co ops/types-of-co-ops/.

University of Wisconsin-Madison Center for Cooperatives, *Utility Cooperatives: Water Cooperatives*, https://uwcc.wisc.edu/resources/utilities/.

Urban Institute, *Nine Charts About Wealth Inequality in America (Updated)* (2017), https://apps.urban.org/features/wealth-inequality-charts/.

U.N. Committee on Economic, Social and Cultural Rights, General Comment No. 15 (2002): The Right to Water (Arts. 11 and 12 of the International Covenant on Economic, Social, and Cultural Rights, U.N. Doc. E/C.12/2002/11 (2003).

U.N. Conference on Environment and Development, *Rio Declaration on Environment and Development*, U.N. Doc. A/CONF.151/26/Rev.1 (Vol. I), annex I (Aug. 12, 1992).

U.N. Conference on Trade and Development, *Climate Change and Trade*, https://unctad.org/en/Pages/DITC/ClimateChange/Climate-Change.aspx.

U.N. Conference on Trade and Development, *Review of Maritime Transport 2019*, U.N. Doc. UNCTAD/RMT/2019/Corr.1 (2020), https://unctad.org/en/PublicationsLibrary/rmt2019_en.pdf.

U.N. Department of Economic and Social Affairs, *Human Right to Water*, https://www.un.org/waterforlifedecade/human_right_to_water.shtml.

U.N. Env't Programme, The Emissions Gap Report 2014: A UNEP Synthesis Report, U.N. Doc. DEW/1833/NA (2014).

U.N.G.A. Res. 64/136, Cooperatives in Social Development (Feb. 11, 2010).

U.N.G.A. Res. 64/292, Human Right to Water and Sanitation (July 28, 2010).

7 U.S.C. § 291 (2012) (United States).

26 U.S.C. § 501 (2018) (United States).

26 U.S.C. § 521 (2004) (United States).

26 U.S.C. §§ 1381–1388 (United States).

26 U.S.C. § 1042 (2018) (United States).

26 U.S.C. § 1042 (c)(2) (2018) (United States).

U.S. Commodity Futures Trading Commission, Climate-Related Market Risk Subcommittee, Market Risk Advisory Committee, *Managing Climate Risk in the U.S. Financial System,* Sep. 2020.

U.S. Energy Info. Admin., *Electricity Explained: Electricity in the United States* (2020), https://www.eia.gov/energyexplained/electricity/electricity-in-the-us.php.

U.S. Energy Info. Admin., *How Much of U.S. Energy Consumption and Electricity Generation Comes from Renewable Energy Sources?* (2020), https://www.eia.gov/tools/faqs/faq.php?id=92&t=4.

U.S. Energy Info. Admin., *Investor-Owned Utilities Served 72% of U.S. Electricity Customers in 2017* (2019), https://www.eia.gov/todayinenergy/detail.php?id=40913.

U.S. Energy Info. Admin., *Renewable Energy Explained* (2019), https://www.eia.gov/energyexplained/renewable-sources/.

U.S. Energy Info. Admin., *U.S. Renewable Energy Consumption Surpasses Coal for the First Time in Over 130 Years* (2020), https://www.eia.gov/todayinenergy/detail.php?id=43895.

U.S. Envtl. Prot. Agency, EPA 430-R-20-002, *Inventory of U.S. Greenhouse Gas Emissions and Sinks: 1990–2018* (2020), https://www.epa.gov/sites/production/files/2020-04/documents/us-ghg-inventory-2020-main-text.pdf.

U.S. Envtl. Prot. Agency, *Global Emissions by Economic Sector,* https://www.epa.gov/ghgemissions/global-greenhouse-gas-emissions-data#Sector.

U.S. Federation of Worker Cooperatives, *About the USFWC,* https://www.usworker.coop/about/.

U.S. Federation of Worker Cooperatives, *Home,* https://www.usworker.coop/home/.

U.S. Federation of Worker Cooperatives, *Worker Ownership,* https://www.usworker.coop/what-is-a-worker-cooperative/.

U.S. Geological Survey, *Water Q&A: How Much Water Do I Use at Home Each Day?,* https://www.usgs.gov/special-topic/water-science-school/science/water-qa-how-much-water-do-i-use-home-each-day?qt-science_center_objects=0#qt-science_center_objects.

Van Allen, Christopher, et al., Benchmarking Air Emissions of the 100 Largest Electric Power Producers in the United States (2018).

Van Allen, Christopher, et al., Benchmarking Air Emissions of the 100 Largest Electric Power Producers in the United States (2019), https://www.mjbradley.com/sites/default/files/Presentation_of_Results_2019.pdf.

Vandenbergh, Michael P., & Jonathan M. Gilligan, *Beyond Politics: The Private Governance Response to Climate Change* (Aseem Prakash ed., Cambridge Univ. Press 2017).

Vermeulen, Sonja J., et al., *Climate Change and Food Systems*, 37 Ann. Rev. Env't & Resources 195 (2012), https://www.annualreviews.org/doi/pdf/10.1146/annurev -environ-020411–130608.

Vermont Electric Coop, *Power Supply*, https://www.vermontelectric.coop/keeping-the -lights-on/power-supply.

Vermont Employee Ownership Center, *Who We Are*, https://www.veoc.org/.

Water & Wastes Digest, *Report Ranks World's 50 Largest Private Water Utilities* (2015), https://www.wwdmag.com/trends-forecasts/report-ranks-world's-50-largest -private-water-utilities.

Watts, Jonathan and Ambrose, Jillian, *Coal Industry Will Never Recover After Coronavirus Pandemic, Say Experts*, The Guardian (May 17, 2020), https://www .theguardian.com/environment/2020/may/17/coal-industry-will-never-recover -after-coronavirus-pandemic-say-experts.

Wilkinson, Richard & Kate Pickett, *The World We Need*, in Co-Operatives in a Post-Growth Era: Creating Co-Operative Economics 61 (Sonja Novkovic & Tom Webb eds., Zed Books 2014).

Williams, Richard C., *The Cooperative Movement: Globalization from Below* (David Crowther ed., Ashgate Pub. 2007).

Willy Street Co-op, *About Our Co-op*, https://www.willystreet.coop/about-us/about -our-coop.

World Fair Trade Organization, *The Big Idea of Fair Trade Is to Challenge the Purpose of Business*, https://wfto.com/news/big-idea-fair-trade-challenge-purpose-business.

World Fair Trade Organization Europe, *The 10 Principles of Fair Trade* (2019), https:// wfto europe.org/the-10-principles-of-fair-trade-2/.

World Health Organization, *Drinking-Water* (2019), https://www.who.int/news -room/fact-sheets/detail/drinking-water.

World Resources Institute, *Green Targets: A Tool to Compare Private Sector Banks' Sustainable Finance Commitments* (2019), https://www.wri.org/finance/banks -sustainable-finance-commitments/.

World Trade Organization, *Climate Change and the Potential Relevance of WTO Rules*, https://www.wto.org/english/tratop_e/envir_e/climate_measures_e.htm.

Zucca, Gary, et al., *Sustainable Viticulture and Winery Practices in California: What Is It, and Do Customers Care?*, 2 Int'l J. Wine Res. 189 (2009).

INDEX